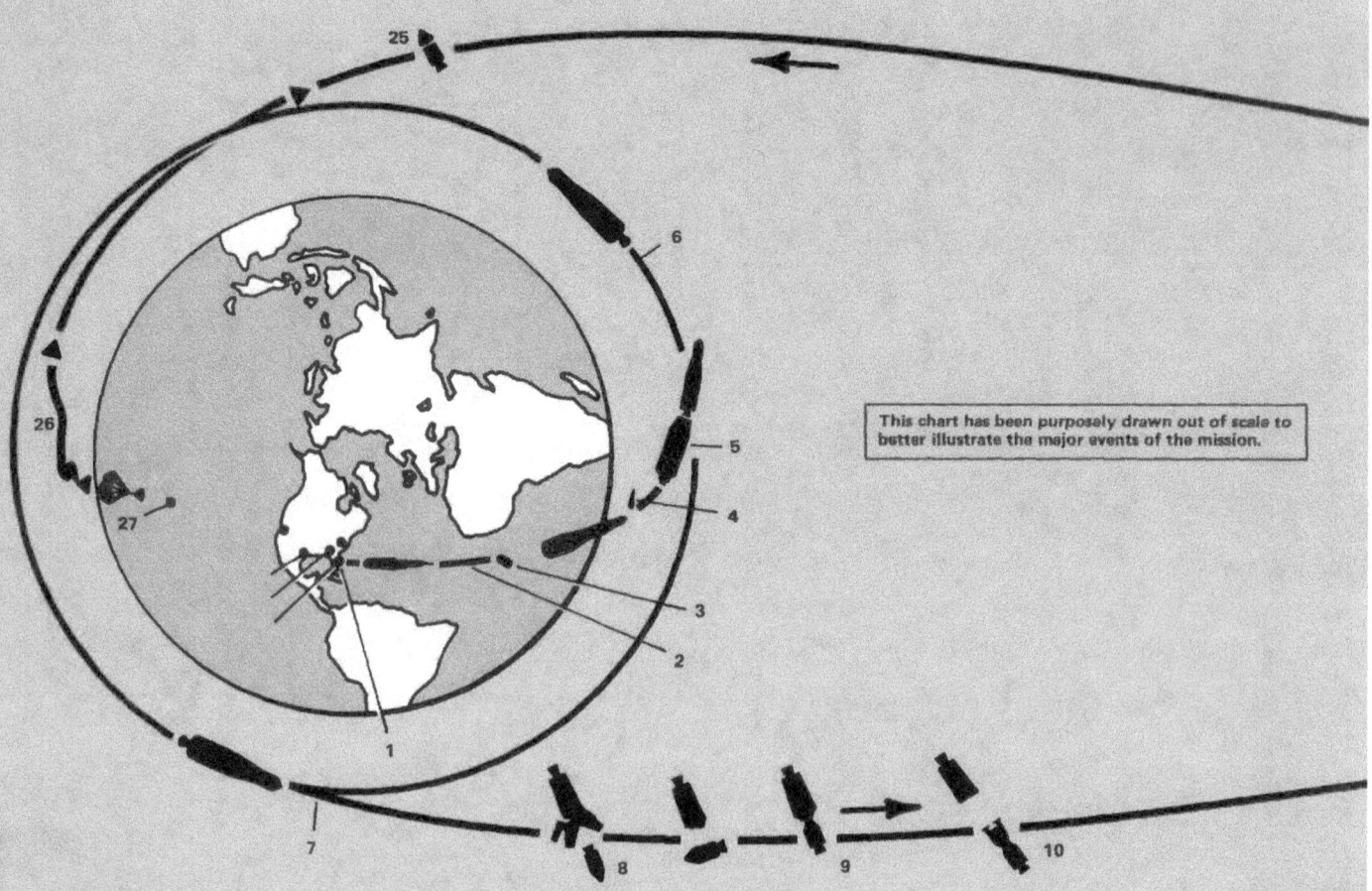

This chart has been purposely drawn out of scale to better illustrate the major events of the mission.

APOLLO MISSION PROFILE

1. Liftoff
2. S-IC powered flight
3. S-IC/S-II separation
4. Launch escape tower jettison
5. S-II/S-IVB separation
6. Earth parking orbit
7. Translunar injection
8. CSM separation from LM adapter
9. CSM docking with LM/S-IVB

10. CSM/LM separation from S-IVB
11. Midcourse correction
12. Lunar orbit insertion
13. Pilot transfer to LM
14. CSM/LM separation
15. LM descent
16. Touchdown
17. Explore surface, set up experiments
18. Liftoff

19. Rendezvous and docking
20. Transfer crew and equipment from LM to CSM
21. CSM/LM separation and LM jettison
22. Transearth injection preparation
23. Transearth injection
24. Midcourse correction
25. CM/SM separation
26. Communication blackout period
27. Splashdown

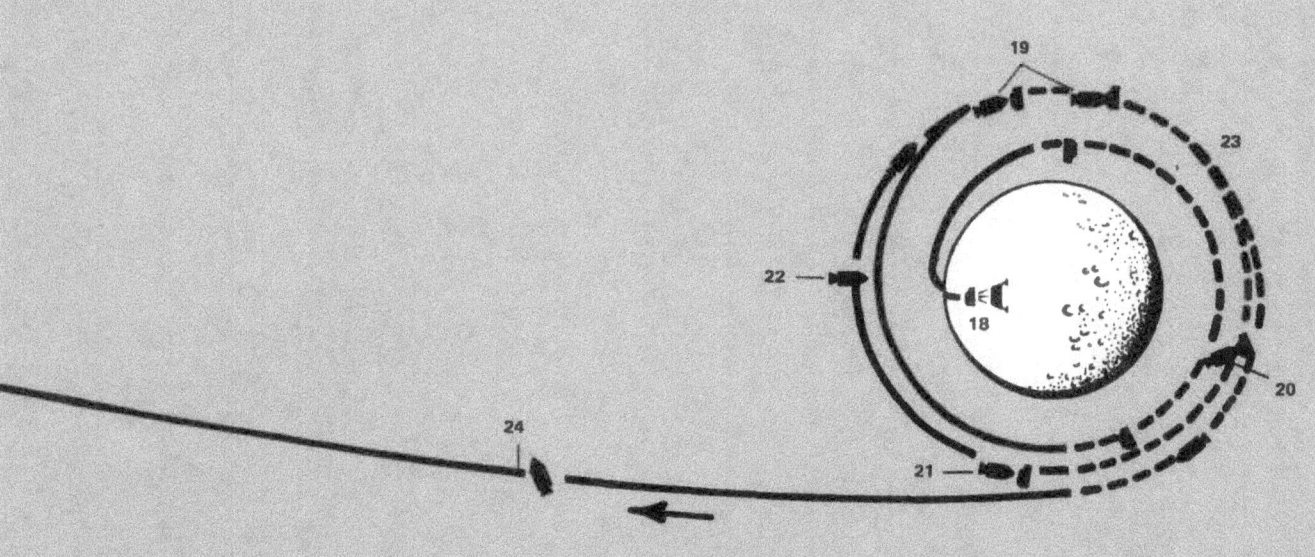

19

23

22 →

18

20

21 —

24 ←

Broken trajectory lines indicate
loss of Earth communications.

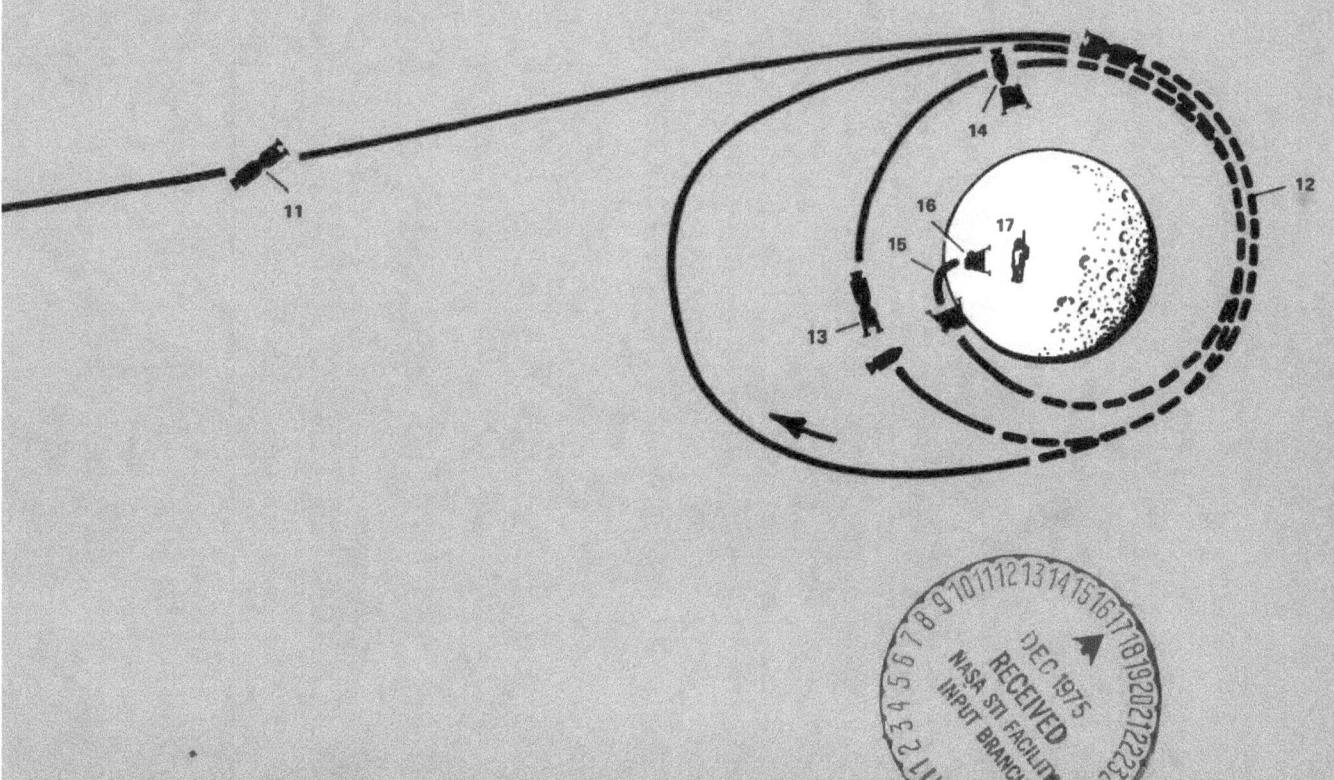

14

16   17

15

13

12

11

HERE MEN FROM THE PLANET EARTH
FIRST SET FOOT UPON THE MOON
JULY 1969 A.D.
WE CAME IN PEACE FOR ALL MANKIND

From the plaque on the Eagle, Apollo 11,
which landed on the Moon, July 20, 1969.

NASA SP–350

Apollo Expeditions

EDITED BY EDGAR M. CORTRIGHT

Scientific and Technical Information Office
NATIONAL AERONAUTICS AND SPACE ADMINISTRATION
Washington, D.C.
1975

to the Moon

Library of Congress Cataloging in Publication Data
Main entry under title:
Apollo expeditions to the moon.
(NASA SP ; 350)
Includes index.
1. Project Apollo—Addresses, essays, lectures. 2. Space flight to the moon—Addresses, essays, lectures. I. Cortright, Edgar M. II. Series: United States. National Aeronautics and Space Administration. NASA SP ; 350.
TL789.8.U6A513   629.45'4   75-600071

Foreword

No nation ever demonstrated its aspirations and abilities as dramatically as did the United States when it landed the first men on the Moon, or as much in public: More people on Earth watched that first small step on a foreign planet than had witnessed any prior event in the ascent of man. While it is still too early to assess the full significance of that remarkable undertaking, I think it is a good time to look back on the total enterprise, while the images are still sharp, and while those concerned are available to give testimony. Historians have observed that ventures into uncharted waters are often illuminated most vividly in the words of those who were there; one thinks of Caesar's *Commentaries,* Bradford's *History of Plymouth Plantation,* Darwin's *Voyage of the Beagle.* An interesting parallel exists between the voyages of H.M.S. *Beagle* and the missions of Apollo: One changed the course of the biological sciences, and the others are reshaping planetary and Earth sciences. In this volume you will find the personal accounts of eighteen men who, like Darwin, were much involved in long and influential voyages.

New scientific insights are an important part of the legacy of Apollo, as well as the worldwide lift to the human spirit that the achievement generated. But there is a third legacy of Apollo that is particularly germane today. This was the demonstration that great and difficult endeavors can be conducted successfully by a steadfast mobilization of national will and resources. Today we face seemingly intractable problems whose resolution may call for similar mobilization of resources and will. Husbanding the planet's finite resources, developing its energy supplies, feeding its billions, protecting its environment, and shackling its weapons are some of these problems. If the zest, drive, and dedication that made Apollo a success can be brought to bear, that may be the most priceless legacy of Apollo.

JAMES C. FLETCHER
Administrator
National Aeronautics and
Space Administration

July 30, 1975

TABLE OF

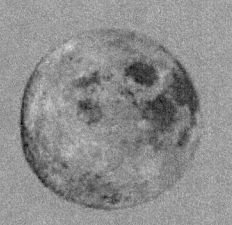

CONTENTS

INTRODUCTION

In looking back at the origins and development of the Apollo program, one word that comes to mind is *action*. From my vantage as Associate Administrator from 1960 to 1965, and then as Deputy Administrator from 1965 to 1968, I had an excellent picture of the intricate action processes that comprised the Apollo program. Disparate and numerous, the actions and their companion reactions came together in a remarkably coordinated and cooperative blending for the goal of placing men on the Moon and bringing them back safely.

The precipitating action was the successful Soviet launching of Sputnik in 1957. This remarkable and surprising achievement was the impetus for NASA's creation in 1958 by President Eisenhower and the Congress. Forged in large part from the widely respected National Advisory Committee for Aeronautics, with its valued research facilities in Virginia, Ohio, and California, NASA also incorporated other research elements. From the Navy came the Vanguard Satellite Project team. From the Army came the ballistic missile team at Redstone Arsenal, to become the nucleus of the Marshall Space Flight Center. The Jet Propulsion Laboratory, operated by the California Institute of Technology for the Defense Department, was also made part of the growing NASA organization.

During the NACA–NASA transition period the elements of Project Mercury for placing a manned capsule in orbit were born. Work also began, and progressed well, on scientific, meteorological, and communication satellites—themselves considerable examples of technological virtuosity—but interest remained high on manned space flight. Estimates of the technical problems and price tag for a manned lunar landing mission were forbidding. The understandable reluctance to make such a major commitment diminished dramatically, however, with Gagarin's successful manned orbital flight in March 1961. Again, as in the period following Sputnik, grave concern about Soviet successes was vocalized in the Congress and President Kennedy asked his administration for plans to make this Nation preeminent in space. Out of this introspection came plans and a favorable response to President Kennedy's special address to Congress in which he stated: "I believe this Nation should commit itself to achieving the goal, before this decade is out, of landing a man on the Moon and returning him safely to Earth." But as we were to learn in carrying out this objective, sustaining the resources meant renewing the commitment annually. The Congress tends to operate with fiscal-year and session perspective; while our horizon was set nine years ahead. We found that an agency's performance as a good steward of public funds is a keystone to its continued support.

As planning for Apollo began, we identified more than 10,000 separate tasks that had to be accomplished to put a man on the Moon. Each task had its particular objectives, its manpower needs, its time schedule, and its complex interrelationship with many other tasks. Which had to be done first? Which could be done concurrently? What were the critical sequences? Vital questions such as these had to be answered in building the network of tasks leading to a lunar landing. The network had to be subdivided into manageable portions, the key ones being: determination of the environment in cislunar space and on the lunar surface; the design and development of the spacecraft and launch vehicles; the conduct of tests and flight missions to prove these components and procedures; and the selection and training of flight crews and ground support to carry out the missions.

Early in the critical planning stages for Apollo, three different approaches to the Moon were considered: direct ascent, rendezvous in Earth orbit, and rendezvous in lunar orbit. The choice of mission mode was a key milestone in our development of Apollo. Like many other decisions, it set us in a direction from which retreat could come only at extreme penalty to the schedule and cost of the program.

Lunar-orbit rendezvous meant considerable payload savings and in turn a reduced propulsion requirement; in fact the reduction was on the order of 50 percent. But in requiring less brute force, we needed more skill and finesse. A module designed especially for landing on and lifting from the lunar surface had to mate with a module orbiting the Moon. Rendezvous and docking, clearly, were of critical importance. The Gemini program was created to provide greater experience than Mercury would in manned operations in space, and especially in perfecting procedures on rendezvous and docking.

While lunar rendezvous was the choice for getting to the Moon, many other fundamental technical, policy, and management questions had to be answered: How and where were major parts to be developed and made? How were they to be shipped? Where were they to be assembled? Where would we site the important supporting facilities and the launch complex? The huge scale of the Apollo operation precluded conventional answers. Facilities that were in themselves major engineering challenges were created, and a separate network of giant deep-space antennas was constructed in Spain, Australia, and California to receive the tremendous volume of data that would flow back from the Moon.

Apollo was an incredible mixture of the large and the small, of huge structures and miniaturized equipment. These astronomical variations in scale had to be dealt with in a thoroughgoing way, with technical competence and managerial assurance. In planning the serial buildup of Apollo missions, we could not take steps so small that the exposure to the ever-present risk outweighed minor gains expected. Yet neither could we take steps so big that we stretched equipment and people dangerously far beyond the capabilities that had already been demonstrated. We followed the fundamental policy of capitalizing on success, always advancing on each mission as rapidly as good judgment dictated.

So step by step, confidence and experience were accumulated. The four manned orbital flights of the Mercury program proved man's ability to survive in space, fly spacecraft, and perform experiments. These abilities were expanded in the Gemini program; in particular, the ability to rendezvous, dock, and conduct extravehicular activity was demonstrated. The Ranger, Surveyor, and Orbiter series contributed necessary cartographic, geologic, and geophysical data about the Moon. All these missions were in preparation for the flights with the powerful Saturn V launch vehicle first flown unmanned in late 1967.

Throughout the program we tried to maintain a flexible posture, keeping as many options open as possible. When difficulties were experienced and delays occurred, alternate plans had to be quickly but carefully evaluated. This meant there was a continual need for steady judgment and cool decision. Apollo managers and astronauts met that need. The Apollo 8 flight was an example of the virtue of schedule flexibility. Lunar-module progress had slowed and the Lunar Orbiter spacecraft had revealed unexpected circumlunar navigational questions. Our action was to send the command and service modules alone into lunar orbit. And so in December 1968 man first flew to the neighborhood of the Moon. Apollo 8 was indeed a significant Christmas gift for the program.

To acquire critical rendezvous and docking experience, Apollo 9 flight-tested the whole system, booster, command and service module, and lunar module in Earth orbit. Could we next put it all together and land on the Moon? Was the step too big? Would it stretch us too far? Yes, it would, we concluded. To resolve the remaining unknowns, astronauts again went to the Moon, not yet to land, but to do as many of the required nonlanding tasks as possible. Apollo 10 was a successful dress rehearsal.

Throughout the testing, both ground and flight, we played deadly serious "what-if" games—designed to anticipate contingencies and cope successfully with them. Computers were invaluable aids to these simulation exercises. Out of these efforts came the experience and team coherence that were the backbone of Apollo's success. Out of these efforts came the ability to adjust to the spurious computer alarms during Apollo 11's descent, to the lightning bolt during the launch of Apollo 12, and to the ruptured oxygen tank on Apollo 13. From the first step of a man onto the Moon in Apollo 11 to the last departing step in Apollo 17, we showed that enormously difficult large endeavors can succeed, given the willingness, discipline, and competence of a dedicated crew of gifted people.

My vantage point in NASA gave me one perspective of the Apollo program's development. As you read this volume you will get other perspectives and insights from key participants in the program. I'm sure you will sense in their writing the exhilaration and pride they justifiably feel in their roles in the Apollo expeditions of the Moon. From my present vantage point as Administrator of the Energy Research and Development Administration I see more clearly that Apollo was as much a triumph of organization as of anything else. It was essential that we had the support of the President and the Congress, the participation of many accomplished scientists and engineers, and the continuing interest of the public at large. No single Government agency nor institution nor corporation can perform alone the tasks associated with reaching major national objectives. Apollo was an outstanding example of how governmental agencies, industrial firms, and universities can work together to reach seemingly impossible goals.

ROBERT C. SEAMANS, JR.
Administrator
Energy Research and
Development Administration

CHAPTER ONE

A Perspective on Apollo

By JAMES E. WEBB

After hundreds of thousands of years of occupancy, and several thousand years of recorded history, man quite suddenly left the planet Earth in 1969 to fly to its nearest neighbor, the Moon. The ten-year span it took to accomplish this task was but a blink of an eye on an evolutionary scale, but the impact of the event will permanently affect man's destiny.

In reflecting on the Apollo program, I am sometimes overwhelmed at the sheer magnitude of the task and the temerity of its undertaking. When Apollo was conceived, a lunar landing was considered so difficult that it could only be accomplished through exceptional large-scale efforts in science, in engineering, and in the development of operational and training systems for long-duration manned flights. These clearly required the application of large resources over a decade.

Industry, universities, and government elements had to be melded into a team of teams. Apollo involved competition for world leadership in the understanding and mastery of rocketry, of spacecraft development and use, and of new departures of international cooperation in science and technology. Like the Bretton Woods monetary agreement, President Truman's Point Four Program, and the Marshall Plan, the Apollo program was a further attempt toward world stability—but with a new thrust.

This chapter will review the origins of this policy and how it was successfully implemented. Subsequent chapters describe how particular problems were solved, how the astronauts and other teams of specialists were trained and performed, how the giant spaceboosters were built and flown, and how all this was joined together in a fully integrated effort. In many of these essays you will find indications of the meaning of the Apollo program to those who devoted much of their lives to it.

In the pre-space years the main defensive shield of the free world against Communist expansion was the preeminence of the United States in aeronautical technology and nuclear weaponry. These were an integral part of a system of mutual-defense treaties with other non-Communist nations.

In the 1950s, when the U.S.S.R. demonstrated rocket engines powerful enough to carry atomic weapons over intercontinental distances, it became clear to United States and free world political and military leaders that we had to add technological strength in rocketry and know-how in the use of space systems to our defense base if we were to play a decisive role in world affairs.

In the United States the first decision was to give this job to our military services. They did it well. Atlas, Titan, Minuteman, and Polaris missiles rapidly added rocket power to the basic air and atomic power that we were pledged to use to support long-held objectives of world stability, peace, and progress.

The establishment of the Atomic Energy Commission as a civilian agency had emphasized in the 1940s our hope that nuclear technology could become a major force for peaceful purposes as well as for defense. In 1958 the establishment of the National Aeronautics and Space Administration, again as a civilian agency, emphasized our hope that space could be developed for peaceful purposes.

NASA was specifically charged with the expansion into space of our high level of aeronautical know-how. It was made responsible for research and development that would both increase our space know-how for military use, if needed, and would enlarge our ability to use space in cooperation with other nations for "peaceful purposes for the benefit of all mankind."

A FERMENT OF DEBATE

The Apollo program grew out of a ferment of imaginative thought and public debate. Long-range goals and priorities within our governmental, quasi-governmental, and private institutions were agreed on. Leaders in political, scientific, engineering, and many other endeavors participated. Debate focused on such questions as which should come first—increasing scientific knowledge or using man-machine combinations to extend both our knowledge of science and lead to advances in engineering? Should we concentrate on purely scientific unmanned missions? Should such practical uses of space as weather observations and communication relay stations have priority? Was it more vital to concentrate on increasing our military strength, or to engage in spectacular prestige-building exploits?

In the turbulent 1960s, Apollo flights proved that man can leave his earthly home with its friendly and protective atmosphere to travel out toward the stars and explore other parts of the solar system. In the 1970s the significance of this new capability is still not clear. Will there be a basic shift of power here on Earth to the nation that first achieves dominance in space? Can we maintain our desired progress toward a prosperous peaceful world if we allow ourselves to be outclassed in this new technology?

Policymakers in Congress, the White House, the State and Defense Departments, the National Science Foundation, the Atomic Energy Commission, NASA, and other agencies agreed in the 1960s that we should develop national competence to operate large space systems repetitively and reliably. It was also agreed that this should be done in full public view in cooperation with all nations desiring to participate. However, this consensus was not unanimous. Critics thought that the Apollo program was too vast and costly, too great a drain on our scientific, engineering, and productive resources,

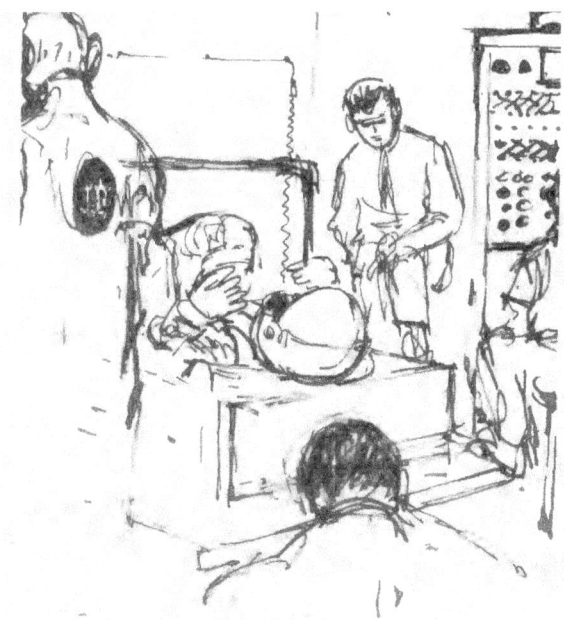

Robert McCall
MERCURY SUIT TEST
felt pen on paper

Paul Calle, SUITING UP, *pencil and wash on paper*

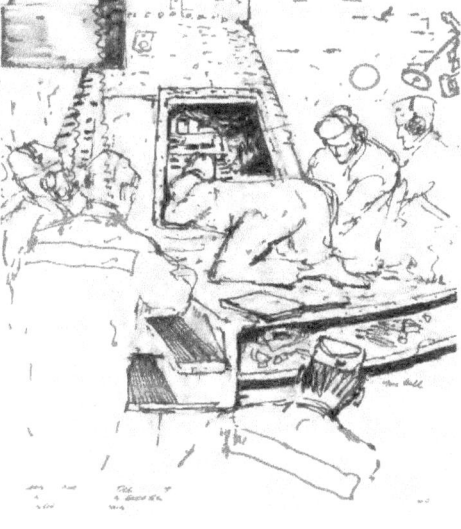

Robert McCall
GANTRY WHITE ROOM
felt pen on paper

Alfred McAdams, RANGE SAFETY, watercolor on paper. From here a straying rocket would be destroyed.

The Artist Looks at Space

Dozens of America's artists were invited by NASA Administrator James E. Webb to record the strange new world of space. Although an intensive use of photography had long characterized NASA's work, the Agency recognized the special ability of the artist's eye to select and interpret what might go unseen by the literal camera lens.

No civilian government agency had ever sponsored as comprehensive and unrestricted an art program before. A sampling of the many paintings and drawings that resulted is presented in this chapter.

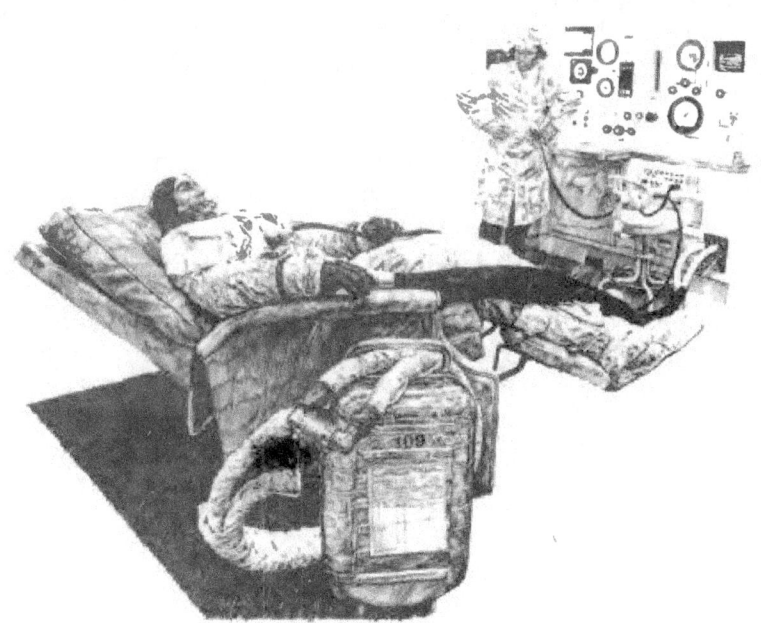

John W. McCoy II, FIRST LIGHT, *watercolor on paper*

Paul Calle
TESTING THE SPACESUIT
pen and ink on paper

too fraught with danger, and contended that automatic unmanned machines could accomplish everything necessary.

Specialized groups frequently overlooked the multiple objectives of developing a means of transporting astronauts to and from the Moon. Some manned spaceflight enthusiasts deplored NASA's simultaneous emphasis on flights to build a solid base of scientific knowledge of space. Some critics failed to recognize the value of having trained men make on-site observations, measurements, and judgments about lunar phenomena, and sending men to place scientific instruments where they could best answer specific questions.

A vast array of government agencies participated in the network of decision-making from which the basic policies that governed the Apollo program evolved. Collaboration between academic and industrial contributors required procedures that often seemed burdensome to scientists and engineers. Even some astronauts failed at times to appreciate the potential benefits of precise knowledge as to the effect of weightlessness and spaceflight stress on their bodies. Fortunately our Nation's most thoughtful leaders recognized the necessity as well as the complexity of the various components of NASA's work and strongly endorsed the Apollo program. It is a tribute to the innate good sense of our citizens that enough of a consensus was obtained to see the effort through to success.

THE GOAL OF APOLLO

The Apollo requirement was to take off from a point on the surface of the Earth that was traveling 1000 miles per hour as the Earth rotated, to go into orbit at 18,000 miles an hour, to speed up at the proper time to 25,000 miles an hour, to travel to a body in space 240,000 miles distant which was itself traveling 2000 miles per hour relative to the Earth, to go into orbit around this body, and to drop a specialized landing vehicle to its surface. There men were to make observations and measurements, collect specimens, leave instruments that would send back data on what was found, and then repeat much of the outward-bound process to get back home. One such expedition would not do the job. NASA had to develop a reliable system capable of doing this time after time.

At the time the decision was made, how to do most of this was not known. But there were people in NASA, in the Department of Defense, in American universities, and in American industry who had the basic scientific knowledge and technical know-how needed to predict realistically that *it could be done.*

Apollo was based on the accumulation of knowledge from years of work in military and civil aviation, on work done to meet our urgent military needs in rocketry, and on a basic pattern of cooperation between government, industry, and universities that had proven successful in NASA's parent organization, the National Advisory Committee for Aeronautics. The space agency built on and expanded the pattern that had yielded success in the past.

Systems engineering and systems management were developed to high efficiency. So was project management. New ways to achieve high reliability in complex machines were worked out. New ways to conduct nondestructive testing were developed. The best of large-scale management theory and doctrine was used to bring together both organizational (or administrative) optimization and join it to responsibility to work within the constraints of accepted organizational behavior.

LARGE ISSUES OF POLICY

In 1961, when President Kennedy asked me to join his administration as head of NASA, I demurred and advised him to appoint a scientist or engineer. The President strongly disagreed. At a time when rockets were becoming so powerful that they could open up "the new ocean of space," he saw this Nation's most important needs as involving many large issues of national and international policy. He pointed to my experience in working with President Truman in the Bureau of the Budget and with Secretary Acheson in the State Department as well as to my experience in aviation and education as his reasons for asking me to take the job. Vice President Johnson also held this view, and emphasized the value of my experience with high-technology companies in the business world.

I could not refuse this challenge, and I found that large issues of policy were indeed to occupy much of my energy. How could NASA, in the Executive Branch, do its work so as to facilitate responsible legislative actions in the Congress? How could public interest in space be made a constructive force? How could other nations' help be assured? In resolving policy and program questions, NASA was fortunate that Dr. Hugh Dryden, as Deputy Administrator, and Dr. Robert Seamans, as Associate Administrator, also had backgrounds of varied experience that could bring great wisdom to the decisions. We early formed a close relationship and stood together in all that was done.

Soon after my appointment, several significant events occurred in rapid succession. The first was a thorough review with Dr. Dryden and Dr. Seamans of what had been learned in both aeronautics and rocketry since NASA had been formed in 1958 to make projections of these advances into the future. We examined the adequacy of NASA's long-range plans and made estimates of the kind of scientific and engineering progress that would be required. We reviewed estimates of cost and found that sufficient priority and funds had not been provided.

The second event was the U.S.S.R.'s successful launch of the first man into Earth orbit, the Gagarin flight on April 12, 1961. A few weeks before this spectacular demonstration of the U.S.S.R.'s competence in rocketry, NASA had appealed to President Kennedy to reverse his earlier decision to postpone the manned spaceflight projects that were planned as a followup to the Mercury program. In his earlier decision, President Kennedy had approved funds for larger rocket engines but not for development of a new generation of man-rated boosters and manned spacecraft. The "talking paper" that I used to urge President Kennedy to support manned flight included the following:

"The U.S. civilian space effort is based on a ten-year plan. When prepared in 1960, this ten-year plan was designed to go hand-in-hand with our military programs. The U.S. procrastination for a number of years had been based in part on a very real skepticism as to the necessity for the large expenditures required, and the validity of the goals sought through the space effort.

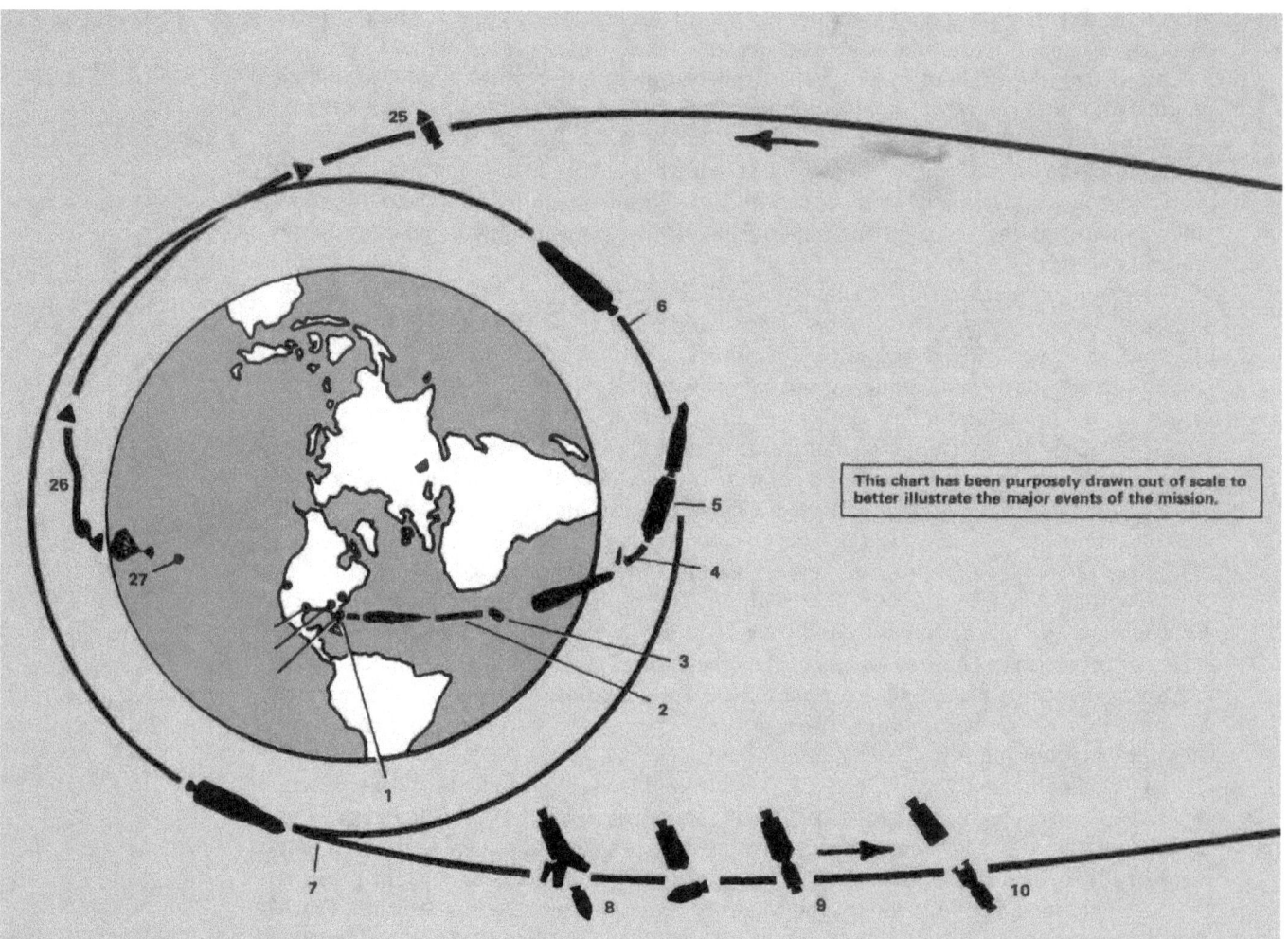

This chart has been purposely drawn out of scale to better illustrate the major events of the mission.

APOLLO MISSION PROFILE

1. Liftoff
2. S-IC powered flight
3. S-IC/S-II separation
4. Launch escape tower jettison
5. S-II/S-IVB separation
6. Earth parking orbit
7. Translunar injection
8. CSM separation from LM adapter
9. CSM docking with LM/S-IVB

10. CSM/LM separation from S-IVB
11. Midcourse correction
12. Lunar orbit insertion
13. Pilot transfer to LM
14. CSM/LM separation
15. LM descent
16. Touchdown
17. Explore surface, set up experiments
18. Liftoff

19. Rendezvous and docking
20. Transfer crew and equipment from LM to CSM
21. CSM/LM separation and LM jettison
22. Transearth injection preparation
23. Transearth injection
24. Midcourse correction
25. CM/SM separation
26. Communication blackout period
27. Splashdown

Robert McCall
ROLLOUT
oil on canvas
Reproduced courtesy of *Life* magazine

Nicholas Solovioff
INSIDE VAB
watercolor on paper

Tom O'Hara
CHECKING THE COMMAND MODULE
acrylic on paper

Billy Morrow Jackson, LUNAR MODULE WHITE ROOM, *watercolor on paper*

Paul Arlt
BIG DISH ANTENNAE, TANANARIVE
acrylic on canvas

John Pike
BOILERPLATE
watercolor on paper

"In the preparation of the 1962 budget, President Eisenhower reduced the $1.35 billion requested by the space agency to the extent of $240 million and specifically eliminated funds to proceed with manned spaceflight beyond Mercury. This decision emasculated the [NASA] ten-year plan before it was even one year old, and, unless reversed, guarantees that the Russians will, for the next five to ten years, beat us to every spectacular exploratory flight. . . .

"The first priority of this country's space effort should be to improve as rapidly as possible our capability for boosting large spacecraft into orbit, since this is our greatest deficiency. . . .

"The funds we have requested for an expanded effort will bring the entire space agency program up to $1.42 billion in FY 1962 and substantially restore the ten-year program. . . .

"The United States space program has already become a positive force in bringing together scientists and engineers of many countries in a wide variety of cooperative endeavors. Ten nations all have in one way or another taken action or expressed their will to become a part of this imaginative effort. We feel there is no better means to reinforce our old alliances and build new ones. . . .

"Looking to the future, it is possible through new technology to bring about whole new areas of international cooperation in meteorological and communication satellite systems. The new systems will be superior to present systems by a large margin and so clearly in the interest of the entire world that there is a possibility all will want to cooperate—even the U.S.S.R."

President Kennedy's March decision had been to proceed cautiously. He had added $126 million to NASA's budget, mostly for engines, but postponed the start on manned spacecraft. In March of 1961, he was not yet ready to move unambiguously toward a resolution of the great national and international policy issues about which he spoke when he asked me to join the administration.

KENNEDY'S DECISION

Gagarin's successful one-orbit flight in Vostok in April 1961 changed Presidential caution into concern and resulted in the Apollo decision.

A thoughtful scholar, Dr. John Logsdon, has described the situation in these words:

"The Soviet Union was quick to capitalize on the propaganda significance of the Gagarin flight. In his first telephone conversation with Gagarin, Nikita Khrushchev boasted, 'Let the capitalist countries catch up with our country.' The Communist Party claimed that in this achievement 'are embodied the genius of the Soviet people and the powerful force of socialism.' . . . Soviet propaganda stressed three themes: (1) The Gagarin flight was evidence of the virtues of 'victorious socialism'; (2) the flight was evidence of the global superiority of the Soviet Union in all aspects of science and technology; (3) the Soviet Union, despite the ability to translate this superiority into powerful military weapons, wants world peace and general disarmament."

"*New York Times* correspondent Harry Schwartz suggested that it appeared likely 'that the Soviet leaders hope their space feat can further alter the atmosphere of international relations so as to create more pressure on Western governments to make concessions on the great world issues of the present day.' "

Logsdon also wrote: ". . . the events of April produced a time of crisis, a time in which a sense of urgency motivated space planners and government policymakers to reexamine our national space goals and space programs. This reexamination resulted in a presidential decision to use the United States space program as an instrument of national strategy, rather than to view it primarily as a program of scientific research. This decision identified, for the world to see, a space achievement as a national goal symbolic of American determination to remain the leading power in the world." [1]

There were, of course, many other elements of national policy and commitment, but it is not easy to relate them to any one event such as the Gagarin flight. Continued Congressional understanding and support was the product of years of work by outstanding legislative leaders, and by devoted committee members and staffs. Cooperation and participation by Department of Defense elements and leaders were essential and are shown throughout this volume.

WORKING WITH INDUSTRIES AND UNIVERSITIES

There is another event, however, that relates to what was done and how NASA proceeded with Apollo. This event was a visit from a sophisticated senior official of a large corporation holding many aerospace contracts. He hit me right between the eyes with the question: "In the award of contracts are you going to follow 100 percent the reports of your technical experts, or are there going to be political influences in these awards?" My answer was just as direct: "In choosing contractors and supervising our industrial partners, we are going to take into account every factor that we should take into account as responsible government officials."

This meant that NASA officials would be required to meet President Kennedy's basic guideline—that we would not limit our decisions to technical factors but would work with American industry in the knowledge that we were together dealing with factors basic to "broad national and international policy." This also became our basic guideline for relations with universities and with scientists in the many disciplines that became so important a part of Apollo.

We constantly endeavored to set our course so that all who participated in Apollo could grow stronger for their own purposes at the same time that they were doing the work to succeed in NASA's projects. As they worked under NASA support, we were determined not to deplete their capability to achieve those goals that were important to them. In essence, our policy was to help them build strength so they could add to the Nation's strength.

Historians will find many lessons in the Apollo program for the managers of future large-scale enterprises. It was a new kind of national venture. Suddenly and dramatically it brought men of action and men of thought into intimate working relationships designed to solve a large number of extremely difficult scientific and technical problems. It was a major challenge to legislators, scientists, and engineers.

[1] Logsdon, John M., "Decision to Go to the Moon," The MIT Press, 1970, pp. 10 and 100.

Robert McCall, APOLLO 8 COMING HOME, *oil on panel*

Paul Calle
INSIDE GEMINI SPACECRAFT
pencil on paper

Paul Calle
GEMINI SPACECRAFT AND GANTRY
pencil on paper

Fred Freeman, SATURN BLOCKHOUSE, *acrylic on canvas*

M. McCaffrey
CHRIS KRAFT AND DR. GILRUTH
mixed media on paper

Robert McCall
GEMINI RECOVERY
watercolor on paper

Leonard Dermott, APOLLO 9 COMING ABOARD, acrylic on canvas

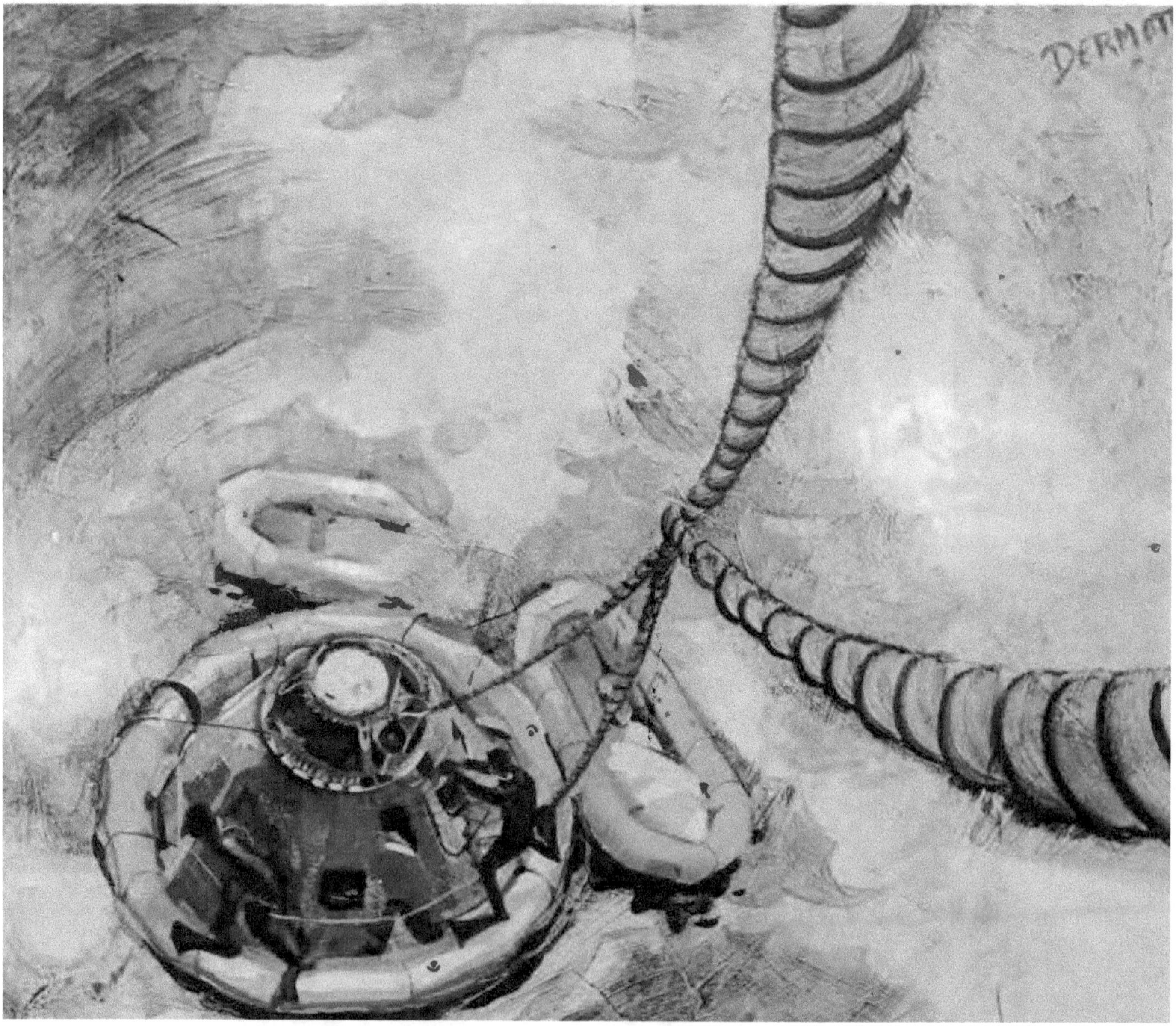

After a careful study of the way we conducted our work, Dr. Leonard R. Sayles and Dr. Margaret K. Chandler of the Graduate School of Business at Columbia University wrote:

"NASA's most significant contribution is in the area of advanced systems design: getting an organizationally complex structure, involving a great variety of people doing a great variety of things in many separate locations, to do what you want, when you want it—and while the decision regarding the best route to your objective is still in the process of being made by you and your collaborators."

Although our goal was clear and unambiguous, to reach it we had to use rapidly developing technology that in turn was based on rapidly increasing scientific knowledge. This required our organization to be highly flexible, and it was altered when unexpected developments made this necessary. As Mercury phased into Gemini, and Apollo reached its peak effort, NASA's work force grew to 390,000 men and women in industry, 33,000 in NASA Centers, and 10,000 in universities. By 1969, the year of the first lunar landing, this total had been reduced by 190,000. By 1974, it was down to 126,000. This is certainly an administrative record.

No more flights to the Moon are scheduled now, and future ones will undoubtedly be made differently, but the Apollo program has not really ended. Instruments placed on the Moon by the American astronauts are still transmitting important data to scientists throughout the world. We know much about how the Moon is bound to the Earth by invisible gravitational fields, and how both are similarly bound to the star we call the Sun. Daily, men and women are learning more about that star, and about the whole universe.

BENEFITS FROM SPACE TECHNOLOGY

No one can yet fully appraise the ultimate benefits from this historic achievement. Many of the technical innovations necessary for men to go to the Moon and back have already been embodied in everyday processes and products. These range from versatile electronic computers to fireproof bedding in hospitals and special equipment for the handicapped. More applications of information acquired from space research are continually being reported.

Possibly more significantly, the idea that if we can go to the Moon we can accomplish other feats long considered impossible has been firmly implanted in people's minds. Confidence that solutions can be found to such urgent problems as an energy shortage, environmental degradation, and strife between nations, has been nourished by this spectacular demonstration in space of man's capabilities.

Apollo was a multidimensional success, triumphant not only as a feat of scientific and engineering precision, but also as a demonstration of our country's spirit and competence. In the worldwide reaction at the time of the lunar landing, it was clear that this great adventure transcended nationality and became a milestone for mankind. Every participant in the Apollo program saw a slightly different facet of it. While reading these personal accounts and studying these pictures, you will possibly perceive more clearly the motives, the hazards faced, and the triumphs that first enabled men to set foot on the Moon.

Speaking to Congress and the Nation, President Kennedy said on May 25, 1961: "I believe that this nation should commit itself to achieving the goal, before this decade is out, of landing a man on the Moon and returning him safely to Earth. No single space project in this period will be more impressive to mankind, or more important in the long-range exploration of space; and none will be so difficult or expensive to accomplish."

CHAPTER TWO

"I Believe We Should Go to the Moon"

By ROBERT R. GILRUTH

President Kennedy's statement, "Fly man to the Moon in this decade," was a beautiful definition of the task. There could be no misunderstanding as to just what was desired, and this clarity of purpose was one factor in the success of Apollo. By definition, it settled the old question of man versus instrument, and it provided a goal so difficult that new concepts, as well as new designs, would be required to accomplish it. Since the landing was geared to the decade of the sixties, the pace of the program was also defined, and a clear test of strength with the Soviets was implicit, if they chose to compete.

The President's decision came after a long series of Russian firsts. They were the first to orbit the Earth, and the first to send instruments to the Moon; only a month earlier, Yuri Gagarin had become the first man in space. Alan Shepard had followed Gagarin into space by only a few weeks to become the first American in space. This feat had given America the feeling that, with a major effort, we might close the gap with the Soviets. The public was ready and willing to expand the space effort, and Congress did not bat an eye at Webb's estimate that $20 to $40 billion could be required to go to the Moon. It was a popular decision, and the vote of the Congress on the Moon program was virtually unanimous.

But could it really be done? Flying man to the Moon required an enormous advance in the science of flight in a very short time. Even the concepts of manned space-flight were only three years old, and voyaging in space over such vast distances was still just a dream. Rendezvous, docking, prolonged weightlessness, radiation, and the meteoroid hazard all involved problems of unknown dimensions. We would need giant new rockets burning high-energy hydrogen; a breakthrough in reliability; new methods of staging and handling; and the ability to launch on time, since going to the Moon required the accurate hitting of launch windows.

Man himself was a great unknown. At the time of the President's decision we had only Alan Shepard's brief 15 minutes of flight on which to base our knowledge. Could man really function on a two-week mission that would involve precise maneuvers, including retrofire into lunar orbit, backing down and landing on the Moon, lunar take-off, midcourse corrections on the way home and, finally, a high-speed reentry into Earth's atmosphere, performed with a precision so far unknown in vehicle guidance? We would have to do intensive work on spacesuit development, since flying to the Moon would be unthinkable without giving astronauts the capability of exploring the Moon on foot, and perhaps later in some roving vehicle. These men would have to be trained in the complex problems of flying and navigating in space, and we would need a cadre of men with space experience before setting out on such a voyage.

On the spacecraft side, we felt that the concepts already under trial in Project Mercury would be applicable for the command and service module, but we would need new sources of onboard power such as the fuel cell. The landing on the Moon would require precision guidance, as well as good visibility from the cabin for the astronauts, and a graduated control of rocket thrust heretofore undeveloped. The state of the art in tracking and communication would be severely tested. Precision navigation techniques using inertial systems would be necessary, and high-speed computing for solving complex navigation problems involving the celestial mechanics of three-body systems. The mass ratios involved in the spacecraft, particularly the lunar lander, would require the ultimate in materials. Safety factors would have to be stretched in order to give the low structural weights required. And in the spacecraft, as well as the launch vehicle, new orders of reliability would be required for so long a mission so far from home.

The Moon itself was a great unknown. Its surface, its mass distribution, and whether the lunar soil would be firm enough to support a landing craft were all open to conjecture. Finally, a master plan had to be evolved. The launch site had to be selected, roles and missions of government centers had to be determined, and a choice had to be made between the various concepts of how to go to the Moon, whether to use Earth-orbit rendezvous, direct ascent, or the most controversial of all, lunar-orbit rendezvous. A government team had to be built that, working closely together with leaders of industry, would manage the development and production of the launch vehicles, the spacecraft, and facilities for tests and operations. Not only would all these things have to be done in the short time available, but many would have to be worked out at the beginning, during what I have called "the year of decisions."

WHAT WE HAD TO BUILD ON

Many of our key people in Apollo, particularly on the spacecraft side, grew out of the old NACA, the National Advisory Committee for Aeronautics. We had the heritage of the airplane to work from, with all its methods of design, test, and operation. For the ballistic-missile programs of the early 1950s we had helped to develop solutions for reentry problems that were to have direct application to spacecraft design. But it was the launching of Sputnik I in October 1957 that put a new sense of value and urgency on the things we were doing. When one month later, the dog Laika was placed in orbit in Sputnik II, I was sure that the Russians were planning for man in space.

It seemed to me that the United States would surely enter into space competition with the Soviet Union, and flying man into space would be a legitimate national goal.

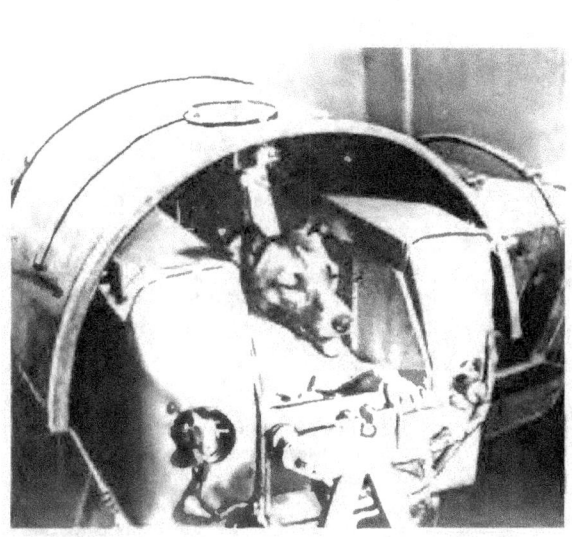

The Russians were ahead. A month after the 184-pound Sputnik I, they sent up the dog Laika (above) in the 1120-pound Sputnik II. Three years later, on April 12, 1961, Yuri Gagarin (right and below) became the first man to orbit the Earth. His Vostok I's payload was 10,417 pounds and his flight lasted 108 minutes. At this time, the United States was in the final stage of preparing for its first manned suborbital flight, on May 5, 1961. The physical hazards of weightlessness were then almost wholly unknown.

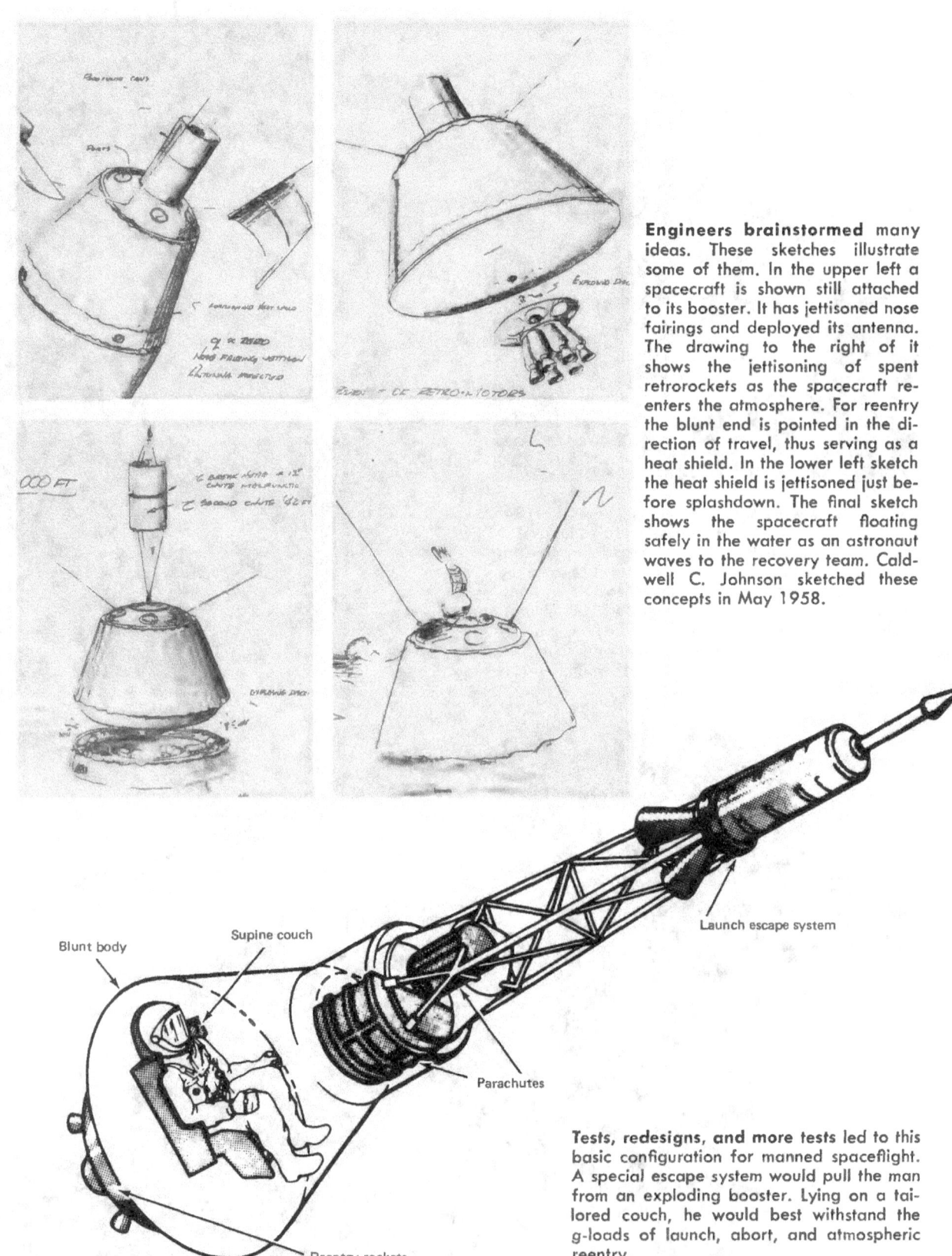

Engineers brainstormed many ideas. These sketches illustrate some of them. In the upper left a spacecraft is shown still attached to its booster. It has jettisoned nose fairings and deployed its antenna. The drawing to the right of it shows the jettisoning of spent retrorockets as the spacecraft reenters the atmosphere. For reentry the blunt end is pointed in the direction of travel, thus serving as a heat shield. In the lower left sketch the heat shield is jettisoned just before splashdown. The final sketch shows the spacecraft floating safely in the water as an astronaut waves to the recovery team. Caldwell C. Johnson sketched these concepts in May 1958.

Blunt body

Supine couch

Launch escape system

Parachutes

Reentry rockets

Tests, redesigns, and more tests led to this basic configuration for manned spaceflight. A special escape system would pull the man from an exploding booster. Lying on a tailored couch, he would best withstand the g-loads of launch, abort, and atmospheric reentry.

Thus it was that our small but creative group in NACA started working intensively on spacecraft-design problems. Most of the effort in those early days had been directed toward hypersonic gliders, or winged vehicles, that would fly at high Mach numbers and perhaps even into orbit. But our views were changing, and Harvey Allen of the Ames Laboratory was the first, to my recollection, to propose a blunt body for flying man in space. He suggested a sphere to enclose the man and said, "you just throw it," meaning, of course, launch it into space with a rocket. In March 1958, Max Faget presented a paper that was to be a milestone in spacecraft design. His paper proposed a simple blunt-body vehicle that would reenter the atmosphere without reaching heating rates or accelerations that would be dangerous to man. He showed that small retrorockets were adequate to initiate reentry from orbit. He suggested the use of parachutes for final descent, and small attitude jets for controlling the capsule in orbit during retrofire and reentry. His paper concluded with a statement that: "As far as reentry and recovery are concerned, the state of the art is sufficiently advanced to proceed confidently with a manned satellite project based upon the use of a blunt-body vehicle."

Starting with the formation of NASA in October 1958, intense efforts were undertaken to create a manned space vehicle and flight organization capable of flying man in orbit around the Earth. The plans for this vehicle were based on the blunt reentry body proposed earlier by Harvey Allen and Max Faget. A special team, the Space Task Group, was formed at Langley Field, Virginia, to manage this effort, and the McDonnell Aircraft Corporation won the competition to build what would be the Mercury spacecraft.

The heat shield was a slightly convex surface constructed of plastic and fiberglass material that would give out gas under intense heat, protecting itself from destruction. The conical afterbody was covered by shingles of high-temperature alloy similar to that used in turbine blades of jet engines. These shingles were insulated from the titanium pressure shell and they dissipated their heat by radiation. Parachutes were by far the lightest and most reliable means of making the final descent to Earth, and the parachute section was protected from heat by shingles of beryllium. Another key factor in the Mercury design was the supine couch for the astronaut. There had been considerable doubt that man could withstand the g-loads associated with rocket launching and reentry, particularly in abort situations. The form-fitting couch gave such well-distributed support that man could withstand over $20g$ without injury or permanent damage.

This concept of the Mercury capsule and, indeed, the whole plan for putting man into space was remarkable in its elegant simplicity. Yet its very daring and unconventional approach made it the subject of considerable controversy. Some people felt that such a means for flying man in space was only a stunt. The blunt body in particular was under fire since it was such a radical departure from the airplane. It was called by its opponents "the man in the can," and the pilot was termed only a medical specimen. Even Dr. Dryden, at the time the Director of NACA, labeled one early ballistic-capsule proposal the same as shooting a young lady from a cannon. However, he approved the Project Mercury design, since it was by then a complete system for orbital flight. The Mercury spacecraft and, in fact, Gemini and Apollo as well, were designed to land on the water because of the large water area which lay east of Cape Canaveral over the South Atlantic. If an abort were required during launch, the spacecraft would have to survive a water landing; and this therefore became the best way to make all landings. It was easier to attenuate landing-impact forces in water landings, although the spacecraft was designed to survive a land impact without harm to flight crews.

The first astronauts were brought onboard the Mercury program in April 1959. They were volunteer military pilots, graduates of test pilot schools. Each was required to have a bachelor's degree in engineering or equivalent, and at least 1500 hours of jet time. Of the first group of 60 candidates called to Washington to hear about the program, more than 80 percent volunteered. All were of such high caliber that selection was difficult. However, I picked seven: three Air Force, three Navy, and one Marine, on the basis that the Mercury program would probably not give more than this number a chance to fly. These men were true pioneers. They volunteered at a time when our plans were only on paper and when no one really knew what the chance of success was. One had to respect their motivation and courage.

We were to have many spectacular successes as well as failures in the Mercury program. However, we were able to learn from each failure, and fortunate in having these failures early in the program so that the astronauts and the animal passengers as well were flown without mishap when their time came. Perhaps our most spectacular failure in Mercury came to be known as the "tower flight." In this sad affair, the escape tower, the parachutes, and the peroxide fuel were all deployed on the launching pad in front of the domestic and international press. A relatively simple ground-circuit defect in the Redstone launch vehicle caused the main rocket engine to ignite and then shut down after a liftoff from the launching pad of about two inches. The capsule events were keyed to the engine shutdown after having been armed by stage liftoff, as this was the normal procedure for sequencing unmanned flights. As you might expect, it was very difficult to explain this spectacular series of events to the working press, and to officials in Washington.

In those days an animal, in our case a chimpanzee, had to precede man into space. The flight of the chimpanzee Ham was a major milestone in our program. Here again we had some problems in the Redstone launch vehicle that resulted in a delayed pickup of the spacecraft, and water entered the spacecraft as a result of landing damage to the pressure shell. However, the animal performed admirably at zero gravity and was picked up unharmed. Ham became quite famous and proved to be a really lovable little fellow as well as a true pioneer.

All the things that were wrong with Ham's flight were corrected by hard work on the ground without further flight tests. We were now ready for our first manned suborbital flight and I recommended to Dr. Dryden and Mr. Webb that we were ready to go ahead. However, the Marshall Center required one more unmanned flight with the Redstone for booster development. It was during this period that the Russians sent Yuri Gagarin aloft in the Vostok spacecraft to become the first man in space.

All of these events were occurring at the time that President Kennedy and his staff were taking over from the outgoing Eisenhower administration. Dr. Glennan, the Administrator of NASA during its first years, gave way to James Webb, who was to be the

Administrator until October 1968. With the change came other events. Project Mercury was examined by the new head of the President's Science Advisory Committee, Dr. Jerome Wiesner, and a staff of medical and physical scientists. Our hearings before the Wiesner Committee went reasonably well until we came to convincing the doctors that it was safe for man to fly at zero gravity. Even though Ham, the chimpanzee, had fared well and was completely normal after his flight, the medical men on the committee were reluctant to accept this evidence that man could stand even 15 minutes of zero gravity. They were even concerned whether man could stand the mental stress of lying on top of a rocket and being blasted into space. However, we were able to convince Mr. Webb and Dr. Dryden that the program was sound, and they, in turn, convinced the President and his staff. It was at this time that Ed Welsh, Executive Secretary of the Space Council, remarked while Mr. Kennedy was pondering the impact of a failure, "Mr. President, can the country stand a success?"

On May 5, 1961, Alan Shepard became the first American to blast off from Cape Canaveral in a flight that was to be of great importance to our future programs. His flight in *Freedom 7* was followed by Gus Grissom in a Mercury capsule called *Liberty Bell 7*. Orbital flights of the Mercury capsule followed with a mechanical man and a chimpanzee named Enos. We were ready for manned flight into orbit. We were extremely fortunate to have six successful Atlas launch vehicles in a row to complete the Mercury program. John Glenn's Mercury-Atlas mission on February 20, 1962, was America's first orbital flight. We were to learn much from the flights of Glenn, Carpenter, Schirra, and Cooper that helped us in planning for the lunar program.

The exposure of man to zero gravity in these early manned flights was perhaps among the greatest medical experiments of all time. All the Mercury astronauts found the weightless state no particular problem. All returned to Earth with no medical difficulties whatever. This finding was so fundamental and straightforward that its importance was missed by many medical critics at the time. It now became simply a question of how long man could withstand weightlessness, and detailed medical measurements were made to cast light on how the body compensated for the new environment. Zero gravity produced some problems in locomotion and habitability, but not in man himself. We believed that even the longest flights of the future would probably require only methods of keeping the human body properly exercised and nourished in order to prevent a different reaction on returning to the gravity of Earth.

THE YEAR OF DECISIONS (JUNE 1961 TO JUNE 1962)

The twelve months following the decision to go to the Moon saw the complete plan unfold. New Centers were created, roles and missions were assigned, and the basic designs for the launch vehicle and spacecraft were agreed upon. In addition, a vital new program, called Gemini, was instituted to explore rendezvous, docking, and the many other factors that were vital before Apollo could set out.

At the onset of the program, there were two government groups eager to participate. There was the Marshall Space Flight Center under Dr. Wernher von Braun, which was a mature Center having more than 5000 people, and there was the small but expert Space Task Group of only a few hundred people, a group already severely

The first seven American astronauts, chosen in April 1959, were (from left, seated) Virgil I. Grissom, M. Scott Carpenter, Donald K. Slayton, and L. Gordon Cooper, Jr.; (standing) Alan B. Shepard, Jr., Walter M. Schirra, and John H. Glenn, Jr. They were test pilots who volunteered to fly a spacecraft—similar to the model shown—that had not yet been built.

Animals flew first, paving the way for man. Chimpanzees were physiologically manlike and easily trained. The Air Force's Aeromedical Field Laboratory provided them.

Tiny one-man spacecraft—then called capsules—orbited the Earth in the Mercury program. They were checked out in a hangar at Cape Canaveral before being hoisted up and mated with a launch vehicle. Heat-resistant shingles covered the afterbody.

Boiler plate models (below) of escape tower and spacecraft were tested with Little Joe boosters at Wallops Station. Little Joe could briefly deliver up to 250,000 pounds of thrust from eight solid-fuel rockets. These launches were in 1959 and 1960.

Maximum public embarrassment in the Mercury program occurred in the inadvertent "tower flight." A circuit quirk cut off the Redstone's engines inches after liftoff. It was just enough to trigger the escape sequence. The enginers could explain it, but the public inevitably wondered if we were really ready to send a man safely into space.

Custom-made couches were provided for the chimpanzees before their flights. Sensors attached to the animals and instruments in the spacecraft greatly reduced uncertainty about the effects of severe g-loads on the body of an astronaut. This medical research resulted in the design of a couch for the Mercury capsule that would distribute the occupant's weight and minimize acceleration loads on his body.

A chimpanzee named Ham, shown hamming it up in the photo below, was "a delightful little fellow." The Mercury-Redstone 2 that carried Ham into space January 31, 1961, over-accelerated and ascended to a higher altitude than planned. Nonetheless, the capsule was recovered with Ham in good shape. He was then retired to a normal lifetime. Mercury researchers, remembering Ham fondly, occasionally visited him afterward.

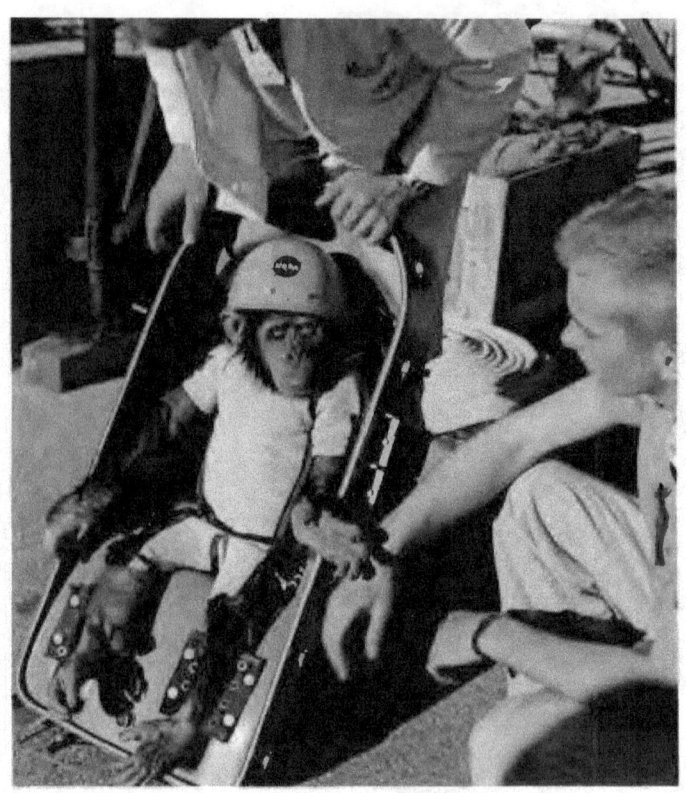

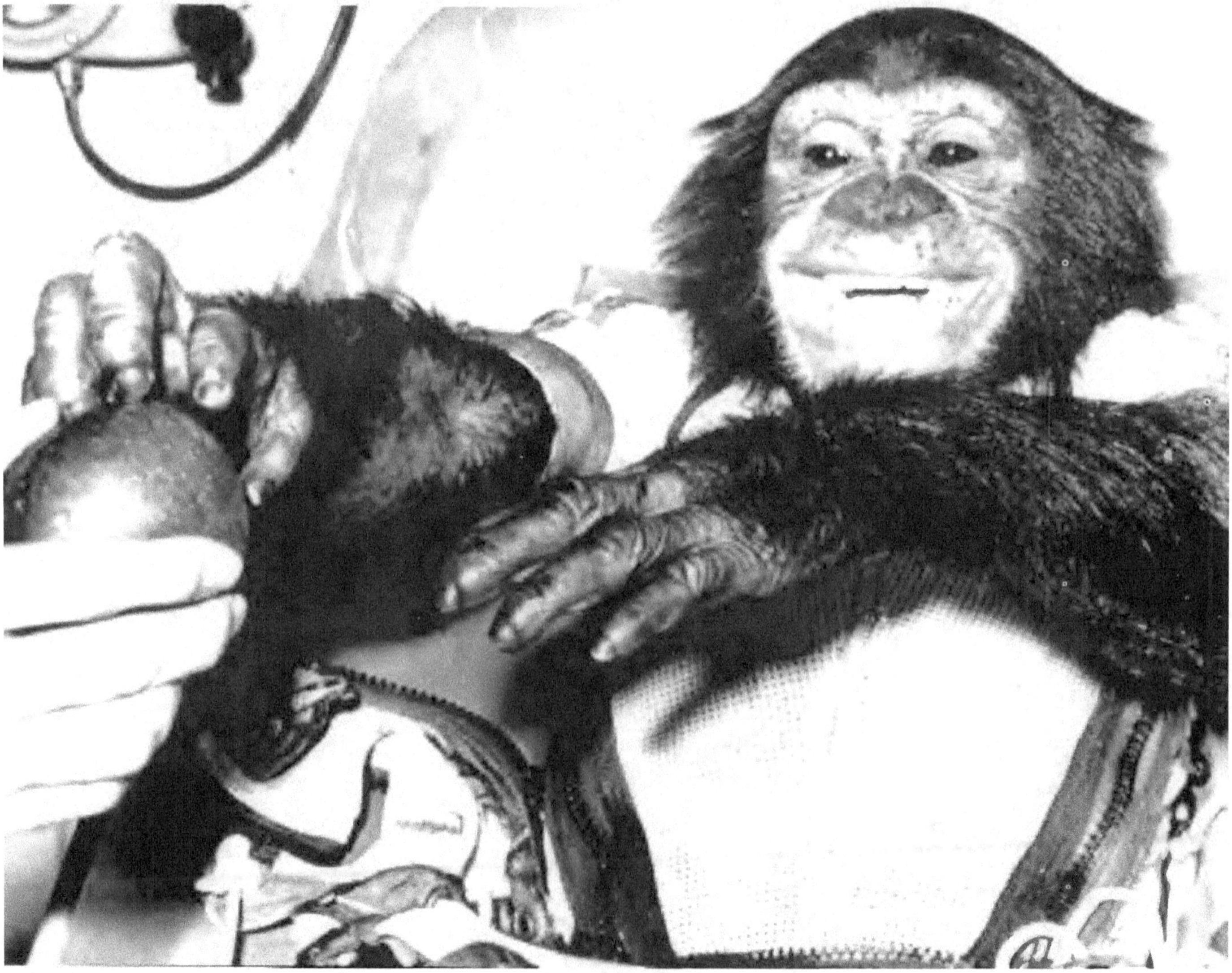

loaded with the Mercury project. Mr. Webb, the Administrator, Dr. Dryden, the Deputy Administrator, and Dr. Seamans, the Associate Administrator, were the top management of the agency, and they brought onboard a young man named Brainerd Holmes to head the manned spaceflight effort in Washington. This group lost little time in deciding roles and missions for the lunar program. A new launch Center was created in Florida, and Dr. Kurt Debus, formerly of Marshall, was named its head. The Marshall Center got the job of developing the huge Moon rockets, although the size of the rocket and the mode for going to the Moon had not yet been decided. The decision was made to expand the Space Task Group into a full NASA Center and assign it the job of developing the spacecraft and of creating a complex of technical facilities for spacecraft research and development, astronaut training, and flight operations. I became head of this new NASA Center in Houston. The Center facilities were authorized by Congress, but they did not yet exist. And so during the first year of Apollo the Space Task Group became the Manned Spacecraft Center and moved from Virginia to Texas. We occupied some 50-odd rented buildings while the new Center was being designed and built. It was a period of growth, organization, and growing pains. We were establishing new contractor relations, moving families and acquiring new homes, as well as conducting the orbital flights of Project Mercury.

Many of the key ideas and designs for going to the Moon were created during this period of upheaval, turmoil, and the stress of major flight activities. Even before the President's decision to land on the Moon, we had been working on designs and guidelines for a manned circumlunar mission. This was done in a series of bull sessions on how we would design the spaceship for this purpose if the opportunity occurred. Our key people would get together evenings, weekends, or whenever we could to discuss such questions as crew size and other fundamental design factors. We believed that we would need three men on the trip to do all the work required, even before the complexity of the landing was added. We believed that man would be able to stand a zero-gravity environment for the time required to go to the Moon and return. We had decided that an oxygen atmosphere of 5 pounds per square inch was the best engineering compromise for a system that would permit extravehicular activity without another module for an airlock. Other basic decisions included the selection of an onboard navigation system as well as the ground-based system, and controlled reentry to reduce g-loads and give pinpoint landings. These original guidelines for lunar flight were presented to all NASA Centers and to the aerospace industry.

The conceptual design of the moonship was done in two phases. The command and service module evolved first as part of our circumlunar studies, and the lunar lander was added later after the mode decision was made. We were extremely fortunate that the design that evolved had such intrinsic merit. We had designed our circumlunar spaceship to have a command module containing the flight crew located on top of the stack, so that the astronauts could escape by means of an escape tower if abort were necessary during launch. The service module containing fuel for space propulsion, electric power equipment, and other stores, was underneath it with its big rocket and its maneuvering systems. In the adapter below the service module, a third element of the spacecraft was located: a mission module to which the crew would transfer for special experiments. Thus when the full landing mission came along, we were able to substitute the lunar lander for this mission module. The turnaround, docking, and tunnel transfers between the command module and the lunar module were then the same ones that we had planned between the command module and the mission module.

The shape of the command module was a refinement of the Mercury capsule, optimized for the higher heating rate and the angles of attack required for controlled reentry. (One must remember that at this early time, reentry was still considered a serious problem.) Reentry from the Moon would generate heating rates twice as great as those in vehicles returning at orbital speed. Experts had warned us that shock-wave radiation would be an additional source of heating. Our studies showed that the blunt body was still the optimum shape, although the afterbody shape should be more highly tapered than in the Mercury capsule. As it turned out, our flights to the Moon showed that the Apollo design was very conservative, particularly on the afterbody, and the margins of safety for the astronauts in returning from the Moon were comfortably large. Max Faget, Caldwell Johnson, and others of the Manned Spacecraft Center were largely responsible for putting down the original lines of the Apollo command module. They also suggested the internal arrangement.

All during the early planning for the lunar missions, I had been greatly concerned about the effects of solar radiation on the astronauts. Experts were not all in agreement as to the amount of radiation that might be received on a mission to the Moon. I remember George Low stating that the normal shielding of the cabin walls, together with the low probability of intense solar activity, would alleviate this hazard. He was right and the radiation experienced by astronauts on trips to the Moon was of no medical significance. Navigation in space might have been a serious problem had not Stark Draper and his group at MIT gotten an early start. They were brought in under contract to devise a system for Apollo back in 1961. Working with their industrial partners, they produced a system that was amazingly accurate.

The pieces of the master plan were now beginning to fit together. In the fall of 1961, North American Aviation had won the contract for the Apollo command and service module. The basic designs of the service propulsion engine, the reaction control system, and the fuel cells were underway, but there were still major technical areas to be settled. One of these was the launch-vehicle design. As a result of many studies, the large rocket originally proposed had lost its backers. Dr. von Braun and the Huntsville team were zeroing in on a rocket of intermediate size. This rocket was to use five of the huge F-1 engines on the first stage and a new hydrogen-oxygen engine in the upper stages. It could easily be sized to send more than 90,000 pounds on a course to the Moon. We in Houston strongly supported this design, which was later called the Saturn V. Only one rocket vehicle of this size would be required to send our spacecraft to the Moon, if we used the lunar-orbit rendezvous technique. Getting official approval for the lunar-orbit rendezvous was, however, to take considerable time and effort. Brainerd Holmes, chosen by Webb to head Apollo in Washington, strongly favored Earth-orbit rendezvous. This mode would use dual launchings of the huge Saturn V rockets, joining them together in orbit and pumping fuel from one to refill the other; and then realigning and lighting off that rocket to the Moon. In this way, much larger payloads

The first suborbital spaceflight of Project Mercury ended happily when a U.S. Marine helicopter recovery team plucked Astronaut Alan B. Shepard, Jr., out of the sea to fly him to the carrier *Champlain* on May 5, 1961.

Shepard reached 116.5-mile altitude on a 15-minute 22-second flight down the Atlantic Missile Range, and found being weightless for 5 minutes pleasant.

Redstone rocket had launched *Freedom 7*. It reached top speed of 5180 miles an hour on 302-mile trip. Astronaut and capsule both landed in fine shape. Three weeks later President Kennedy proposed that U.S. astronauts go to the Moon in next decade.

John Glenn rounded the Earth three times in *Friendship 7,* enjoyed his February 20, 1962, encounter with zero g, and wished his capsule were glass so that he could see more. Operations men on ground feared the heat shield was not locked in place. Glenn, too, had nervous moments and splashed his spacecraft into Atlantic 40 miles short of the projected area. Difficulties in controlling his vehicle intensified engineers' drive for perfect performances.

A destroyer picked up Glenn's capsule in 17 minutes. He skinned his knuckles blowing the hatch, and said that "it was hot in there." He had lost weight but doctors' exhaustive tests showed no adverse effects from his 4-hour 22-minute flight. City after city feted Glenn, and his capsule was put in the Smithsonian Institution near the Wright brothers' airplane.

could be flown to the Moon than by a single rocket, but the technical and operational problems seemed to me to be overwhelming.

In contrast, I believed in and supported lunar-orbit rendezvous. In this mode, the lander leaves the mother ship in lunar orbit and goes down to the Moon's surface. Upon returning to lunar orbit, it links up with the mother ship and the astronauts transfer to the command module and return to Earth for reentry and landing. Lunar-orbit rendezvous was espoused by John Houbolt, chairman of the group that studied this plan at the Langley Research Center. When I heard of this plan, I was convinced that this was the way to go. It required far less weight injected toward the Moon, but even more important, in my view, was the fact that one spacecraft could be designed specifically for lunar landing and takeoff, while the other could be designed for flying to and from the Moon and specifically for reentry and Earth landing. An additional bonus was that it allowed the tremendous industrial job to be divided between two major contractors since there would be two spacecraft, thereby giving each one a more manageable task.

By the late fall of 1961, all of us at the new Manned Spacecraft Center were unified in support of lunar-orbit rendezvous and were working tooth and nail to find out all we could about lunar landers, rendezvousing, and the tradeoffs to be made. In December of 1961 we made an earnest appeal to Brainerd Holmes to approve lunar-orbit rendezvous. He could not be convinced at that time, however, and only six months later was the final decision made. Much of the credit for selling the lunar-orbit mode must be given to the Houston people. Charles Frick, who was our Apollo Spacecraft Manager at that time, was particularly effective. Studies conducted by Frick's people converted first the key engineers at the Marshall Space Flight Center, including Drs. Rees and von Braun, and, finally, Brainerd Holmes. Dr. Joe Shea, Holmes' assistant, then carried the decision on to higher echelons of the Government. Mr. Webb approved the lunar-orbit plan and only Dr. Wiesner and a few others of the President's Science Advisors remained unconvinced. However, the White House accepted Mr. Webb's decision.

We were extremely fortunate during this period to have Brainerd Holmes in charge of the Apollo program. He encouraged the key Center leaders to work together by establishing a management council with regular meetings. During these meetings, we argued out our different opinions and developed into a management team. A less skillful leader might have forced an early arbitrary decision that would have made the whole task of getting to the Moon virtually impossible.

Our Administrator, Mr. Webb, now had a master plan. It consisted of the giant three-stage launch vehicle, the Saturn V. There would be a command module with three astronauts onboard. The command module would be a blunt body, properly shaped and ballasted for controlled gliding reentry. It would use ablative material for the heat shield and would land at sea with parachutes. A separate service module would carry the space-propulsion engine, attitude-control jets, the fuel cells for electric power, together with supplies of fuel and oxygen. There would be a lunar-landing stage designed specifically for the job of landing on the Moon. It would carry two men down to the Moon's surface and back to rendezvous with the mother ship in orbit. In simple terms, this was the technical plan for Apollo, and it was to need no change as it went forward in development. All of this had been decided within one year after the President's announcement. Less than six months later, Grumman had won the contract to build the lunar lander.

One major element of basic program planning was still missing. How were we to bridge the tremendous gap between the simple Mercury Earth-orbital program and the Apollo voyage to the Moon? We needed a chance to train our men in many new elements of spaceflight, and we needed an engineering prototype for our ideas as well. The answer was Project Gemini.

GEMINI PROGRAM

The Gemini program was designed to investigate in actual flight many of the critical situations which we would face later in the voyage of Apollo. The spacecraft carried an onboard propulsion system for maneuvering in Earth orbit. A guidance and navigation system and a rendezvous radar were provided to permit astronauts to try out various techniques of rendezvous and docking with an Agena target vehicle. After docking, the astronauts could light off the Agena rocket for large changes in orbit, simulating the entry-into-lunar-orbit and the return-to-Earth burns of Apollo. Gemini was the first to use the controlled reentry system that was required for Apollo in returning from the Moon. It had hatches that could be opened and closed in space to permit extravehicular activity by astronauts, and fuel cells similar in purpose to those of Apollo to permit flights of long duration. The spacecraft was small by Apollo standards, carrying only two men in close quarters. However, the Titan II launch vehicle, which was the best available at that time, could not manage a larger payload.

A total of 10 manned flights were made in the Gemini program between March 1965 and November 1966. They gave us nearly 2000 man-hours in space and developed the rendezvous and docking techniques essential to Apollo. By burning the Agena rockets after docking, we were able to go to altitudes of more than 800 nautical miles and prove the feasibility of the precise space maneuvers essential to Apollo. Our first experience in EVA was obtained with Gemini and difficulties here early in the program paved the way for the smoothly working EVA systems used later on the Moon. The Borman and Lovell flight, Gemini VII, showed us that durations up to two weeks were possible without serious medical problems, and the later flights showed the importance of neutral buoyancy training in preparation of zero-gravity operations outside the spacecraft.

Gemini gave us the confidence we needed in complex space operations, and it was during this period that Chris Kraft and his team really made spaceflight operational. They devised superb techniques for flight management, and Mission Control developed to where it was really ready for the complex Apollo missions. Chris Kraft, Deke Slayton, head of the astronauts, and Dr. Berry, our head of Medical Operations, learned to work together as a team. Finally, the success of these operations and the high spaceflight activity kept public interest at a peak, giving our national leaders the broad supporting interest and general approval that made it possible to press ahead with a program of the scale of Apollo.

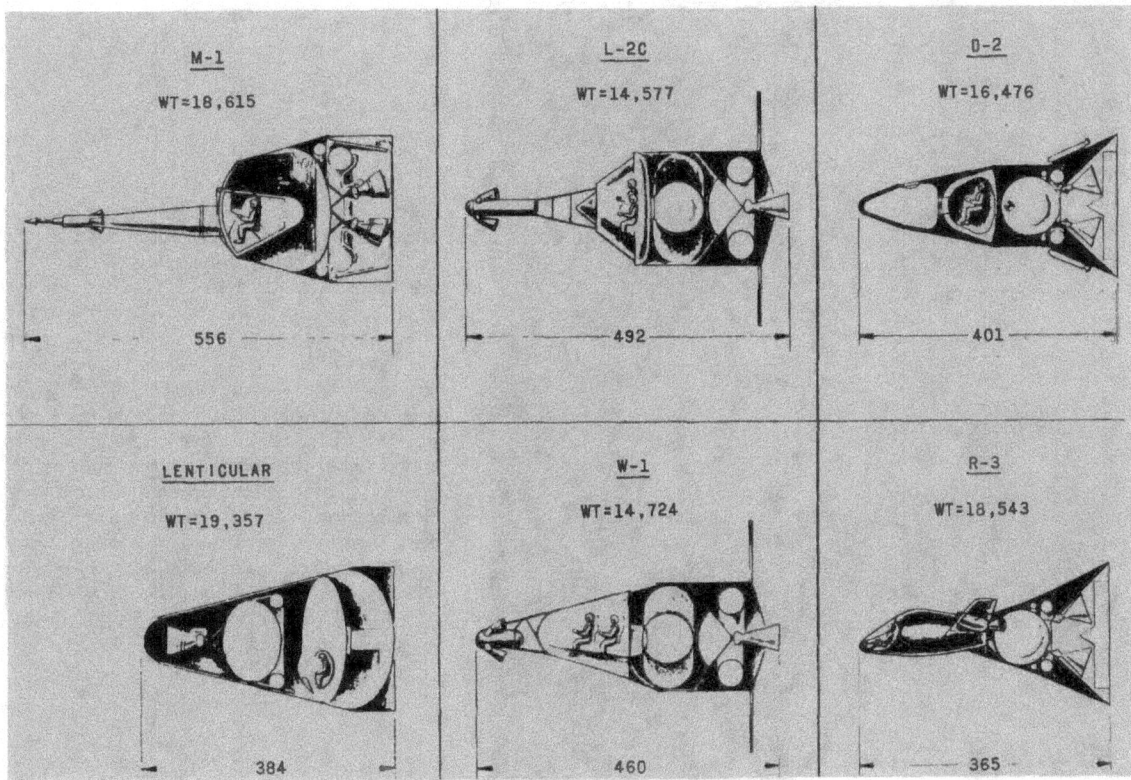

M-1	L-2C	O-2
WT=18,615	WT=14,577	WT=16,476
556	492	401

LENTICULAR	W-1	R-3
WT=19,357	WT=14,724	WT=18,543
384	460	365

Early proposals for manned space vehicles varied greatly in configuration and weight. In some, the men within faced one way during launch and another during reentry; in others, the vehicle was turned around, not the seats. Different approaches to the problem of escape from launching disaster were shown in these six industrial proposals. Environmental control, thermal and radiation shielding, and protection against meteorite impact were all unknowns facing early spacecraft designers.

A one-man lunar lander weighing 5000 pounds was envisioned as early as 1961 by a pair of Space Task Group engineers, James A. Chamberlin and James T. Ross, and here drawn by Harry A. Shoaf. It was seen as part of a 35,000-pound payload that might be carried by a post-Mercury spacecraft. The other extreme in early ideas to send men to the Moon called for a direct-ascent manned lunar vehicle weighing some 150,000 pounds. It would have been launched by Nova, a giant booster capable (on paper) of approximately 12 million pounds of thrust.

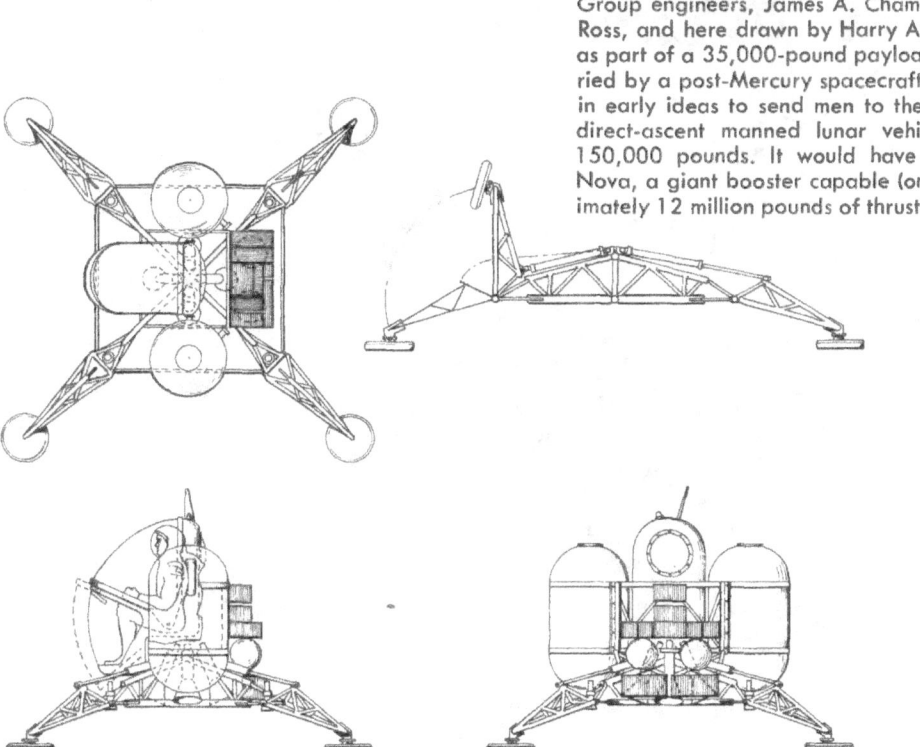

This Gemini spacecraft, in preparation in the Pyrotechnical Installation Building at the Cape, was to climb to a record altitude of 853 miles in September 1966. It docked in space with an Agena, and then used the big Agena rocket for the energy needed to reach the larger orbit. Gemini flights provided priceless experience in the tricky business of rendezvousing two craft in space with the minimum expenditure of energy. They also supplied practice at docking and in extravehicular activity, both needed for future Moon voyages. Finally, they helped build up experience with the mission-control system developing on the ground to support manned space-flight.

The two-man Gemini seemed capacious after tiny Mercury but it was actually very cramped. The astronauts rubbed elbows, and the man in the right seat, returning after EVA with his bulky spacesuit and tether, had to jam himself in to close the hatch over his head.

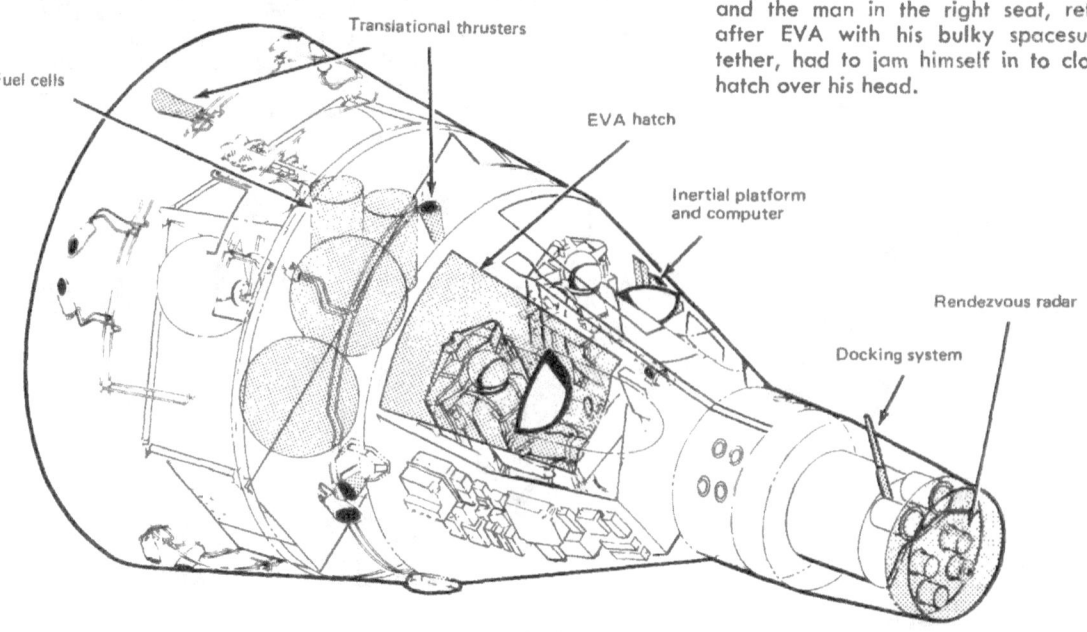

Fuel cells

Transiational thrusters

EVA hatch

Inertial platform and computer

Rendezvous radar

Docking system

Gemini launches drew hundreds of thousands of spectators, awed by the roar, flame, and smoke of the big Titan II booster. Viewers clogged the highways and camped by roadsides. Millions of others watched launchings on television, and the astronauts received tumultuous welcomes on their return. The launch at left is Gemini V, which carried Astronauts Cooper and Conrad for 120 revolutions of the Earth during August 1965. Fuel cells had their first space test on this flight.

First U.S. rendezvous in space occurred on December 15, 1965, when Gemini VI found and came within 6 feet of Gemini VII, which had been launched 11 days earlier. Picture below was shot by Tom Stafford, aboard Gemini VI with Wally Schirra. The other spacecraft, shown here at a range of 37 feet, was flown by Astronauts Borman and Lovell in a flight lasting more than 330 hours. Rendezvous proved entirely feasible but tricky to manage with minimum fuel use.

Astronaut Edward H. White was the first American to step outside in space. Jim McDivitt, Gemini IV's command pilot, took this picture on June 3, 1965. A 25-foot umbilical line and tether linked White to the spacecraft. In his left hand is an experimental personal propulsion unit. His chest pack contained an eight-minute emergency oxygen supply, as a backstop.

An Agena target vehicle was docked with by Gemini VIII on March 16, 1966. A short-circuited thruster set the two craft spinning dangerously, forcing Astronauts Armstrong and Scott to end the mission.

THE PLAN IN RETROSPECT

In thinking back over the flights of Apollo, I am impressed at the intrinsic excellence of the plan that had evolved. I have, of course, somewhat oversimplified its evolution, and there were times when we became discouraged, and when it seemed that the sheer scope of the task would overwhelm us in some areas there were surprises and other areas proceeded quite naturally and smoothly.

The most cruel surprise in the program was the loss of three astronauts in the Apollo fire, which occurred before our first manned flight. It was difficult for the country to understand how this could have occurred, and it seemed for a time that the program might not survive. I believe that the self-imposed discipline that resulted, and the ever-greater efforts on quality, enhanced our chances for success, coming as they did while the spacecraft was being rebuilt and final plans formulated.

The pogo problem was another surprise. Like the fire, it showed how difficult it was to conquer this new ocean of space. Fortunately, intensive and brilliant work with the big Saturns solved the problem with the launch rocket, permitting the flights to proceed without mishap.

We had planned a buildup of our flights, starting with a simple Earth-orbit flight of the command and service modules (Apollo 7), to be followed by similar trials with the lunar module (LM) added, for tests of rendezvous and docking and various burns of the LM engines (Apollo 9). These tests would have then been followed by flights to lunar orbit with the LM scouting the landing but not going all the way in (Apollo 10), and then the landing (Apollo 11).

After Apollo 7, however, the LM was not yet ready and the opportunity occurred to fly to the Moon with command and service module (CSM) only. This flight (Apollo 8) was to give us many benefits early in the program. Technically, it gave us information on our communication and tracking equipment for later missions, a close view of our landing sites, and experience in cislunar space with a simplified mission. Politically, it may have assured us of being first to the Moon, since the stepped-up schedule precluded the Russians from flying a man around the Moon with their Zond before we reached the Moon following our previously scheduled missions.

The flights came off almost routinely following Apollo 8 on through the first lunar landing and the flight to the Surveyor crater. But Apollo 13 was to see our first major inflight emergency when an explosion in the service module cut off the oxygen supply to the command module. Fortunately, the LM was docked to the CSM, and its oxygen and electric power, as well as its propulsion rocket, were available. During the 4-day ordeal of Apollo 13, the world watched breathlessly while the LM pushed the stricken command module around the Moon and back to Earth. Precarious though it was, Apollo 13 showed the merit of having separate spacecraft modules, and of training of flight and ground crews to adapt to emergency. The ability of the flight directors on the ground to read out the status of flight equipment, and the training of astronauts to meet emergencies, paid off on this mission.

Apollo surely is a prototype for explorations of the future when we again send men into space to build a base on the Moon or to explore even farther away from Earth.

By dawn's first light, a giant Apollo/Saturn V aboard its mobile launcher trundles toward its rendezvous with the Moon. Riding its crawler past spaceport marshes, the rocket moves at about one mph. (During its voyage in space, a part of it containing men will travel at 24,300 mph.) Nothing of the size and power of this formidable creation had ever been built before.

CHAPTER THREE

Saturn the Giant

By WERNHER VON BRAUN

With the *beep—beep—beep* of Sputnik on October 4, 1957, the Soviet Union had inaugurated the Space Age. It had also presented American planners with the painful realization that there was no launch vehicle in the U.S. stable capable of orbiting anything approaching Sputnik's weight.

Responding to a proposal submitted by the Army Ballistic Missile Agency, the Department of Defense was in just the right mood to authorize ABMA to develop a 1,500,000-pound-thrust booster. That unprecedented thrust was to be generated by clustering eight S-3D Rocketdyne engines used in the Jupiter and Thor missiles. The tankage for the kerosene and liquid oxygen was also to be clustered to make best use of tools and fixtures available from the Redstone and Jupiter programs. The program was named "Saturn" simply because Saturn was the next outer planet after Jupiter in the solar system.

Gen. John B. Medaris, commander of ABMA and my boss, felt that for a good design job on the booster it was necessary for us also to study suitable upper stages for the Saturn. On November 18, 1959, Saturn was transferred to the new National Aeronautics and Space Administration. NASA promptly appointed a committee to settle the upper-stage selection for Saturn. It was chaired by Dr. Abe Silverstein who, as associate director of NASA's Lewis Center in Cleveland, had spent years exploring liquid hydrogen as a rocket fuel. As a result of this work the Air Force had let a contract with Pratt & Whitney for the development of a small 15,000-pound-thrust liquid hydrogen/liquid oxygen engine, two of which were to power a new "Centaur" top stage for the Air Force's Atlas. Abe was on solid ground when he succeeded in persuading his committee to swallow its scruples about the risks of the new fuel and go to high-power liquid hydrogen for the upper stage of Saturn.

In the wake of Gagarin's first orbital flight on April 12, 1961, Saturn gained increased importance. Nevertheless, when the first static test of the booster with all eight engines was about to begin, at least one skeptical witness predicted a tragic ending of

"Cluster's last stand." Doubts about the feasibility of clustering eight highly complex engines had indeed motivated funding for two new engine developments. One was in essence an uprating and simplification effort on the S-3D, and it led to the 188,000-pound-thrust H-1 engine. The other aimed at a very powerful new engine called F-1, which was to produce a full 1.5-million-pound thrust in a single barrel. Both contracts went to Rocketdyne.

Following up on the recommendation of the Silverstein committee, NASA awarded a contract to the Douglas Aircraft Company for the development of a second stage for Saturn that became known as S-IV. It was to be powered by six Centaur engines. On September 8, 1960, President Eisenhower came to Huntsville to dedicate the new Center, named after Gen. George C. Marshall. It was to become the focal point for NASA's new large launch vehicles, and I was appointed as its first director.

DETERMINING SATURN'S CONFIGURATION

The first launch of the Saturn booster was still five months away when, on May 25, 1961, President John F. Kennedy proposed that the United States commit itself to land a man on the Moon "in this decade." For this ambitious task a launch vehicle far more powerful than our eight-engine Saturn would be needed. To determine its exact power requirements, a selection had to be made from among three operational concepts for a manned voyage to the Moon: direct ascent, Earth orbit rendezvous (EOR), and lunar orbit rendezvous (LOR).

With direct ascent, the entire spacecraft would soft-land on the Moon carrying enough propellants to fly back to Earth. Weight and performance studies showed that this would require a launch vehicle of a lift-off thrust of 12 million pounds, furnished by eight F-1 engines. We called this hypothetical launch vehicle Nova. The EOR mode envisioned two somewhat smaller rockets that were to rendezvous in Earth orbit where their payloads would be combined. In the LOR mode a single rocket would launch a payload consisting of a separable spacecraft toward the Moon, where an onboard propulsion unit would ease it into orbit. A two-stage lunar module (LM) would then detach itself from the orbiting section and descend to the lunar surface. Its upper stage would return to the circumlunar orbit for rendezvous with the orbiting section. In a second burst of power, the propulsion unit would finally drive the reentry element with its crew out of lunar orbit and back to Earth.

As all the world knows, the LOR mode was ultimately selected. But even after its adoption, the number of F-1 engines to be used in the first stage of the Moon rocket remained unresolved for quite a while. H. H. Koelle, who ran our Project Planning Group at Marshall, had worked out detailed studies of a configuration called Saturn IV with four F-1's, and another called Saturn V with five F-1's in its first stage. Uncertainty about LM weight and about propulsion performance of the still untested F-1 and upper-stage engines, combined with a desire to leave a margin for growth, finally led us to the choice of the Saturn V configuration.

Despite the higher power offered by liquid hydrogen, Koelle's studies indicated that little would be gained by using it in the first stage also, where it would have needed disproportionately large tanks. (Liquid hydrogen is only one twelfth as dense as kero-

Dr. von Braun, standing next to one of the five engines at the after end of the Saturn V vehicle's first stage (on display at the Alabama Space Museum), provides a scale reference for the human figures shown alongside two of the rockets in the drawing below. The sequence of launch vehicles of ever-increasing size and power that led from the 46-foot-high V-2 rocket through the Mercury and Gemini boosters to the 363-foot Saturn V is drawn here at a single scale.

Saturn V

Saturn IB

Saturn I

Gemini Titan

Atlas Agena

Atlas Mercury

Mercury Redstone

V2

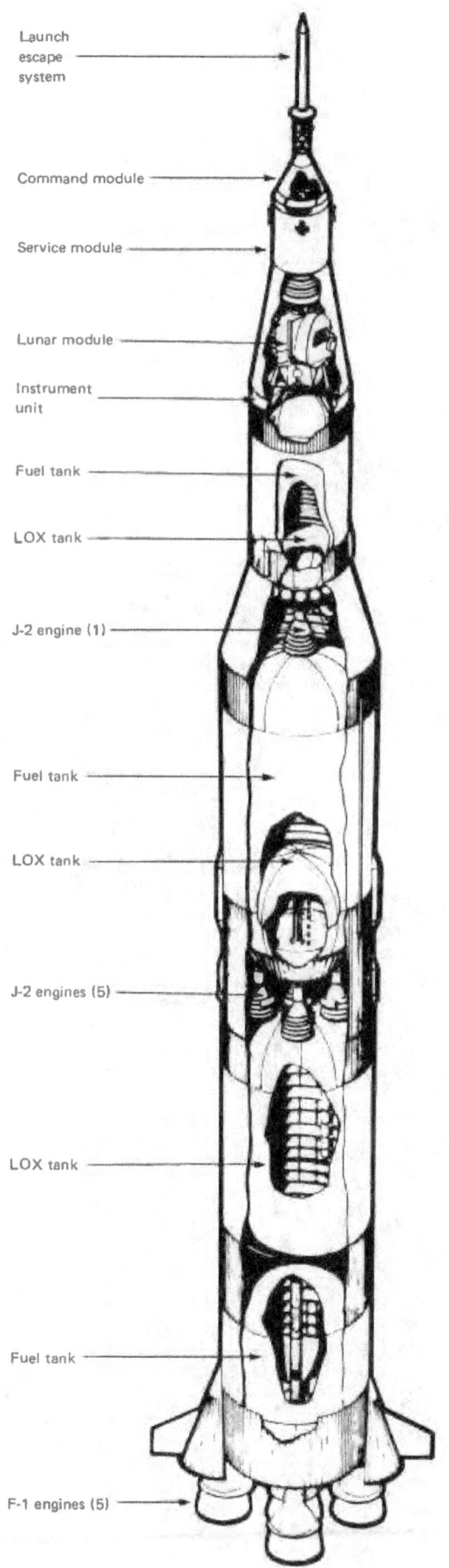

Launch escape system

Command module

Service module

Lunar module

Instrument unit

Fuel tank

LOX tank

J-2 engine (1)

Fuel tank

LOX tank

J-2 engines (5)

LOX tank

Fuel tank

F-1 engines (5)

The stack: the three-stage launch vehicle, Saturn V, surmounted by its payload, the Apollo spacecraft. The greater part of the launch vehicle consists of tankage for the fuel and for the oxidant, LOX (liquid oxygen), used in all three Saturn V stages. The powerful F-1 engines of the first stage burn kerosene to produce a combined thrust of 7.5 million pounds. The fuel for the J-2 engines of the two upper stages is liquid hydrogen. The combined thrust of the second stage's five engines is just over a million pounds, or five times that of the third stage's single J-2 engine. Development of the original hydrogen tanks was difficult because the low boiling point of hydrogen (−253 °C) required insulation sufficient to prevent transfer of heat from the outside and the comparatively warm (−183 °C) liquid oxygen.

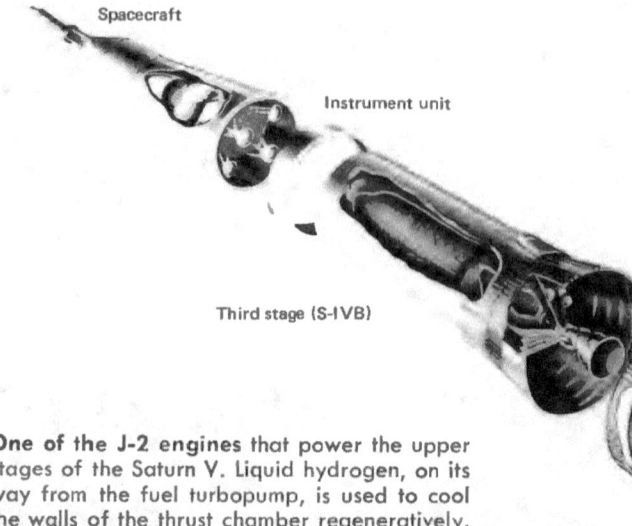

Spacecraft

Instrument unit

Third stage (S-IVB)

One of the J-2 engines that power the upper stages of the Saturn V. Liquid hydrogen, on its way from the fuel turbopump, is used to cool the walls of the thrust chamber regeneratively.

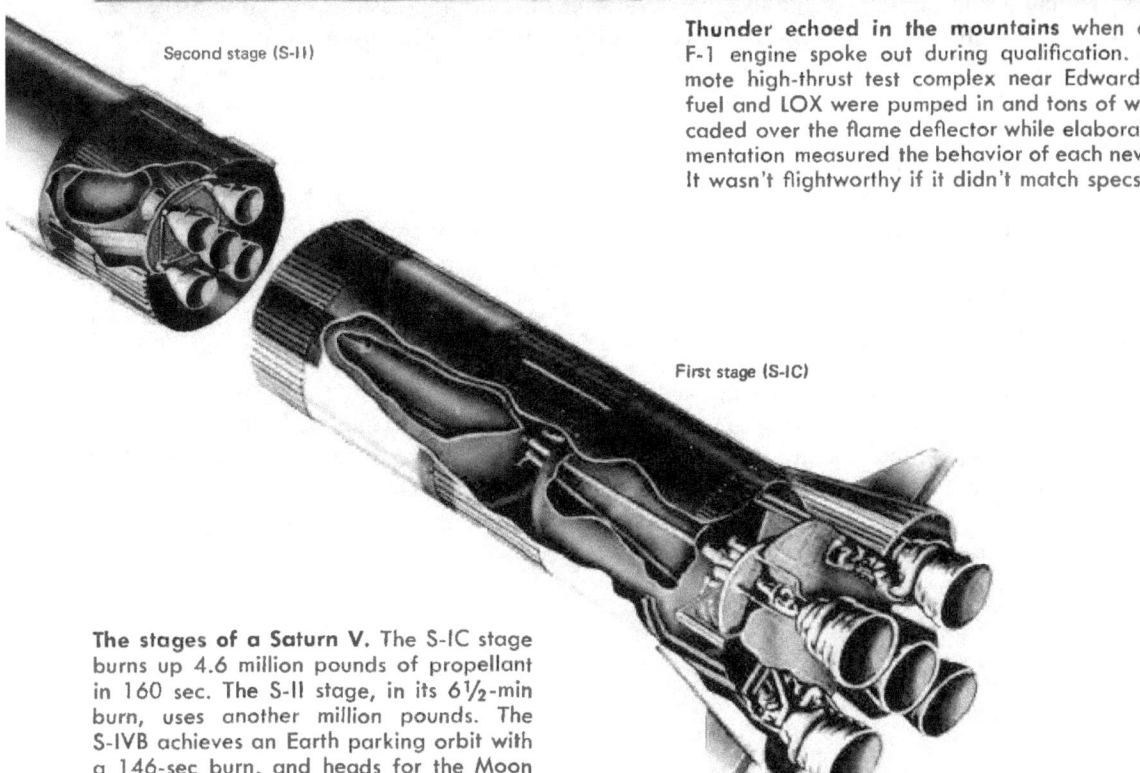

Second stage (S-II)

Thunder echoed in the mountains when a mighty F-1 engine spoke out during qualification. At a remote high-thrust test complex near Edwards, Calif., fuel and LOX were pumped in and tons of water cascaded over the flame deflector while elaborate instrumentation measured the behavior of each new engine. It wasn't flightworthy if it didn't match specs.

First stage (S-IC)

The stages of a Saturn V. The S-IC stage burns up 4.6 million pounds of propellant in 160 sec. The S-II stage, in its 6½-min burn, uses another million pounds. The S-IVB achieves an Earth parking orbit with a 146-sec burn, and heads for the Moon with a second burn of 345 sec.

sene, so a much larger tank volume would have been required.) In all multistage rockets the upper stages are lighter than the lower ones. Thus heavier but less energetic kerosene in the first stage, in combination with lighter but more powerful hydrogen in the upper stages, made possible a better launch-vehicle configuration.

Saturn V, as it emerged from the studies, would consist of three stages—all brand new. The first one, burning kerosene and oxygen, would be powered by five F-1 engines. We called it S-IC. The second stage, S-II, would need about a million pounds of thrust and, if also powered by five engines, would call for the development of new 200,000-pound hydrogen-oxygen engines. A single engine of this thrust would just be right to power the third stage. The Saturn I's S-IV second stage was clearly not powerful enough to serve as the Saturn's third one. A much larger tankage and at least thirteen of Pratt & Whitney's little LR-10 engines would be required; this did not appear very attractive.

When bids for the new J-2 engine were solicited, Pratt & Whitney with its ample liquid-hydrogen experience was a strong contender. But when all the points in the sternly controlled bidding procedure were counted, North American's Rocketdyne Division won again.

North American had been involved in the development of liquid fuel rocket engines since the immediate postwar years and the Navajo long range ramjet program. The engines it developed for the Navajo booster and their offspring later found their way into the Atlas, Redstone, Thor, and Jupiter programs. For the testing of these engines NAA's Rocketdyne Division had acquired a boulder-strewn area high in the Santa Susana mountains, north of Los Angeles, that had previously served as rugged background for many a Western movie. The Santa Susana facility would henceforth serve not only for the development of the new J-2 engine, but also for short duration "battleship" testing of the five-engine cluster of these engines powering the S-II stage. (Safety and noise considerations ruled out the use of Santa Susana for the 1.5-million-pound-thrust F-1 engine. Test stands for its development were therefore set up in the Mojave desert, adjacent to Edwards Air Force Base.)

CHOOSING THE BUILDERS

How many prime contractors, we wondered, should NASA bring in for the development of the Saturn V? Just one, or one per stage? How about the Instrument Unit that was to house the rocket's inertial-guidance system, its digital computer, and an assortment of radio command and telemetry functions? Who would do the overall systems engineering and monitor the intricate interface between the huge rocket and the complex propellant-loading and launching facilities at Cape Canaveral? Where would the various stages be static-tested?

Understandably, the entire aerospace industry was attracted by both the financial value and the technological challenge of Saturn V. To give the entire plum to a single contractor would have left all others unhappy. More important, Saturn V needed the very best engineering and management talent the industry could muster. By breaking up the parcel into several pieces, more top people could be brought to bear on the program.

The first (S-IC) stage of the Saturn V launch vehicle being hoisted into the static test stand at Marshall Space Flight Center. This was the "battleship," or developmental test version of the stage, built heavily to permit repeated testing of flight-version working components. The first three flight S-IC stages were assembled at MSFC and tested in this stand. The massively reinforced construction of the 300-foot-tall stand was essential to withstand the 7.5 million pounds of thrust developed by the stage's engines during static testing.

The **"pogo problem,"** a lengthwise mode of vibration recognized in the second Saturn V launch, was speedily solved through mathematical analysis supported by data collected in shake tests. To supplement shake tests in Marshall's Dynamic Test Tower, Boeing quickly erected this tower for special pressure tests at the Michoud Assembly Facility.

A test at the "Arm Farm." Just to the man's left a skin section representing the S-II stage is mounted to the Random Motion/Lift-Off Simulator, which can simulate at ground level the swaying of the space vehicle in a Florida storm. A duplicate of the Mobile Launch Tower's S-II Forward Swingarm projects from the left, carrying the umbilicals that are connected to the skin section.

The Boeing Company was the successful bidder on the first stage (S-IC); North American Aviation won the second stage (S-II); and Douglas Aircraft fell heir to the Saturn V's third stage (S-IVB). Systems engineering and overall responsibility for the Saturn V development was assigned to the Marshall Space Flight Center. The inertial-guidance system had emerged from a Marshall in-house development, and as it had to be located close to other elements of the big rocket's central nervous system, it was only logical to develop the Instrument Unit (IU) to house this electronic gear as a Marshall in-house project. IU flight units were subsequently produced by IBM, which had developed the launch-vehicle computer.

Uniquely tight procurement procedures introduced by NASA Administrator Jim Webb made it possible to acquire billions of dollars' worth of exotic hardware and facilities without overrunning initial cost estimates and without the slightest hint of procurement irregularity. Before it could issue a request for bids, the contracting NASA Center had to prepare a detailed procurement plan that required the Administrator's personal approval, and that could not be changed thereafter. It had to include a point-scoring system in which evaluation criteria—technical merits, cost, skill availability, prior experience, etc.—were given specific weighting factors. Business and technical criteria were evaluated by separate teams not permitted to know the other's rankings. The total matrix was then assembled by a Source Evaluation Board that gave a complete presentation of all bids and their scoring results to the three top men in the agency, who themselves chose the winner. There was simply no room for arbitrariness or irregularity in such a system.

The tremendous increase in contracts needed for the Saturn V program required a reorganization of the Marshall Space Flight Center. Most of our resources had been spent in-house, and our contracts had either been let to support contractors or to producers of our developed products. Now 90 percent of our budget was spent in industry, much of it on complicated assignments which included design, manufacture, and testing. So on September 1, 1963, I announced that Marshall would henceforth consist of two major elements, one to be called Research and Development Operations, the other Industrial Operations. Most of my old R&D associates then became a sort of architect's staff keeping an eye on the integrity of the structure called Saturn V, and the other group funded and supervised the industrial contractors.

That same year Dr. George Mueller had taken over as NASA's Associate Administrator for Manned Space Flight. He brought with him Air Force Maj. Gen. Samuel Phillips, who had served as program manager for Minuteman, and now became Apollo Program Director at NASA Headquarters. Both men successfully shaped the three NASA Centers involved in the lunar-landing program into a team. I was particularly fortunate in that Sam Phillips persuaded his old friend and associate Col. (later Maj. Gen.) Edwin O'Connor to assume the directorship of Marshall's Industrial Operations.

On September 7, 1961, NASA had taken over the Michoud Ordnance plant at New Orleans. The cavernous plant—46 acres under one roof—was assigned to Chrysler and Boeing to set up production for the first stages of Saturn I and Saturn V. In October 1961 an area of 13,350 acres in Hancock County, Miss., was acquired. Huge test stands were erected there for the static testing of Saturn V's first and second stages.

Shipment of the oversize stages between Huntsville, Michoud, the Mississippi Test Facility, the two California contractors, and the Kennedy Space Center in Florida required barges and seagoing ships. Soon Marshall found itself running a small fleet that included the barges *Palaemon*, *Orion*, and *Promise*. For shipments through the Panama Canal we used the USNS *Point Barrow* and the SS *Steel Executive*. For rapid transport we had two converted Stratocruisers at our disposal with the descriptive names "Pregnant Guppy" and "Super Guppy." Their bulbous bodies could accommodate cargo up to the size of an S-IVB stage.

AN ALL-UP TEST FOR THE FIRST FLIGHT

In 1964 George Mueller visited Marshall and casually introduced us to his philosophy of "all-up" testing. To the conservative breed of old rocketeers who had learned the hard way that it never seemed to pay to introduce more than one major change between flight tests, George's ideas had an unrealistic ring. Instead of beginning with a ballasted first-stage flight as in the Saturn I program, adding a live second stage only after the first stage had proven its flightworthiness, his "all-up" concept was startling. It meant nothing less than that the very first flight would be conducted with all three live stages of the giant Saturn V. Moreover, in order to maximize the payoff of that first flight, George said it should carry a live Apollo command and service module as payload. The entire flight should be carried through a sophisticated trajectory that would permit the command module to reenter the atmosphere under conditions simulating a return from the Moon.

It sounded reckless, but George Mueller's reasoning was impeccable. Water ballast in lieu of a second and third stage would require much less tank volume than liquid-hydrogen-fueled stages, so that a rocket tested with only a live first stage would be much shorter than the final configuration. Its aerodynamic shape and its body dynamics would thus not be representative. Filling the ballast tanks with liquid hydrogen? Fine, but then why not burn it as a bonus experiment? And so the arguments went on until George in the end prevailed.

In retrospect it is clear that without all-up testing the first manned lunar landing could not have taken place as early as 1969. Before Mueller joined the program, it had been decided that a total of about 20 sets of Apollo spacecraft and Saturn V rockets would be needed. Clearly, at least ten unmanned flights with the huge new rocket would be required before anyone would muster the courage to launch a crew with it. (Even ten would be a far smaller number than the unmanned launches of Redstones, Atlases, and Titans that had preceded the first manned Mercury and Gemini flights.) The first manned Apollo flights would be limited to low Earth orbits. Gradually we would inch our way closer to the Moon, and flight no. 17, perhaps, would bring the first lunar landing. That would give us a reserve of three flights, just in case things did not work as planned.

Mueller changed all this, and his bold telescoping of the overall plan bore magnificent fruit: With the third Saturn V ever to be launched, Frank Borman's Apollo 8 crew orbited the Moon on Christmas 1968, and the sixth Saturn V carried Neil Armstrong's Apollo 11 to the first lunar landing. Even though production was

An Instrument Unit being readied for checkout at the IBM facility in Huntsville. A cylinder 22.7 feet across and 3 feet high, the structure consists of 24 panels with stiffening rings at the top and bottom. The units that perform the guidance, control, and telemetry functions for the Saturn are mounted to the inside of the cylinder. The foam rubber pads at the top, and the plastic strips around the outside, are for protection during manufacture.

Pressure test during predelivery checkout of an Instrument Unit at the IBM facility. This IU was destined for Saturn vehicle 505, which launched the Apollo 10 mission. On the launch pad, the IU, which weighs two tons, sits atop the third (S-IVB) stage, with the Apollo spacecraft directly above it.

whittled back to fifteen units, Saturn V's launched a total of two unmanned and ten manned Apollo missions, plus one Skylab space station. Two uncommitted rockets went into mothballs.

But let us go back to 1962. To develop and manufacture the large S-II and S-IVB stages, two West Coast contractors required special facilities. A new Government plant was built at Seal Beach where North American was to build the S-II. S-IVB development and manufacture was moved into a new Douglas center at Huntington Beach, while static testing went to Sacramento. The Marshall Center in Huntsville was also substantially enlarged. A huge new shop building was erected for assembly of the first three S-IC stages. A large stand was built to static-test the huge stage under the full 7,500,000-pound-thrust of its five F-1 engines. These engines generated no less than 180 million horsepower. As about 1 percent of that energy was converted into noise, neighborhood windows could be expected to break and plaster rain from ceilings if the wind was blowing from the wrong direction or the clouds were hanging low. A careful meteorological monitoring program had to be instituted to permit test runs only under favorable weather conditions.

Although the most visible and audible signs of Marshall's involvement in Saturn V development were the monstrous and noisy S-IC engines, equally important work was done in its Astrionics Laboratory. The Saturn V's airbearing-supported inertial guidance platform was born there, along with a host of other highly sophisticated electronic devices. In the Astrionics Simulator Facility, guidance and control aspects of a complete three-stage flight of the great rocket could be electronically simulated under all sorts of operating conditions. The supersonic passage of the rocket through a high-altitude jet stream could be duplicated, for instance, or the sudden failure of one of the S-II stage's five engines. The simulator would faithfully display the excursions of the swivel-mounted rocket engines in response to external wind forces or unsymmetrical loss of thrust, establishing the dynamic response of the entire rocket and the resulting structural loads.

The Saturn V's own guidance system would guide the Apollo flights not only to an interim parking orbit but all the way to translunar injection. It fed position data to the onboard digital computer, which in turn prepared and sent control signals to the hydraulic actuators that swiveled the big engines for flight-path control. As propellant consumption lightened the rocket, and as it traversed the atmosphere at subsonic and supersonic speeds, the gain settings of these control signals had to vary continuously, for proper control damping. Serving as the core of the Saturn V's central nervous system, the computer did many other things too. It served in the computerized prelaunch checkout procedure of the great rocket, helped calibrate the telemetry transmissions, activated staging procedures, turned equipment on and off as the flight proceeded through various speed regimes, and even watched over the cooling system that stabilized the temperatures of the array of sensitive blackboxes within the IU. So although the working flight lifetime of the Saturn computer was measured in minutes, it performed many exacting duties during its short and busy life.

In planning the lunar mission, why did we plan to stop over in a parking orbit? The reason was twofold: For one, in case of a malfunction it is much easier and safer

Super Guppy, bigger sister of the aptly named *Pregnant Guppy*, was the only airplane in the world capable of carrying a complete S-IVB stage. Both aircraft were built by John M. Conroy, who started with the fuselages of two surplus Boeing C-97 Stratocruisers, ballooned out the upper decks enormously, and hinged the front sections so that they could be folded back 110 degrees. *Super Guppy* flew smoothly at a 250-mph cruising speed, and its cargo deck provided a 25-foot clear diameter. Below, a finished S-IVB stage is being unloaded onto a cargo lift trailer at Kennedy Space Center.

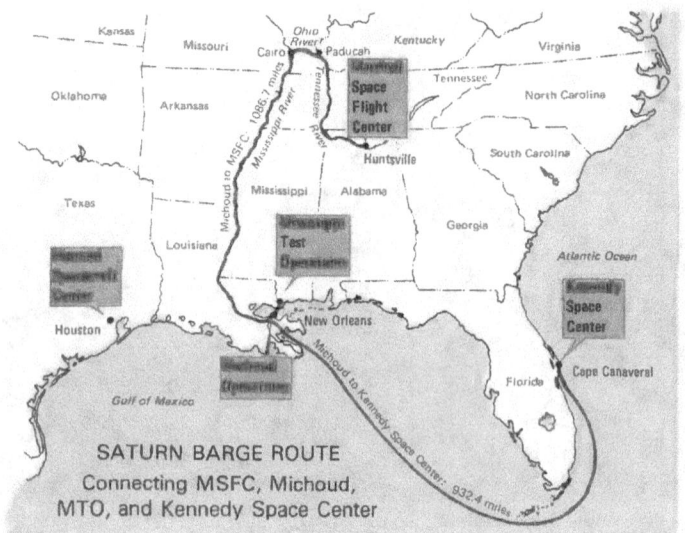

SATURN BARGE ROUTE
Connecting MSFC, Michoud,
MTO, and Kennedy Space Center

Bulky but fragile, huge launch-vehicle stages could not travel by rail or public road; tunnels, bridges, and low wires would have dictated endless detours. Right, even a comparatively small Saturn I wasn't easy to back into its barge at Huntsville. Below, a rocket-laden barge is escorted by two tugs through Sun-bronzed waters. The biggest California-built parts of Saturn V voyaged by ship through the Panama Canal and across the Gulf of Mexico.

for astronauts to return from Earth orbit than from a high-speed trajectory carrying them toward the Moon. A parking orbit offers both crew and ground controllers an opportunity to give the vehicle a thorough once-over before committing it to the long voyage. Second, there is the consideration of operational flexibility. If the launch came off at precisely the right instant, only one trajectory from the launch pad to the Moon had to be considered. But as there was always the possibility of a last-minute delay it appeared highly desirable to provide a launch window of reasonable duration. This meant not only that the launch azimuth had to be changed, but due to Earth rotation and to orbital motion the Moon would move to a different position in the sky. A parking orbit permitted an ideal way to take up the slack: the longer a launch delay, the shorter the stay in the parking orbit. Restart of the third stage in parking orbit for translunar injection would take place at almost the same time of day regardless of launch delays. (As it happened, all but two of the manned Apollo-Saturns lifted off within tiny fractions of a second of being *precisely* on time. One was held for weather and the other was held because of a faulty diode in the ground-support equipment.)

Why was the big rocket so reliable? Saturn V was not overdesigned in the sense that everything was made needlessly strong and heavy. But great care was devoted to identifying the real environment in which each part was to work—and "environment" included accelerations, vibrations, stresses, fatigue loads, pressures, temperatures, humidity, corrosion, and test cycles prior to launch. Test programs were then conducted against somewhat more severe conditions than were expected. A methodology was created to assess each part with a demonstrated reliability figure, such as 0.9999998. Total rocket reliability would then be the product of all these parts reliabilities, and had to remain above the figure of 0.990, or 99 percent. Redundant parts were used whenever necessary to attain this reliability goal.

Marshall built an overall systems simulator on which all major subsystems of the three-stage rocket could be exercised together. This facility featured replicas of propellant tanks that could be loaded or unloaded, pressurized or vented, and that duplicated the pneumatic and hydraulic dynamics involved. Electrically, it simulated the complete network of the launch vehicle and its interfacing ground support equipment.

THE PERILS OF POGO

An important Marshall facility was the Dynamic Test Tower, the only place outside the Cape where the entire Saturn V vehicle could be assembled. Electrically powered shakers induced various vibrational modes in the vehicle, so that its elastic deformations and structural damping characteristics could be determined. The Dynamic Test Tower played a vital role in the speedy remedy of a problem that unexpectedly struck in the second flight of a Saturn V. Telemetry indicated that during the powered phases of all three stages a longitudinal vibration occurred, under which the rocket alternately contracted and expanded like a concertina. This "pogo" oscillation (the name derived from the child's toy) would be felt particularly strongly in the command module.

Analysis, supported by data collected in engine tests, confirmed that the oscillation was caused by resonance coupling between the springlike elastic structure of the

tankage, and the rocket engines' propellant-feed systems. Susceptibility to pogo (a phenomenon not unknown to missile designers) had been thoroughly investigated by the Saturn stage contractors, who had certified that their respective designs would be pogo-free. It turned out that these mathematical analyses had been conducted on an inadequate data base.

Once the problem was understood, a fix was quickly found. "In sync" with the pogo oscillations, pressures in the fuel and oxidizer feed lines fluctuated wildly. If these fluctuations could be damped by gas-filled cavities attached to the propellant lines, which would act as shock absorbers, the unacceptable oscillation excursions should be drastically reduced. Such cavities were readily available in the liquid-oxygen prevalves, whose back sides were now filled with pressurized helium gas tapped off the high-pressure control system. After a few weeks of hectic activity, a pogo-free Saturn flight no. 3 successfully boosted the Apollo 8 crew to their Christmas flight in lunar orbit.

ARTIFICIAL STORMS AT THE ARM FARM

The connections between the ground and the towering space vehicle posed a tricky problem. An umbilical tower, even higher than the vehicle itself, was required to support an array of swingarms that at various levels would carry the cables and the pneumatic, fueling, and venting lines to the rocket stages and to the spacecraft. The swingarms had to be in place during final countdown, but in the last moments they had to be turned out of the way to permit the rocket to rise. There was always the possibility, however, of some trouble after the swingarms had been disconnected. For instance, the holddown mechanism would release the rocket only after all five engines of the first stage produced full power. If this condition was not attained within a few seconds, all engines would shut down. In such a situation, unless special provisions were made for reattachment of some swingarms, Launch Control would be unable to "safe" the vehicle and remove the flight crew from its precarious perch atop a potential bomb.

These considerations led to the establishment, at Marshall, of a special Swingarm Test Facility, where detachment and reconnection of various arms was tested under brutally realistic conditions. On the "Arm Farm" extreme conditions (such as a launch scrub during an approaching Florida thunderstorm) could be simulated. Artificial rain was blown by aircraft propellers against the swingarms and their interconnect plugs, while the vehicle portion was moved back and forth, left and right, simulating the swaying motions that the towering rocket would display during a storm.

Throughout Saturn V's operational life, its developers felt a relentless pressure to increase its payload capability. At first, the continually growing weight of the LM (resulting mainly from additional operational features and redundancy) was the prime reason. Later, after the first successful lunar landing, the appetite for longer lunar stay times grew. Scientists wanted landing sites at higher lunar latitudes, and astronauts like tourists everywhere wanted a rental car at their destination. How well this growth demand was met is shown by a pair of numbers: The Saturn V that carried Apollo 8 to the Moon had a total payload above the IU of less than 80,000 pounds; in comparison, the Saturn that launched the last lunar mission, Apollo 17, had a payload of 116,000 pounds.

Silhouetted by the glare of the first Saturn V launch, a flock of birds calmly conducts its dawn patrol of the lagoon. As the vehicle begins to clear the launch pad, several more seconds will pass before the crashing roar reaches the flock. This is Apollo 4, the first "all up" test of the launch vehicle and spacecraft, proving out their flight compatibility in an unmanned Earth-orbiting mission.

Circling the Moon once every two hours in the CSM, one lunar explorer awaits his colleagues from the lunar surface. At the nose of the craft is the extended docking probe, ready to *receive* the LM. The bell-shaped rocket engine at the rear *must* work one more time for return.

58 APOLLO

Like a spider dancing upside down, the lunar module makes its first solo flight in Earth orbit. The rods protruding from the footpads are to give first indication of contact with the lunar surface. The ladder on the front leg would soon serve Neil Armstrong to take that "small step for a man. . . ."

CHAPTER FOUR

The Spaceships

By GEORGE M. LOW

On April 3, 1967, NASA 2, a Grumman Gulfstream, was taxiing for takeoff at Washington National Airport. Bob Gilruth, Director of NASA's Manned Spacecraft Center and I (his Deputy at that time) were about to return to Houston after a series of meetings in Washington. But just before starting down the runway, the pilot received a cryptic message from the tower: return to the terminal and ask the passengers to wait in the pilot's lounge. Soon arrived Administrator Jim Webb, his Deputy Bob Seamans, George Mueller, the head of Manned Space Flight, and Apollo Program

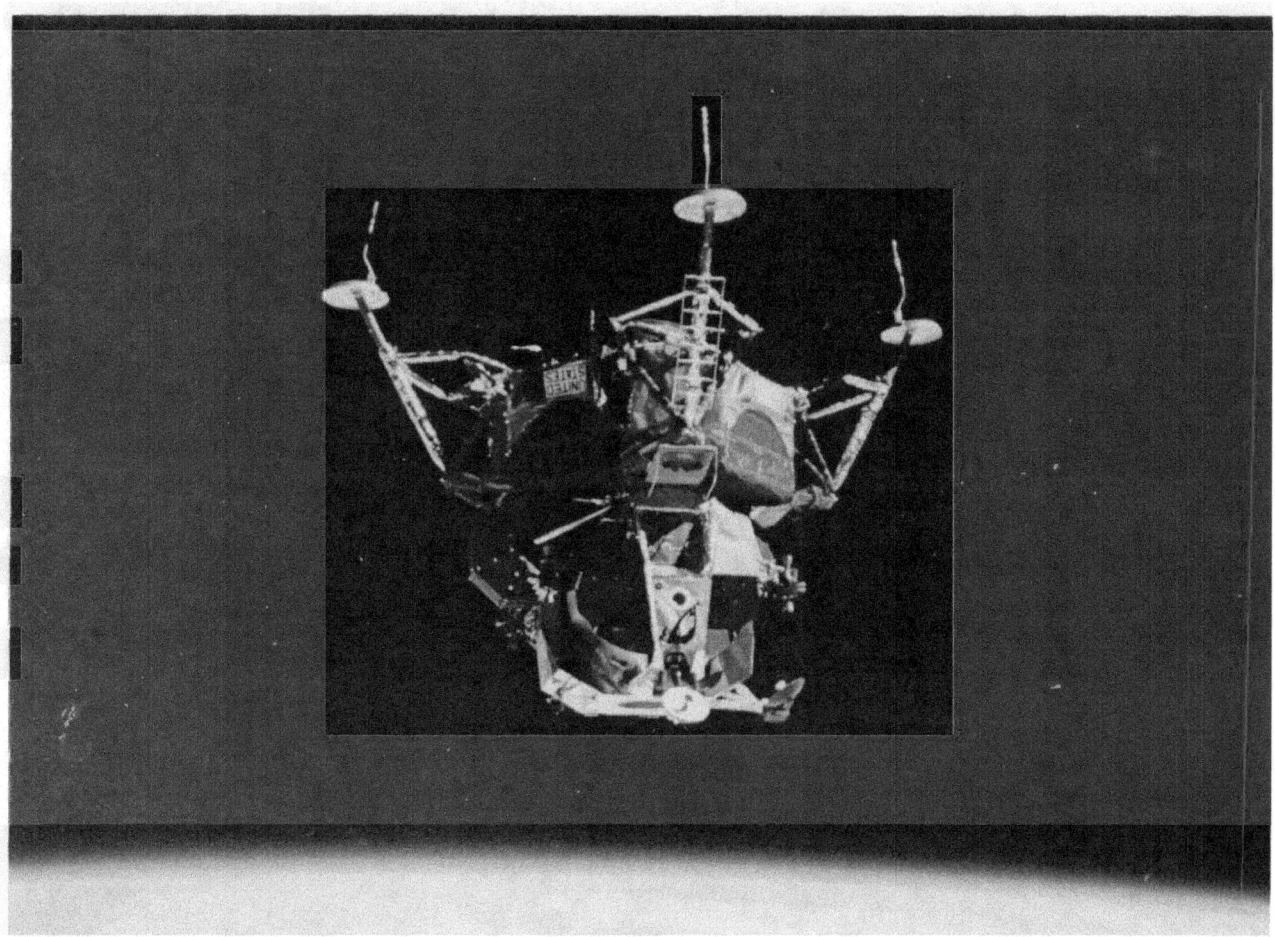

Director Sam Phillips. Counting Bob Gilruth, everybody in the NASA hierarchy between me and the President was there.

Jim Webb, using fewer words than usual, came right to the point: Apollo was faltering; the catastrophic fire on January 27 that had taken the lives of three astronauts had been a major setback. All its consequences were not yet known; time was running out on the Nation's commitment to land on the Moon before the end of the decade. Then the punch line: NASA wanted me to take on the task of rebuilding the Apollo spacecraft, and to see to it that we met the commitment.

Thus began the most exciting, most demanding, sometimes most frustrating, and always most challenging 27 months in my career as an engineer. Not that Apollo was completely new to me. Six years earlier I had chaired the NASA committee that recommended a manned lunar landing and provided the background work for President Kennedy's decision to go to the Moon. In the intervening years I had not been involved in the day-to-day engineering details of the Apollo spaceships—yet 27 months later, sitting at a console in the Launch Control Center during the final seconds of the countdown for Apollo 11, I had come to know and understand two of the most complex flying machines ever built by man.

TWO MAGNIFICENT FLYING MACHINES

But in April 1967 these machines were essentially strangers to me. How were they designed? How were they built and tested? What were their strengths and their weaknesses? Above all, what flaw in their design had caused the fire, and what other flaws lurked in their complexity?

First there was the command and service module—the CSM—collectively a single spacecraft, but separable into two components (the command module and the service module) for the final minutes of reentry. It was built by North American Rockwell in Downey, California, a place which would become one of my many "homes" for the next 27 months. The command module was compact, solid, and sturdy, designed with one overriding consideration: to survive the fiery heat of reentry as it abandoned the service module and slammed back into the atmosphere at the tremendous speed of 25,000 miles an hour. It was a descendant of Mercury and Gemini, but its task was much more difficult. The speed of reentry from the Moon is nearly one and one-half times as fast as returning from Earth orbit; to slow down from that speed required the dissipation of great amounts of energy. In fact, there is enough energy at reentry to melt and vaporize all the material in the command module several times over, so the spacecraft had to be protected by an ablative heat shield that charred and slowly burned away, thereby protecting all that it surrounded. The command module was also crammed with equipment and subsystems; and of course three men lived in it for most of the lunar journey, and one of them for all of it. It was cone-shaped, with a blunt face for reentry; it was 11 feet long, 13 feet in diameter, and weighed 6 tons.

The service module was the quartermaster of the pair. It carried most of the stores needed for the journey through space; oxygen, power-generation equipment, and water as a byproduct of power generation. More than that, it had a propulsion system bigger and more powerful than many upper stages of present launch vehicles. It made all the maneuvers needed to navigate to the Moon, to push itself and the lunar module into lunar orbit, and to eject itself out of orbit to return to Earth. The service module was a cylinder 13 feet in diameter and 24 feet long. Fully loaded it weighed 26 tons.

Then there was the lunar module (LM, pronounced LEM, which had actually been its designation—for lunar excursion module—until someone decided that the word "excursion" might lend a frivolous note to Apollo). LM was the first true spaceship; it was hidden in a cocoon during the launch through the atmosphere because it could operate only in the vacuum of space. Built by Grumman in Bethpage, New York (another of my many homes away from home), it was somewhat flimsy, with paper-thin walls and spindly legs. Its mission was to carry two explorers from lunar orbit to the surface of the Moon, provide a base for them on the Moon, and then send its upper half back into lunar orbit to a rendezvous with its mother ship, the CSM. Designed by aeronautical engineers who for once did not have to worry about airflow and stream-lining, it looked like a spider, a gargantuan, other-world insect that stood 23 feet tall and weighed 16 tons. When Jim McDivitt returned from Apollo 9, LM's first manned flight, he gave me a photograph of his *Spider* in space, with this caption: "Many thanks for the funny-looking spacecraft. It sure flies better than it looks."

These were the Apollo spacecraft: two machines, 17 tons of aluminum, steel, copper, titanium, and synthetic materials; 33 tons of propellant; 4 million parts, 40 miles of wire, 100,000 drawings, 26 subsystems, 678 switches, 410 circuit breakers.

To look after them there was a brand-new program manager who would have to leap upon this fast-moving train, learn all about it, decide what was good enough and what wasn't, what to accept, and what to change. In the meanwhile, the clock ticked away, bringing the end of the decade ever closer.

COMPLEX SUBSYSTEMS PERFORMED VITAL FUNCTIONS

At the heart of each spacecraft were its subsystems. "Subsystem" is space-age jargon for a mechanical or electronic device that performs a specific function such as providing oxygen, electric power, and even bathroom facilities. CSM and LM subsystems performed similar functions, but differed in their design because each had to be adapted to the peculiarities of the spacecraft and its environment.

Begin with the environmental control system—the life-support system for man and his machine. It was a marvel of efficiency and reliability, with weight and volume at a premium. A scuba diver uses a tank of air in 60 minutes; in Apollo an equivalent amount of oxygen lasted 15 hours. Oxygen was not simply inhaled once and then discarded: the exhaled gas was scrubbed to eliminate its CO_2, recycled, and reused. At the same time, its temperature was maintained at a comfortable level, moisture was removed, and odors were eliminated. That's not all: the same life-support system also maintained the cabin at the right pressure, provided hot and cold water, and a circulating coolant to keep all the electronic gear at the proper temperature. (In the weightless environment of space, there are no convective currents, and equipment must be cooled by means of circulating fluids.) Because astronauts' lives depended on this system, most of the functions were provided with redundancy—and yet the entire unit was not much bigger than a window air conditioner.

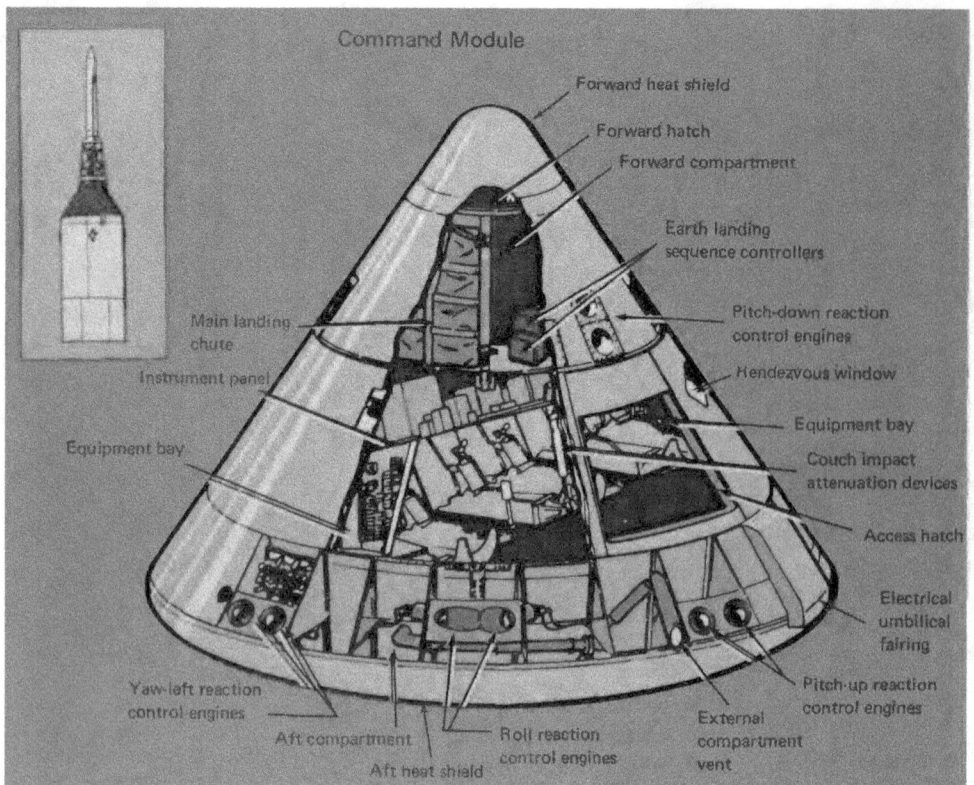

Command Module

Forward heat shield
Forward hatch
Forward compartment
Earth landing sequence controllers
Pitch-down reaction control engines
Rendezvous window
Equipment bay
Couch impact attenuation devices
Access hatch
Electrical umbilical fairing
Pitch-up reaction control engines
External compartment vent
Aft heat shield
Roll reaction control engines
Aft compartment
Yaw-left reaction control engines
Equipment bay
Instrument panel
Main landing chute

Service Module

Fitting pad (6 places)
Fuel tank
Helium tanks
Reaction control propellant tank
Pressure system panel
Fuel cells
Oxidizer tank (2 places)
Fuel tank
High gain antenna (retracted)
Service propulsion engine
Umbilical connector
Reaction control engine (4 places)
LO₂ tank
LH₂ tank
LO₂ tank

Looking like a huge toy top, the conical command module was crammed with some of the most complex equipment ever sent into space. The three astronaut couches were surrounded by instrument panels, navigation gear, radios, life-support systems, and small engines to keep it stable during reentry. The entire cone, 11 feet long and 13 feet in diameter, was protected by a charring heat shield. The 6.5-ton CM was all that was finally left of the 3000-ton Saturn V stack that lifted off on the journey to the Moon.

Packed with plumbing and tanks, the service module was the CM's constant companion until just before reentry. So all components not needed during the last few minutes of flight, and therefore requiring no protection against reentry heat, were transported in this module. It carried oxygen for most of the trip; fuel cells to generate electricity (along with the oxygen and hydrogen to run them); small engines to control pitch, roll, and yaw; and a large engine to propel the spacecraft into —and out of—lunar orbit.

The lunar module (facing page) was also a two-part spacecraft. Its lower or descent stage had the landing gear and engines and fuel needed for the landing. When the LM blasted off the Moon, the descent stage served as the launching pad for its companion ascent stage, which was also home for the two explorers on the surface. In function if not in looks the LM was like the CM, full of gear to communicate, navigate, and rendezvous. But it also had its own propulsion system, an engine to lift it off the Moon and send it on a course toward the command module orbiting above.

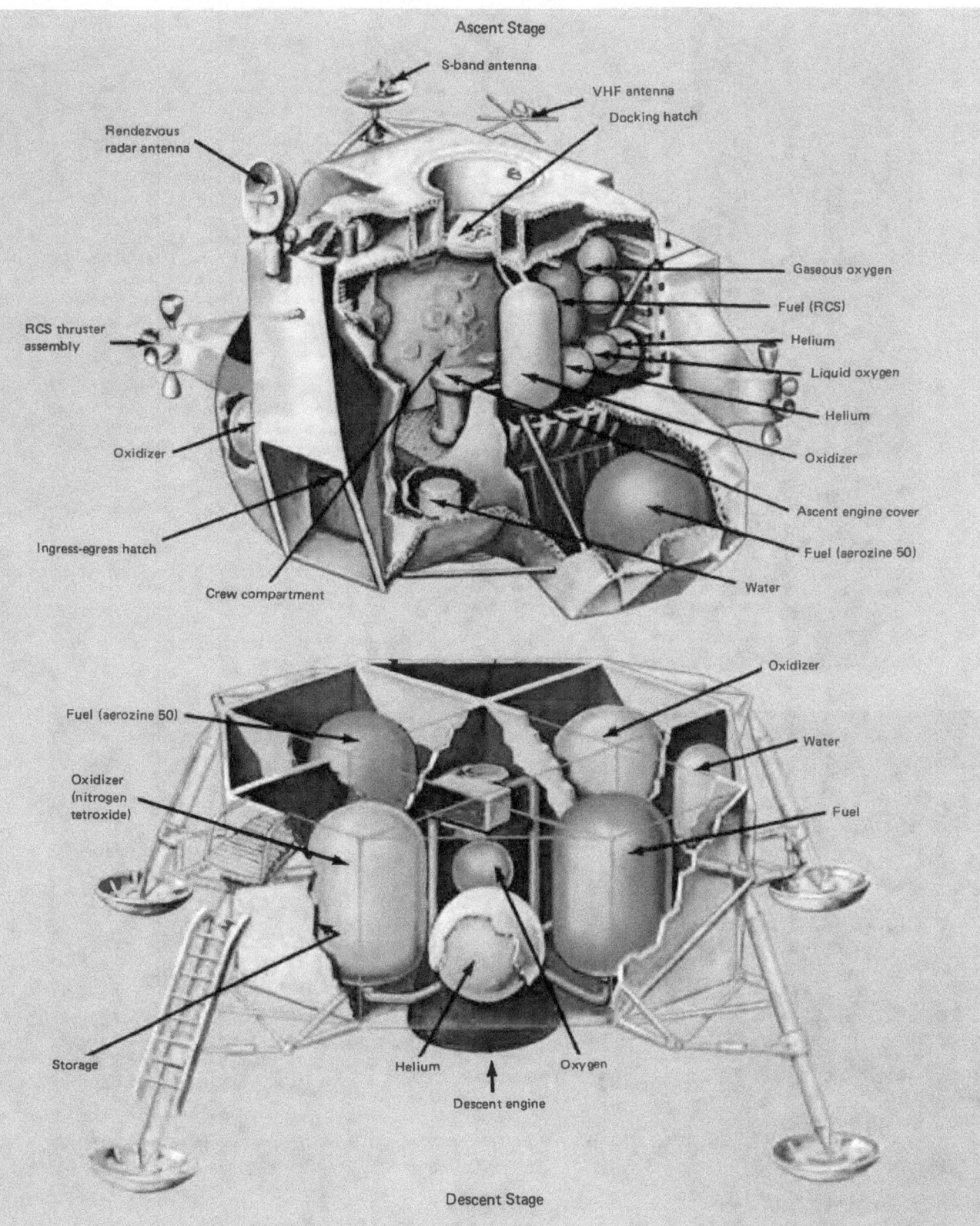

Ascent Stage

S-band antenna

VHF antenna

Docking hatch

Rendezvous radar antenna

Gaseous oxygen

Fuel (RCS)

Helium

Liquid oxygen

RCS thruster assembly

Helium

Oxidizer

Oxidizer

Ascent engine cover

Ingress-egress hatch

Fuel (aerozine 50)

Crew compartment

Water

Oxidizer

Fuel (aerozine 50)

Water

Oxidizer (nitrogen tetroxide)

Fuel

Storage

Helium

Oxygen

Descent engine

Descent Stage

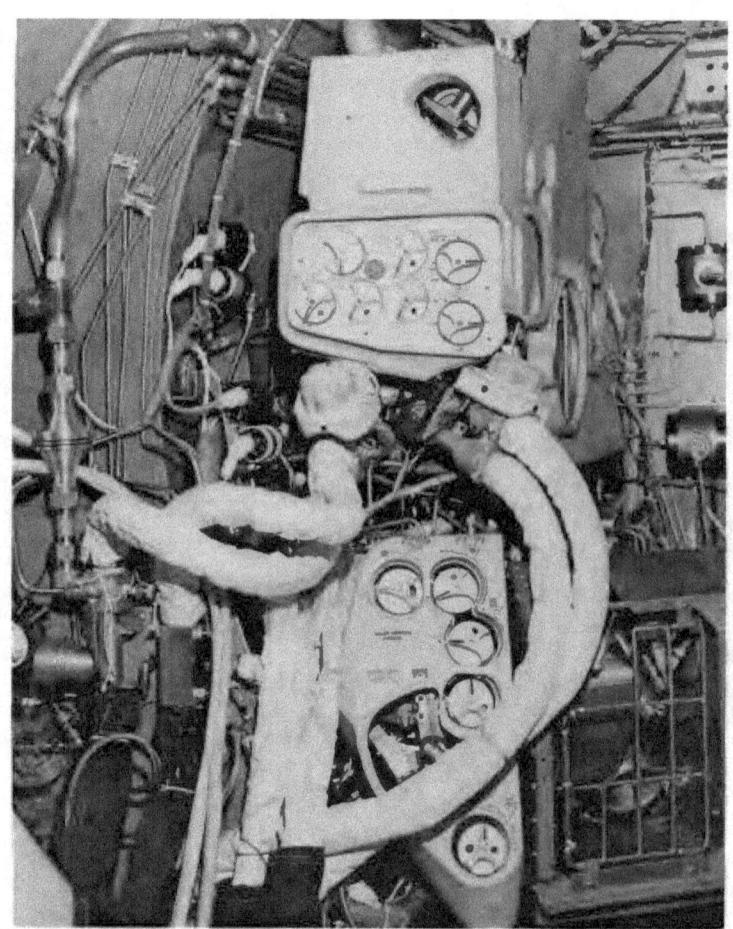

Like a plumber's dream, the LM's environmental control system nestled in a corner of the ascent stage. Those hoses provided pure oxygen to two astronauts at a pressure one-third that of normal atmosphere, and at a comfortable temperature. The unit recirculated the gas, scrubbed out CO_2 and moisture exhaled, and replenished oxygen as it was used up.

Sound is deadened and not an echo can be heard in this anechoic test chamber. Used to simulate reflection-free space, its floor, walls, and ceiling are completely covered with foam pyramids that absorb stray radiation, so that an antenna's patterns can be accurately measured. Here two NASA engineers inspect a test setup of an astronaut's backpack. Any interference between the astronaut and his small antenna could be detected and fixed before a real astronaut set foot on the Moon.

How do you generate enough electric power to run a ship in space? In the CSM, the answer was fuel cells; in the LM, storage batteries. Apollo fuel cells used oxygen and hydrogen—stored as liquids at extremely cold temperatures—that when combined chemically yielded electric power and, as a byproduct, water for drinking. (In early flights the water contained entrapped bubbles of hydrogen, which caused the astronauts no real harm but engendered major gastronomical discomfort. This led to loud complaints, and the problem was finally solved by installing special diaphragms in the system.) The fuel-cell power system was efficient, clean, and absolutely pollution-free. Storing oxygen and hydrogen required new advances in leakproof insulated containers. If an Apollo hydrogen tank were filled with ice and placed in a room at 70 F, it would take 8.5 years for the ice to melt. If an automobile tire leaked at the same rate as these tanks, it would take 30 million years to go flat.

"Houston, this is Tranquility." These words soon would be heard from another world, coming from an astronaut walking on the Moon, relayed to the LM, then to a tracking station in Australia or Spain or California, and on to Mission Control in Houston with only two seconds' delay. Communications from the Moon were clearer and certainly more reliable than they were from my home in Nassau Bay (a stone's throw from the Manned Spacecraft Center) to downtown Houston. At the same time, a tiny instrument would register a reading in the astronauts' life-support system, and a few seconds later an engineer in Mission Control would see a variation in oxygen pressure, or a doctor a change in heart rate; and around the world people would watch on their home television sets. Behind all of this would be the Apollo communications system—designed to be the astronauts' life line back to Earth, to be compact and lightweight, and yet to function with absolute reliability; an array of receivers, transmitters, power supplies and antennas, all tuned to perfection, that allowed the men and equipment on the ground to extend the capabilities of the astronauts and their ships. (Later on, when the computer on Apollo 11's LM was overloaded during the critical final seconds of the landing, it was this communications system that enabled a highly skilled flight controller named Steve Bales to tell Neil Armstrong that it was safe to disregard the overload alarms and to go ahead with the lunar landing.)

If you had to single out one subsystem as being most important, most complex, and yet most demanding in performance and precision, it would be Guidance and Navigation. Its function: to guide Apollo across 250,000 miles of empty space; achieve a precise orbit around the Moon; land on its surface within a few yards of a predesignated spot; guide LM from the surface to a rendezvous in lunar orbit; guide the CM to hit the Earth's atmosphere within a 27-mile "corridor" where the air was thick enough to capture the spacecraft, and yet thin enough so as not to burn it up; and finally land it close to a recovery ship in the middle of the Pacific Ocean. Designed by the Massachusetts Institute of Technology under Stark Draper's leadership, G&N consisted of a miniature computer with an incredible amount of information in its memory; an array of gyroscopes and accelerometers called the inertial-measurement unit; and a space sextant to enable the navigator to take star sightings. Together they determined precisely the spacecraft location between Earth and Moon, and how best to burn the engines to correct the ship's course or to land at the right spot on the Moon with a minimum expenditure of fuel. Precision was of utmost importance; there was no margin for error, and there were no reserves for a missed approach to the Moon. In Apollo 11, *Eagle* landed at Tranquility Base, after burning its descent engine for 12 minutes, with only 20 seconds of landing fuel remaining.

But the guidance system only told us where the spacecraft was and how to correct its course. It provided the brain, while the propulsion system provided the brawn in the form of rocket engines, propellant tanks, valves, and plumbing. There were 50 engines on the spacecraft, smaller but much more numerous than those on the combined three stages of the Saturn that provided the launch toward the Moon. Most of them—16 on the LM, 16 on the SM, and 12 on the CM—furnished only 100 pounds of thrust apiece; they oriented the craft in any desired direction just as an aircraft's elevators, ailerons, and rudder control pitch, roll, and yaw.

Three of the engines were much larger. On the service module a 20,500-pound-thrust engine injected Apollo into lunar orbit, and later brought it back home; on the LM there was a 10,500-pound-thrust engine for descent, and a 3500 pounder for ascent. All three *had* to work: a failure would have stranded astronauts on the Moon or in lunar orbit. They were designed with reliability as the number one consideration. They used hypergolic propellants that burned spontaneously on contact and required no spark plugs; the propellants were pressure-fed into the thrust chamber by bottled helium, eliminating complex pumps; and the rocket nozzles were coated with an ablative material for heat protection, avoiding the need for intricate cooling systems.

Three other engines could provide instant thrust at launch to get the spacecraft away from the Saturn if it should inadvertently tumble or explode. The largest of these produced 160,000 pounds of thrust, considerably more than the Redstone booster which propelled Alan Shepard on America's first manned spaceflight. (Since we never had an abort at launch, these three were never used.)

There were other subsystems, each with its own intricacies of design, and, more often than not, with its share of problems. There were displays and controls, backup guidance systems, a lunar landing gear on the LM and an Earth landing system (parachutes) on the CM, and a docking system designed with the precision of a Swiss watch, yet strong enough to stop a freight car. There were also those things that fell between the subsystems: wires, tubes, plumbing, valves, switches, relays, circuit breakers, and explosive charges that started, stopped, ejected, separated, or otherwise activated various sequences.

A TRAGIC FIRE TAKES THREE LIVES

Apollo in January 1967 was adjudged almost ready for its first manned flight in Earth orbit. And then disaster. A routine test of Apollo on the launching pad at Cape Kennedy. Three astronauts—Grissom, White, and Chaffee—in their spacesuits in a 100-percent oxygen environment. A tiny spark, perhaps a short circuit in the wiring. It was all over in a matter of seconds. Yet it would be 21 months before Apollo would again be ready to fly.

By April 1967, when I was given the Apollo spacecraft job, an investigation board had completed most of its work. The board was not able to pinpoint the exact cause

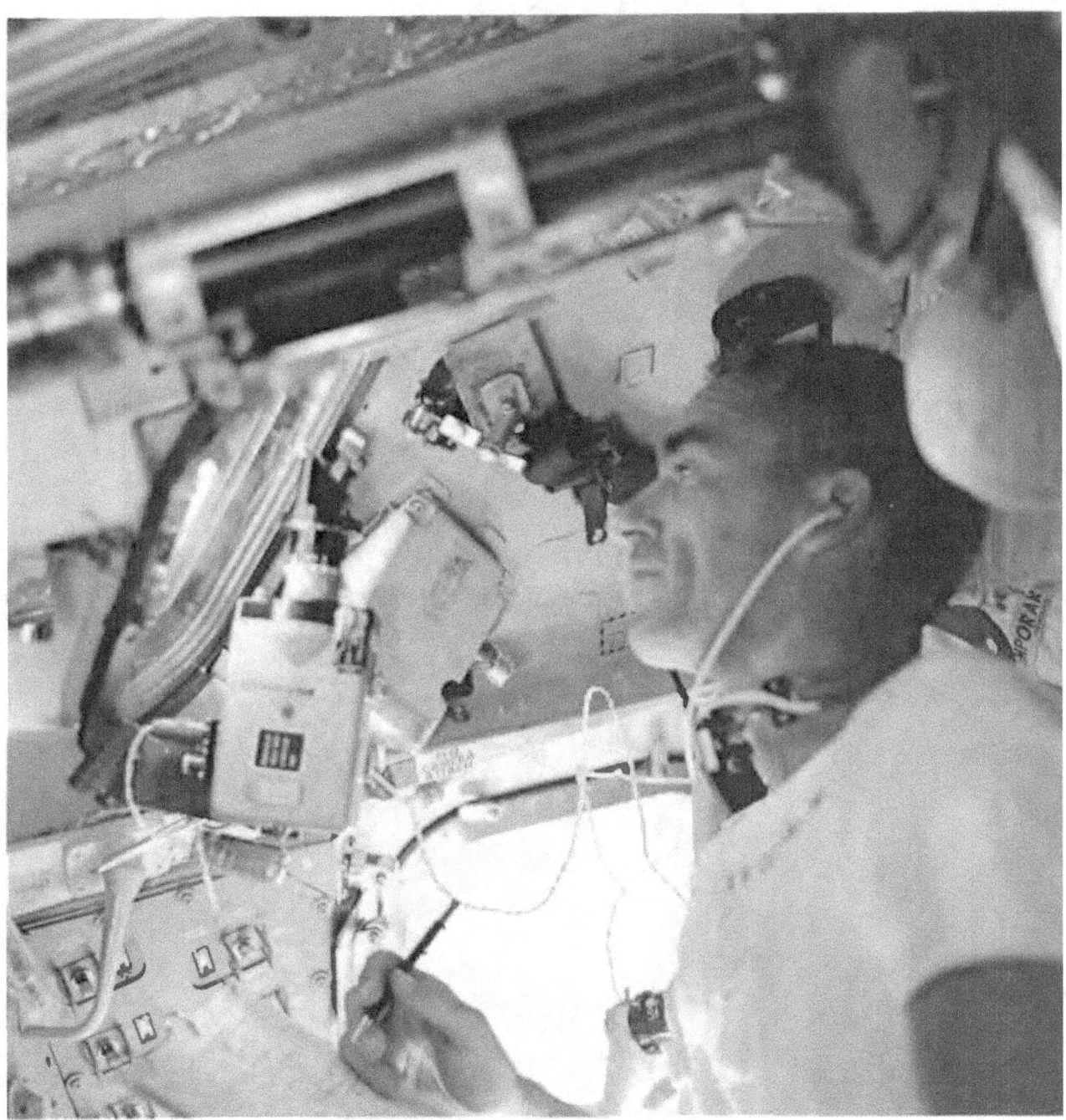

Like new Magellans, astronauts learned to navigate in space. Here Walt Cunningham makes his observations through a spacecraft window. The tools of a space navigator included a sextant to sight on the stars, a gyroscopically stabilized platform to hold a constant reference in space, and a computer to link the data and make the most complex and precise calculations.

Because there is no air to deflect, a spacecraft lacks rudders or ailerons. Instead, it has small rocket engines to pitch it up or down, to yaw it left or right, or to roll it about one axis. Sixteen of them were mounted on the service module, in "quads" of four. Here one quad is tested to make sure that hot rocket exhaust will not burn a hole in the spacecraft's thin skin.

Similar in shape but not size were the three big engines aboard Apollo spacecraft. Two of them had no backup, so they were designed to be the most reliable engines ever built. If the service-propulsion engine failed in lunar orbit, three astronauts would be unable to return home; if the ascent engine failed on the Moon, it would leave two explorers stranded. (A descent-engine failure would not be as critical, because the ascent engine might be used to save the crew members.)

Apollo Spacecraft Engines

Service propulsion engine
thrust: 20,500 lb

LM descent engine
thrust: 10,500 lb

LM ascent engine
thrust: 3,500 lb

of the fire, but this only made matters worse because it meant that there were probably flaws in several areas of the spacecraft. These included the cabin environment on the launch pad, the amount of combustible material in the spacecraft, and perhaps most important, the control (or lack of control) of changes.

Apollo would fly in space with a pure oxygen atmosphere at 5 psi (pounds per square inch), about one-third the pressure of the air we breathe. But on the launching pad, Apollo used pure oxygen at 16 psi, slightly above the pressure of the outside air. Now it happens that in oxygen at 5 psi things will generally burn pretty much as they do in air at normal pressures. But in 16 psi oxygen most nonmetallic materials will burn explosively; even steel can be set on fire. Mistake number one: Incredible as it may sound in hindsight, we had all been blind to this problem. In spite of all the care, all the checks and balances, all the "what happens if's," we had overlooked the hazard on the launching pad.

Most nonmetallic things will burn—even in air or 5 psi oxygen—unless they are specially formulated or treated. Somehow, over the years of development and test, too many nonmetals had crept into Apollo. The cabin was full of velcro cloth, a sort of space-age baling wire, to help astronauts store and attach their gear and checklists. There were paper books and checklists, a special kind of plastic netting to provide more storage space, and the spacesuits themselves, made of rubber and fabric and plastic. Behind the panels there were wires with nonmetallic insulation, and switches and circuit breakers in plastic cases. There were also gobs of insulating material called RTV. (In Gordon Cooper's Mercury flight, some important electronic gear had malfunctioned because moisture condensed on its uninsulated terminals. The solution for Apollo had been to coat all electronic connections with RTV, which performed admirably as an insulator, but, as we found out later, burned in an oxygen environment.) Mistake number two: Far too much nonmetallic material had been incorporated in the construction of the spacecraft.

There is an old saying that airplanes and spacecraft won't fly until the paper equals their weight. There was a time when two men named Orville and Wilbur Wright could, unaided, design and build an entire airplane, and even make its engine. But those days are long gone. When machinery gets as complex as the Apollo spacecraft, no single person can keep all of its details in his head. Paper, therefore, becomes of paramount importance: paper to record the exact configuration; paper to list every nut and bolt and tube and wire; paper to record the precise size, shape, constitution, history, and pedigree of every piece and every part. The paper tells where it was made, who made it, which batch of raw materials was used, how it was tested, and how it performed. Paper becomes particularly important when a change is made, and changes must be made whenever design, engineering, and development proceed simultaneously as they did in Apollo. There are changes to make things work, and changes to replace a component that failed in a test, and changes to ease an astronaut's workload or to make it difficult to flip the wrong switch.

Mistake number three: In the rush to prepare Apollo for flight, the control of changes had not been as rigorous as it should have been, and the investigation board was unable to determine the precise detailed configuration of the spacecraft, how it was

The **pedigree** of just one Apollo spacecraft took this many books. A mind-numbing degree of documentation contributed to reliability, safety, and success. If one batch of one alloy in one part was found to be faulty, for example, a search could show if the bad material had found its way into other spacecraft, to lie in wait there.

Inspecting the new hatch, Wally Schirra makes sure his crew cannot be trapped as was the crew that died in the terrible Apollo spacecraft fire. Opening outward (to swing freely if pressure built up inside), the new hatch had to be much sturdier than the old inward-opening one. The complicated latch sealed against tiny leaks but allowed very rapid release.

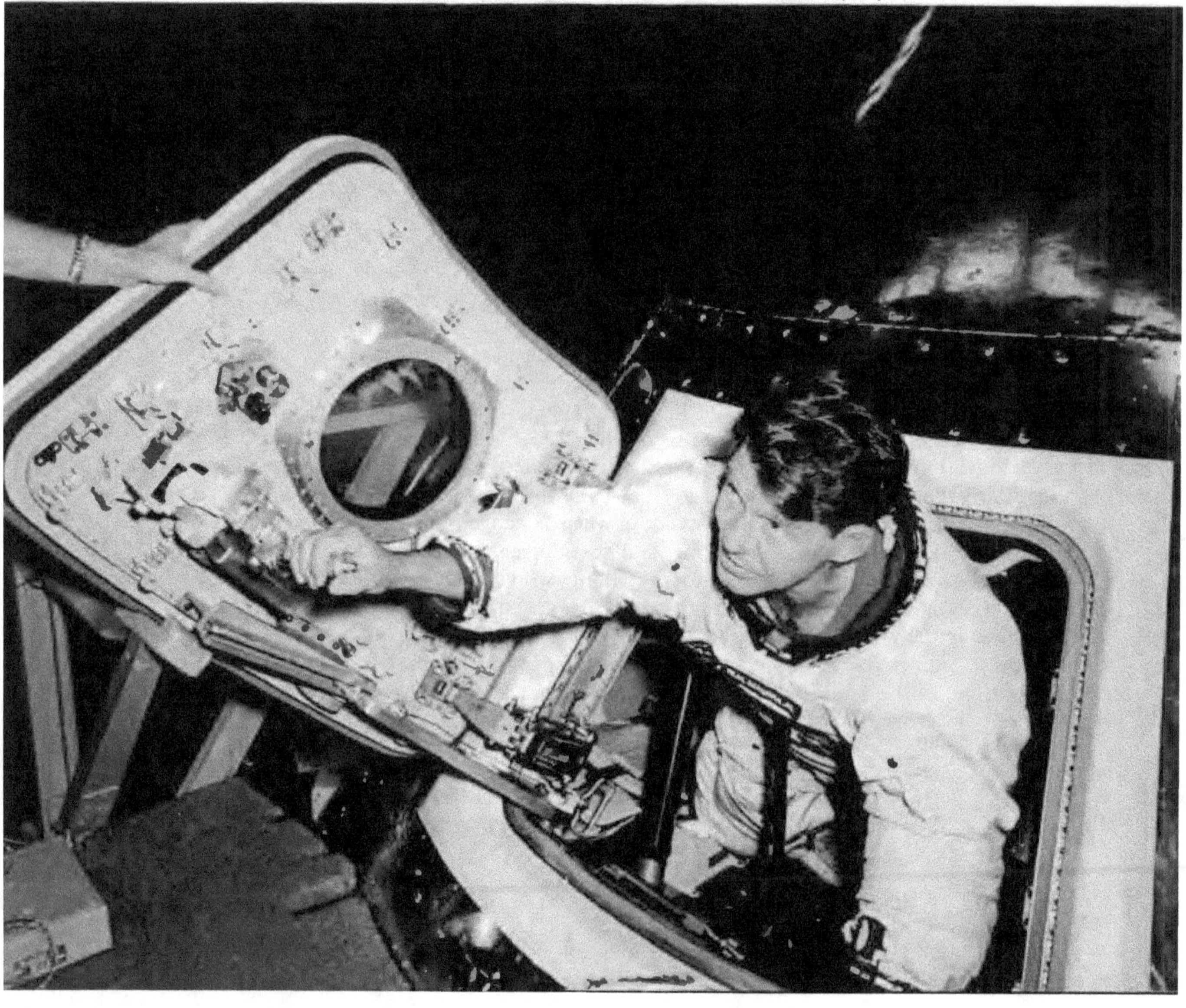

After the fire, flammability and self-extinguishment were key concerns. In the test setup at right a wiring bundle is purposely ignited, using the white flammable material within the coil near the bottom to simulate a short circuit. Picture at far right shows the aftermath: a fire that initially propagated but soon extinguished itself. It took great effort and ingenuity to devise materials that would not burn violently in the pure-oxygen atmosphere. If a test was not satisfactory and a fire did not put itself out, the material or wire routing was redesigned and then retested.

Seared at temperatures hotter than the surface of the Sun, a sample of heat-shield material survives the blast from a space-age furnace. Machines used to check out Apollo components were as demanding as those in the mission itself, because a mistake or miscalibration during preflight trials could easily lay the groundwork for disaster out in unforgiving space.

Meant to fly in a vacuum, and to survive fiery re-entry, the command module had also to serve as a boat. Although its parachutes appeared to lower it gently, its final impact velocity was still a jarring 20 mph. Tests like this one established its resistance to the mechanical and thermal shocks of impact, and its ability to float afterward.

Hitting land was possible, even though water was the expected landing surface. For this, a shock-absorbing honeycomb between the heat shield and the inner shell was one protection, along with shock absorbers on the couch supports. A third defense against impact was the way each couch was molded to its astronaut's size and shape, to provide him with the maximum support.

Through the portal of a huge test chamber, the command and service modules can be seen in preparation for a critical test: a simulated run in the entire space environment except for weightlessness. In this vacuum chamber one side of the craft can be cooled to the temperature of black night in space while the opposite side is broiled by an artificial Sun. Will coolant lines freeze or boil? Will the cabin stay habitable?

made, and what was in it at the time of the accident. Three mistakes, and perhaps more, added up to a spark, fuel for a fire, and an environment to make the fire explosive in its nature. And three fine men died.

AND THEN THE REBUILDING BEGAN

Now time was running out. The race against time began, with only 33 months remaining from April 1967 and the end of the decade. The work to be done appeared to be overwhelming and dictated 18-hour days, seven days a week. My briefcase was my office, my suitcase my home, as I moved from Houston to Downey, to Bethpage, to Cape Kennedy, and back to Houston again. At Tranquility Base, the Sun would only rise 33 more times before 1970.

Rebuilding meant changes and changes meant trouble if they were not kept under perfect control. Our solution was the CCB, the Configuration Control Board. On it were some of the best engineers in the world: my two deputies, Ken Kleinknecht and Rip Bolender; Apollo's Assistant for Flight Safety, Scott Simpkinson; Max Faget, Houston's Chief Engineer; Chris Kraft, the Chief of Flight Operations; Deke Slayton, the head of the astronauts; Dale Myers for North American Rockwell; and Joe Gavin for Grumman. The Board was rounded out with Chuck Berry for medical inputs and Bill Hess for science. It was organized by my technical assistant, George Abbey, who knew everything about everybody on Apollo, and who was always able to get things done. I was its chairman and made all decisions. Arguments sometimes got pretty hot as technical alternatives were explored. In the end I would decide, usually on the spot, always explaining my decision openly and in front of those who liked it the least. To me, this was the true test of a decision—to look straight into the eyes of the person most affected by it, knowing full well that months later on the morning of a flight, I would look into the eyes of the men whose lives would depend on that decision. One could not make any mistakes.

When I wasn't sure of myself or when I didn't trust my judgment, I knew where to go to get help—Bob Gilruth, my boss, who himself had been through every problem in Mercury. An extremely able engineer, Bob had acquired great wisdom over the years dealing with men and their flying machines. Bob was always there when I needed him.

The CCB met every Friday, promptly at noon, and often well into the night. From June 1967 to July 1969 the Board met 90 times, considered 1697 changes and approved 1341. We dealt with changes large and small, discussed them in every technical detail, and reviewed their cost and schedule impact. Was the change really necessary? What were its effects on other parts of the machine, on computer programs, on the astronauts, and on the ground tracking systems? Was it worth the cost, how long would it take, and how much would it weigh?

We redesigned the command module hatch to open out instead of in, because the old hatch had been a factor in trapping Grissom, White, and Chaffee inside their burning craft. This may sound simple, but it wasn't. An inward-opening hatch was much easier to build, because when it was closed it tended to be self-sealing since the pressure inside the spacecraft forced it shut. The opposite was true for an outward-opening hatch, which had to be much sturdier, and hence heavier, with complicated latches.

Escape from disaster was the objective of this spectacular test. The peril occurs in the early moments of launch, when the Saturn V contains thousands of tons of propellant: if things go wrong, the manned command module must be pulled away to a safe distance by the launch-escape rocket. Above, the launch-escape rocket is fitted to a test CM atop a Little Joe II booster. This booster, far cheaper than a Saturn, can duplicate its initial flight phases.

Up and away goes the command module, right, when the solid-fuel escape rocket—a single rocket firing through three nozzles—lights off. The sequence is begun only when the booster has accelerated the command module to "worst-case" speeds and heights. As it happened, the escape system was never needed during any of the Apollo launches.

We rewired the spacecraft, rerouted wire bundles, and used better insulation on the wires. We looked at every ounce of nonmetallic material, removed much of it, and concocted new materials for insulation and for pressure suits. We invented an insulating coating that would not burn, only to find that it would absorb moisture and become a conductor, so we had to invent another one. Pressure suits had to shed their nylon outer layer, to be replaced with a glass cloth; but the glass would wear away quickly, and shed fine particles which contaminated the spacecraft and caused the astronauts to itch. The solution was a coating for the glass cloth. We solved the problem of fire in the space atmosphere of 5 psi oxygen; but try as we might, we could not make the ship fireproof in the launch-pad atmosphere of 16 psi oxygen. Then Max Faget came up with an idea: Launch with an atmosphere that was 60 percent oxygen and 40 percent nitrogen, and then slowly convert to pure oxygen after orbit had been reached and the pressure was 5 psi. The 60–40 mixture was a delicate balance between medical requirements on the one hand (too much nitrogen would have caused the bends as the pressure decreased) and flammability problems on the other. It worked.

Weight is a problem in the design of any flying machine. Apollo, with its many changes, was anything but an exception. Problems are always easier to solve if one can afford a little leeway for making a change, but difficult and expensive if there is no weight margin. In the command module, we found a way to gain an extra 1000 pound margin by redesigning the parachute to handle a heavier CM. This margin made other CM changes relatively simple, and certainly less costly and time consuming.

For LM there was no such solution. We had to shave an ounce here, another there, to make room for the changes that had to be made. It was difficult, lengthy, and expensive.

TESTING AND RETESTING TO GET READY FOR FLIGHT

We tested for "sneak circuits" (inadvertent electrical paths), discovered some, and made changes. We ran a "failure mode and effects analysis"—a search for all the "what happens if's"—and made more changes. We tested, and retested, and changed and fixed and tested again. We set off small explosive charges inside the burning rocket engines, and to our horror found the all-important LM ascent engine was prone to catastrophic instability—a way of burning that could destroy the engine on takeoff and leave the astronauts stranded on the Moon. Much to the consternation of my bosses in Washington, we sent out new bids and selected a different contractor who built a new engine faster than anyone believed possible. But it worked.

No detail was too small to consider. We asked questions, received answers, asked more questions. We woke up in the middle of the night, remembering questions we should have asked, and jotted them down so we could ask them in the morning. If we made a mistake, it was not because of any lack of candor between NASA and contractor, or between engineer and astronaut; it was only because we weren't smart enough to ask all the right questions. Every question was answered, every failure understood, every problem solved.

We built mockups of the entire spacecraft, and tried to set them on fire. If they burned, we redesigned, rebuilt, and tried again. By vibration we tried to shake things apart; we tested in chambers simulating the vacuum of space, the heat of the Sun, and the cold of the lunar night. We subjected all systems to humidity and salt spray, to the noise of the booster, and the shock of a hard landing. We dropped the command module into water to simulate normal landings and on land to test for emergency landings; we plopped the lunar module on simulated lunar terrain. We overstressed and overloaded until things broke, and if they broke too soon, we redesigned and rebuilt and tested again.

The final exam came in flight. First the command module was tested with only the launch-escape tower, against the possibility of a Saturn exploding on the launch pad. Then we launched the CSM on a special booster, the Little Joe II, to see whether it would survive if the Saturn should fail in the atmosphere, when air loads are at their peak. (There is a big difference between manned and unmanned flight. If the launch vehicle should stray off course while lifting an automated payload, the range safety officer could press a button and destroy booster and payload together; in manned flight the spacecraft would first be separated from the errant booster, which would then be blown up before it wandered off, leaving the CM to be carried to safety by the launch escape tower. This separation maneuver demanded the utmost in speed and power.)

The CSM, unmanned, was flown twice on the Saturn 1B (1,600,000 pounds of thrust). Then, on November 9, 1967, came the most critical test of all: Apollo 4, the first flight of the Saturn V (7,500,000 pounds), would subject the CSM to the lunar return speed of 25,000 mph. After achieving an altitude of 10,000 miles, the spacecraft's engines drove Apollo back down into the atmosphere at unprecedented speed. Temperatures on the heat shield reached 5000 F, more than half the surface temperature of the Sun. The heat shield charred as expected, but the inside of the cabin remained at a comfortable 70 F. A major milestone had been passed.

Apollo 5 on January 22, 1968, was the first flight test of LM—an unmanned flight in Earth orbit that put the lunar module through its paces. There were problems. The computer shut down the LM's descent engine prematurely on its first burn. But then the flight controllers on the ground took over and continued the flight with an alternate mission. Now another question arose: Should we repeat this flight? Grumman felt we should; I disagreed. After considerable technical debate, we decided that the next flight with LM would be manned—which it was, 14 months later.

Apollo 6, three months after Apollo 5, was to be a simple repeat of Apollo 4, but it wasn't. The Saturn had problems, and so did the spacecraft adapter—that long conical section which joined the CSM to the booster, and which also served as LM's cocoon. (The spacecraft itself did a beautiful job.) After a fantastic piece of detective work by Don Arabian, our chief test engineer, we found a flaw in the manufacturing of the honeycomb structure of the adapter, and how to fix it.

October 11, 1968. Eighteen months since that day in the pilot's lounge at Washington Airport when I said yes, I would take on Apollo. Eighteen of the greatest months an engineer could ask for. In that time 150,000 Americans had worked around the clock, dedicating their skills and their lives to forge two of the most magnificent flying machines yet devised: CSM and LM. It was a beautiful morning in Florida, just the kind of morning for another launch. This time Apollo was ready for its men.

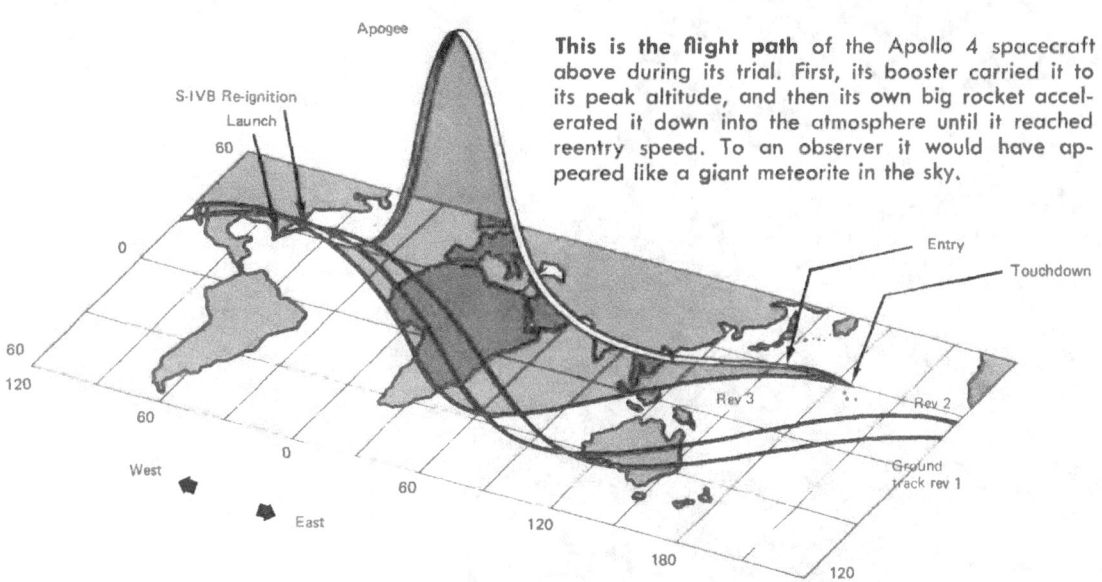

Charred but perfectly intact, the CM here had passed its most severe test of reentry at a speed of 25,000 mph. From left, Ralph Ruud, Dale Myers, George Low, and Robert Gilruth.

Apogee

S-IVB Re-ignition

Launch

60

0

60

120

West

East

60

0

60

120

180

120

Entry

Touchdown

Rev 3

Rev 2

Ground track rev 1

This is the flight path of the Apollo 4 spacecraft above during its trial. First, its booster carried it to its peak altitude, and then its own big rocket accelerated it down into the atmosphere until it reached reentry speed. To an observer it would have appeared like a giant meteorite in the sky.

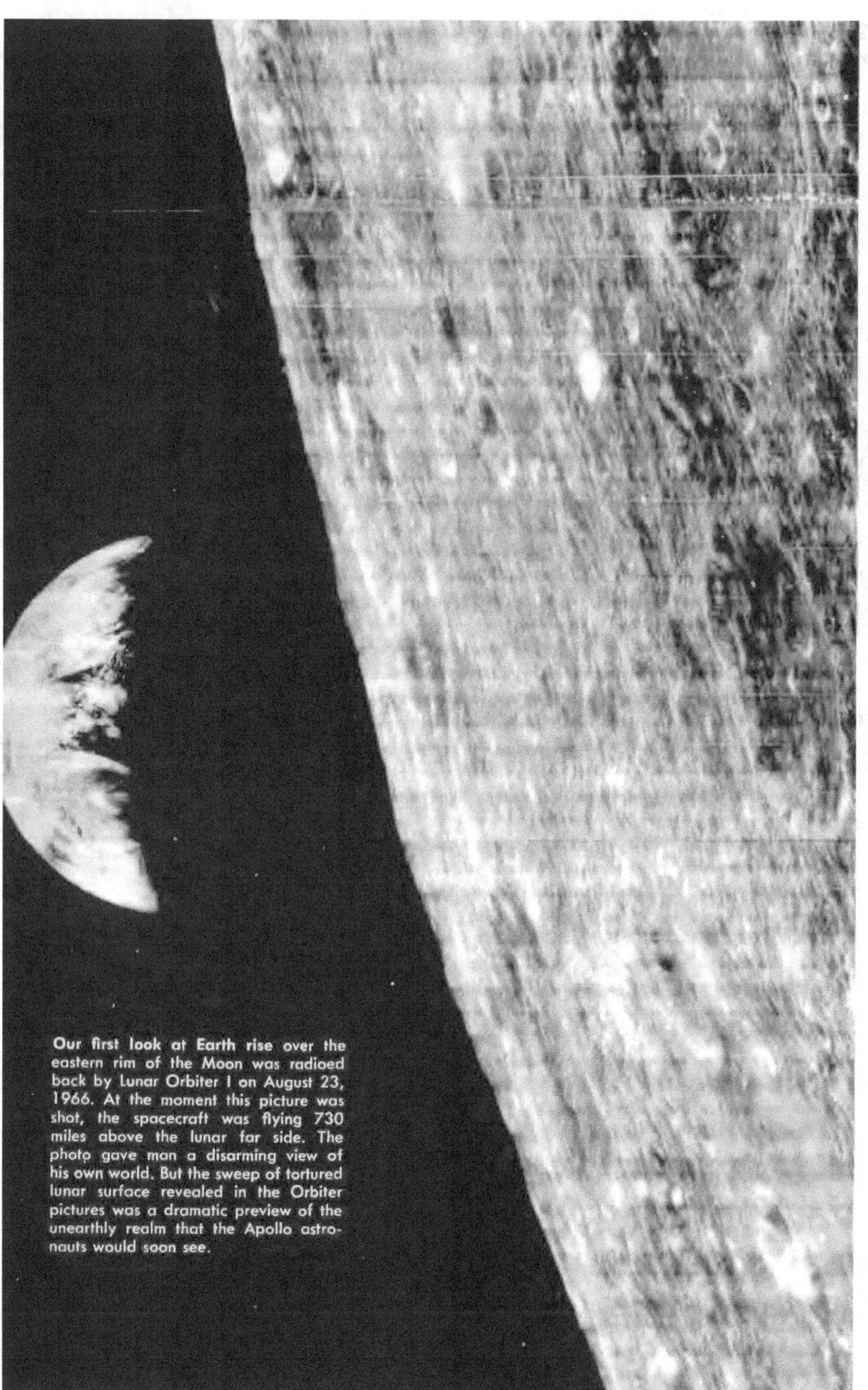

Our first look at Earth rise over the eastern rim of the Moon was radioed back by Lunar Orbiter I on August 23, 1966. At the moment this picture was shot, the spacecraft was flying 730 miles above the lunar far side. The photo gave man a disarming view of his own world. But the sweep of tortured lunar surface revealed in the Orbiter pictures was a dramatic preview of the unearthly realm that the Apollo astronauts would soon see.

CHAPTER FIVE

Scouting the Moon

By EDGAR M. CORTRIGHT

After centuries of studying the Moon and its motions, most astronomers—faced with diminishing returns—had abandoned it to lovers and poets by the time that Sputnik ushered in the space age. The hardy few who had not been wooed away to greener astronomical pastures were soon to be richly rewarded for their patience.

Before the invention of the telescope in 1608, astronomers had to be content with two good eyes and a fertile imagination to surmise the nature of the lunar surface. As a consequence they mainly devoted themselves to the mathematics of the Moon's motions relative to the Earth and Sun. The early telescopes that first revealed the crater-pocked face of the Moon touched off several centuries of speculation about the lunar surface by scientists and fiction writers alike—it often being unclear who was writing the fiction. But telescopes peering through the turbulent atmosphere of Earth have severe limitations. By 1956 the very best terrestrial telescope images of the Moon were only able to resolve objects about the size of the U.S. Capitol. Anything smaller was a mystery.

So the question remained: What was the lunar surface really like? While few people really believed the Moon to be made of green cheese, many scientific hypotheses

Galileo drew the Moon in 1610 and described its surface as "uneven, rough, replete with cavities and packed with protruding eminences." While a correct description, his details are unrecognizable now.

Harold Urey enhanced our view of the lunar surface by creating this montage made from segments of Lick Observatory photos taken when the Sun angle was low so that shadows emphasized relief.

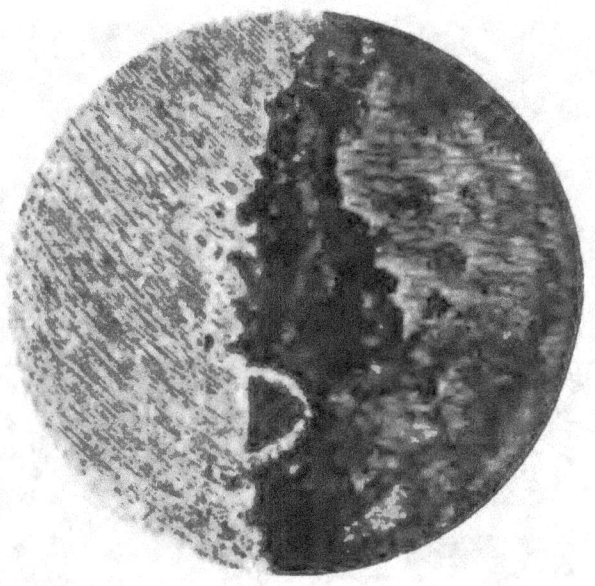

A **Juno II launch** vehicle was made up of an Army Ballistic Missile Agency's modified Jupiter first stage with a spin-stabilized solid-rocket upper stage developed by the Jet Propulsion Laboratory. JPL also developed the 13.4-lb Pioneer IV payload shown below. Launched March 3, 1959, it helped detect and measure the second of the Earth's great radiation belts.

The Thor-Able, with more advanced upper stages, propelled the 94.4-lb Pioneer V spacecraft to escape velocity on March 11, 1960. The instrumentation of this payload (below) measured the Earth's outer magnetosphere, detected an interplanetary magnetic field, helped explain the effect of solar flares on cosmic rays, and first detected the plasma clouds emitted by the Sun during solar storms. But it told nothing new about the Moon.

An unsuccessful effort to orbit the Moon was made with the Atlas-Able launch vehicle (above) and the Pioneer P—31 spacecraft (below). The launch vehicle was capable of injecting 380 lb into a translunar trajectory. The 39-in. spacecraft carried two 18.5-lb-thrust liquid-propellant engines for flight-path corrections and injection into lunar orbit.

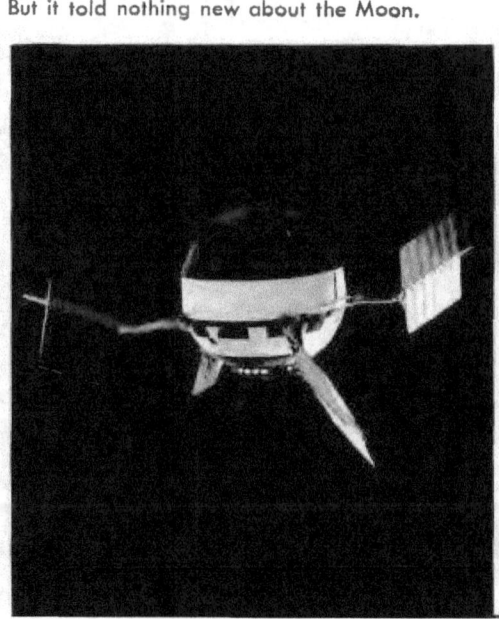

A sophisticated craft for its day, the 800-lb Ranger or its launch vehicle failed in its first six tries. Then it behaved beautifully, returning thousands of pictures in its last three flights, most of them far superior to the best that could be obtained from telescopes on Earth. Rangers crashed on the Moon at nonsurvivable velocity; their work was done in the few short moments from camera turn-on to impact.

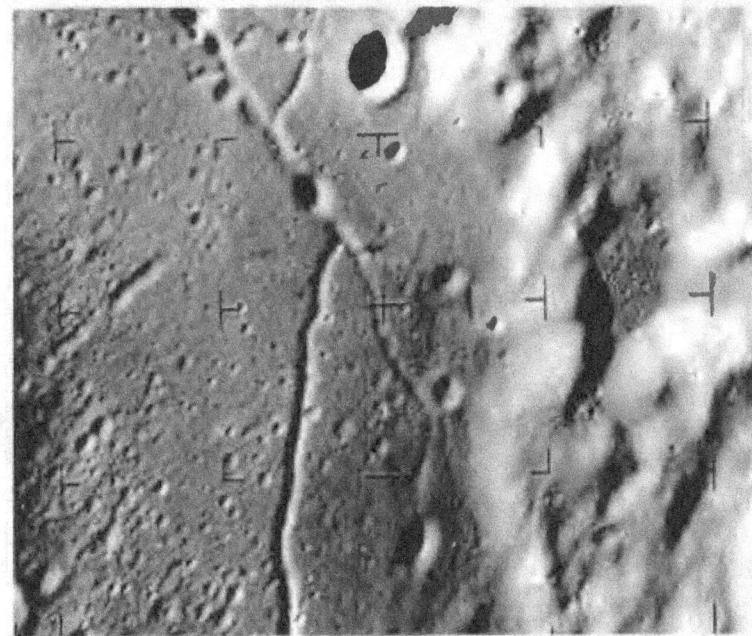

Heading in toward Alphonsus, a lunar crater of high scientific interest, Ranger IX sent back 5814 pictures of the surface before it crashed. The one at left, taken several score miles away, shows part of the crater floor and slumped wall of Alphonsus, a rille structure, and a varied population of craters. Ranger pictures were exciting in the wholly new details of the Moon that they provided.

The last instant before it smashed, Ranger IX radioed back this historic image, taken at a spacecraft altitude of one-third mile about a quarter of a second before impact. The area pictured is about 200 by 240 feet, and details about one foot in size are shown. The Ranger pictures revealed nothing that discouraged Apollo planners, although they did indicate that choosing an ideally smooth site for a manned landing was not going to be an easy task.

cherished not long ago were equally strange and rather more ominous. They included deep fields of dust into which a spacecraft might sink; a labyrinth of "fairy castles" such as children build by dripping wet sand at the beach; electrostatic dust that might spring up and engulf an alien object; and treacherously covered crevasses into which an unwary astronaut might fall. What proved to be the most accurate prediction, however, likened the Moon to a World War I battlefield, bombarded by a rain of meteoroids throughout the millennia, and churned into a wasteland of craters and debris. The absence of an atmosphere and the low gravitational field would allow small secondary particles to be blasted from the surface by a primary meteoroid impact and thrown unimpeded halfway around the Moon. This led to the concept of a uniform blanket of ejecta over the entire Moon.

But our story is getting ahead of itself. The surface properties of the Moon were largely unknown in 1958, a matter which assumed great practical importance when man's first journeys to the Moon began to take shape. How much weight would the surface support? What were the slopes? Were there many rocks—and of what size? Would the dust or dirt cling? What was the intensity of primary and secondary meteoroid bombardment? What was the exact size and shape of the Moon, and what were the details of the lunar gravity field into which our spaceships would one day plunge?

A SHAKY START

The military rockets developed in the 1950s provided a basic tool with which it became possible to send rudimentary spacecraft to the Moon. Both the Army and the Air Force were quick to initiate efforts to be the first to the Moon with a manmade object. (The Russians, as it proved, were equally quick, or quicker.) These first U.S. projects, which were transferred in 1958 to the newly formed National Aeronautics and

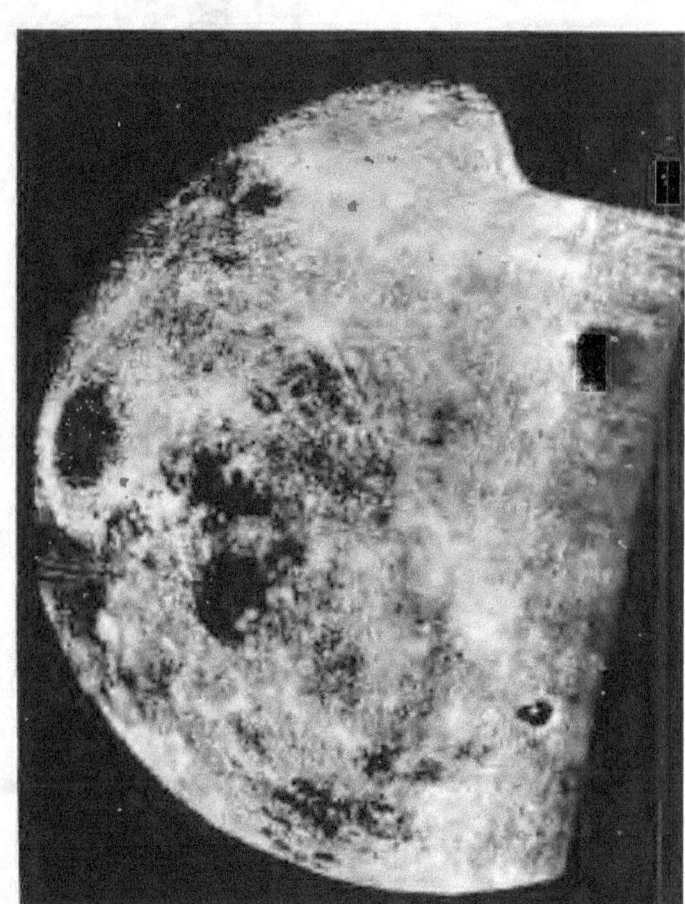

Mankind's first glimpse of the far side of the Moon came in October 1959, provided by the Soviet spacecraft Luna 3. Although crude compared with later views, its pictures showed a number of lunar features for the first time. One of these was the crater Tsiolkovsky, named for the famed Russian mathematician, which appears here in the lower right as a small sea with an island in it. The images from Luna 3 indicated that the Moon's far side lacked the large mare areas on the side facing Earth.

Space Administration, consisted of four Air Force Thor-Able rockets, and two Army Juno II rockets, each with tiny payloads, designed to measure radiation and magnetic fields near the Moon and, in some cases, to obtain rudimentary pictures. NASA and the Air Force then added three Atlas-Able rockets, which could carry heavier payloads, in an attempt to bolster these early high-risk efforts. Of these nine early missions launched between August 1958 and December 1960, none really succeeded. Two Thor-Able and all three Atlas-Able vehicles were destroyed during launch. One Thor-Able and one of the Juno II's did not attain sufficient velocity to reach the Moon and fell back to Earth. Two rockets were left.

The Soviets were also having problems. But on January 4, 1959, Luna 1, the first space vehicle to reach escape velocity, passed the Moon within about 3700 miles and went into orbit about the Sun. Two months later the United States repeated the feat with the last Juno II, although its miss distance was 37,300 miles. A year later the last Thor-Able payload flew past the Moon, but like its predecessors it yielded no new information about the surface. On October 7, 1959, the Soviet Luna 3 became the first spacecraft to photograph another celestial body, radioing to Earth crude pictures of the previously unseen far side of the Moon. The Moon was not a "billboard in the sky" with slatted back and props. Its far side was found to be cratered, as might be expected, but unlike the front there were no large mare basins. The primitive imagery that Luna 3 returned was the first milepost in automated scientific exploration of other celestial bodies.

Undaunted by initial failures, and certainly spurred on by Soviet efforts, a NASA team began to plan a long-term program of lunar exploration that would embody all necessary ingredients for success. The National Academy of Sciences was enlisted to help draw the university community into the effort. The Jet Propulsion Laboratory, a California Institute of Technology affiliate that had been transferred from the Army to NASA in 1958, was selected to carry out the program. JPL was already experienced in rocketry and had participated in the Explorer and Pioneer IV projects.

OUR FIRST CLOSE LOOK

The first project to emerge from this government/university team was named Ranger, to connote the exploration of new frontiers. Subsequently Surveyor and Prospector echoed this naming theme. (Planetary missions adopted nautical names such as Mariner, Voyager, and Viking.) The guideline instructions furnished JPL for Ranger read in part: "The lunar reconnaissance mission has been selected with the major objective . . . being the collection of data for use in an integrated lunar-exploration program. . . . The [photographic] system should have an overall resolution of sufficient capability for it to be possible to detect lunar details whose characteristic dimension is as little as 10 feet." Achieving this goal did not come about easily.

The initial choice of launch vehicle for the Ranger was the USAF Atlas, mated with a new upper stage to be developed by JPL, the Vega. Subsequently NASA cancelled the Vega in favor of an equivalent vehicle already under development by the Air Force, the Agena. This left JPL free to concentrate on the Ranger. The spacecraft design that evolved was very ambitious for its day, incorporating solar power, full

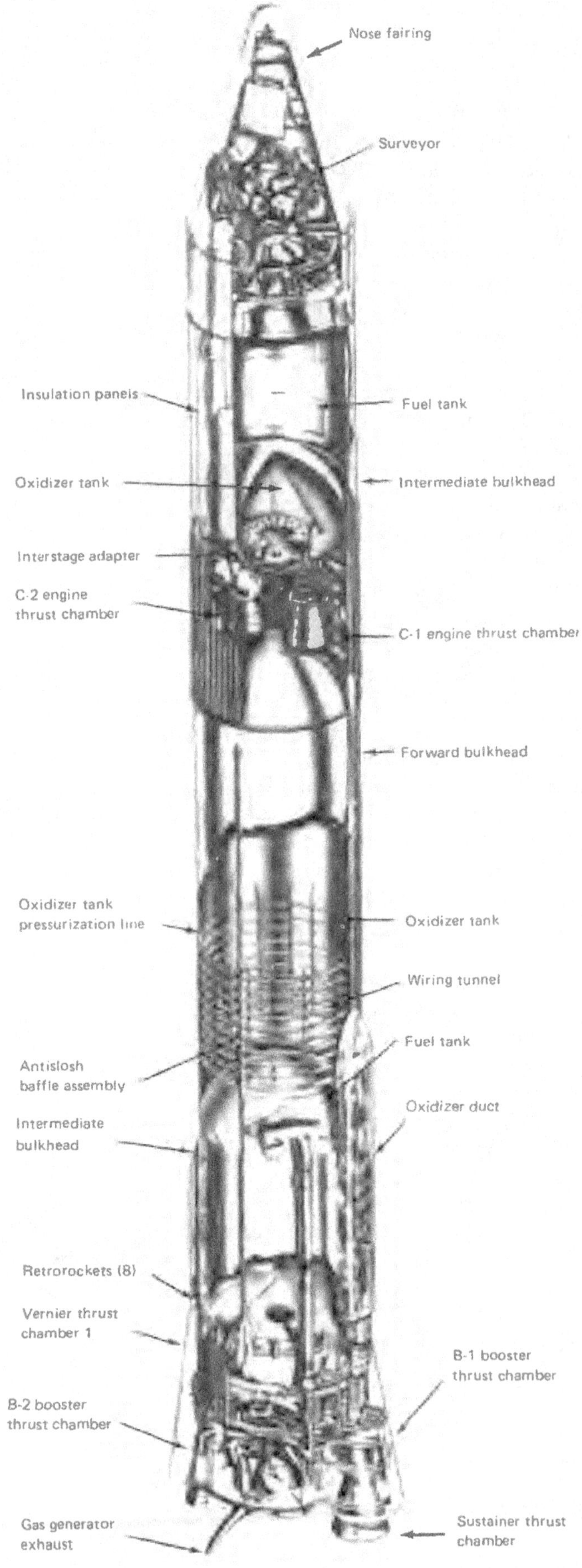

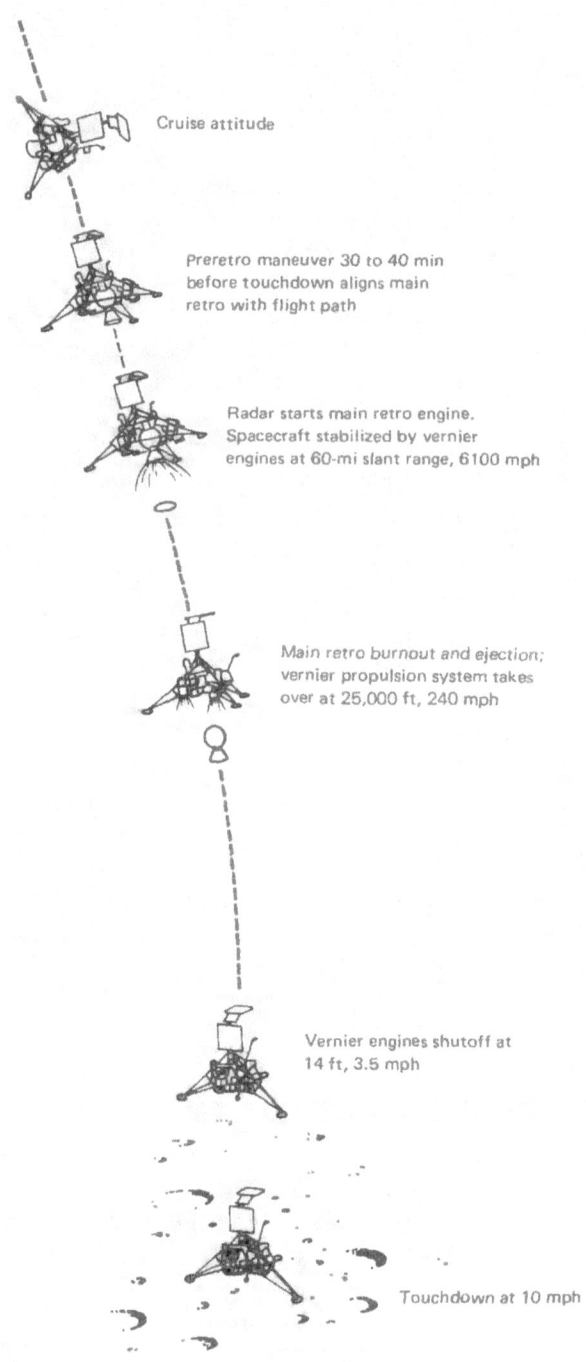

Nose fairing

Surveyor

Insulation panels

Fuel tank

Oxidizer tank

Intermediate bulkhead

Interstage adapter

C-2 engine thrust chamber

C-1 engine thrust chamber

Forward bulkhead

Oxidizer tank pressurization line

Oxidizer tank

Wiring tunnel

Fuel tank

Antislosh baffle assembly

Oxidizer duct

Intermediate bulkhead

Retrorockets (8)

Vernier thrust chamber 1

B-1 booster thrust chamber

B-2 booster thrust chamber

Gas generator exhaust

Sustainer thrust chamber

Cruise attitude

Preretro maneuver 30 to 40 min before touchdown aligns main retro with flight path

Radar starts main retro engine. Spacecraft stabilized by vernier engines at 60-mi slant range, 6100 mph

Main retro burnout and ejection; vernier propulsion system takes over at 25,000 ft, 240 mph

Vernier engines shutoff at 14 ft, 3.5 mph

Touchdown at 10 mph

The Surveyor mission had been conceived in 1959 as a scheme to soft-land scientific instruments on the Moon's surface. It was a highly ambitious plan that required both development of a radical new launch vehicle and the new technology of a closed-loop, radar-controlled automated landing. The cutaway drawing at left shows the Atlas-Centaur launch vehicle, the photo at the top of the facing page the Surveyor as flown, and the drawing above the main events in a successful landing sequence. The Atlas-Centaur, a major step forward in rocket propulsion, was the first launch vehicle to use the high-energy propellant combination of hydrogen and oxygen. Its new Centaur upper stage, built by General Dynamics, had two Pratt & Whitney RL—10 engines of 15,000-lb thrust each. The first stage was a modified Atlas D having enlarged tanks and increased thrust.

The spidery Surveyor consisted of a tubular framework perched on three shock-absorbing footpads. Despite its queer appearance, it incorporated some of the most sophisticated automatic systems man had ever hurled into space (see specifications below). The first one launched made a perfect soft landing on the Moon, radioing back to Earth a rich trove of imagery and data. Seven were launched in all; one tumbled during course correction, one went mysteriously mute during landing, and the remaining five were unqualified successes.

TYPICAL SURVEYOR SPECIFICATIONS WERE:	GUIDANCE AND CONTROL	PROPULSION
Weight at launch ....2193 lb	Inertial reference ....3-axis gyros	Main retrorocket ....9000-lb solid fuel
Landed weight ....625 lb	Celestial reference ....Sun and Canopus	Vernier retrorackets ......throttlable between
POWER	sensors	30- and 102-lb thrust each
Solar panel ....90 watts	Attitude control ....cold gas jets	TV CAMERA
Botteries ....230 ampere-hours	Terminal landing ....automated closed loop,	Focal length ....25 or 100 mm
COMMUNICATIONS	with radar altimeter and	Aperture ....f/4 to f/22
Dual transmitters ....10 watts each	doppler velocity sensor	Resolution ....1 mm at 4 m

Its insectlike shadow was photographed by Surveyor I on the desolate surface of Oceanus Procellarum. During the long lunar day it shot 10,386 pictures, including the 52 in this mosaic. The noon temperature of 235° F dropped to 250° below zero an hour after the Sun went down.

three-axis stabilization, and advanced communications. Clearly JPL also had its eye on the planets in formulating this design.

Of a total of nine Rangers launched between 1961 and 1965, only the last three succeeded. From the six failures we learned many lessons the hard way. Early in the program, an attempt was made to protect the Moon from earthly contamination by sterilizing the spacecraft in an oven. This technique, which is now being used on the Mars/Viking spacecraft, had to be abandoned at that time when it wreaked havoc with Ranger's electronic subsystems.

In the first two launches in 1961 the new Agena B upper stage failed to propel the Ranger out of Earth orbit. Failures in both the launch vehicle and spacecraft misdirected the third flight. On the fourth flight the spacecraft computer and sequencer malfunctioned. And on the fifth flight a failure occurred in the Ranger power system. The U.S. string of lunar missions with little or no success had reached fourteen. Critics were clamoring that Ranger was a "shoot and hope" project. NASA convened a failure review board, and its studies uncovered weaknesses in both the design and testing of Ranger. Redundancy was added to electronic circuits and test procedures were tightened. As payload Ranger VI carried a battery of six television cameras to record surface details during the final moments before impact. When it was launched on January 30, 1964, we had high confidence of success. Everything seemed to work perfectly. But when the spacecraft plunged to the lunar surface, precisely on target, its cameras failed to turn on. I will never forget the feeling of dismay in the JPL control room that day.

But we all knew we were finally close. Careful detective work with the telemetry records identified the most probable cause as inadvertent turn-on of the TV transmitter while Ranger was still in the Earth's atmosphere, whereupon arcing destroyed the system. The fix was relatively simple, although it delayed the program for three months. On July 28, 1964, Ranger VII was launched on what proved to be a perfect mission. Eighteen minutes before impact in Oceanus Procellarum, or Ocean of Storms, the cameras began transmitting the first of 4316 excellent pictures of the surface. The final frame was taken only 1400 feet above the surface and revealed details down to about 3 feet in size. It was a breathless group of men that waited the arrival of the first quick prints in the office of Bill Pickering, JPL's Director. The prints had not been enhanced and it was hard to see the detail because of lack of contrast. But those muddy little pictures with their ubiquitous craters seemed breathtakingly beautiful to us.

By the time of the Ranger VII launch, the Apollo program had already been underway for three years, and Ranger had been configured and targeted to scout possible landing sites. Thus Ranger VIII was flown to a flat area in the Sea of Tranquility where it found terrain similar to that in the Ocean of Storms: gently sloping plains but craters everywhere. It began to look as if the early Apollo requirement of a relatively large craterless area would be difficult to find. As far as surface properties were concerned, the Ranger could contribute little to the scientific controversy raging over whether the Moon would support the weight of a machine—or a man.

To get maximum resolution of surface details, it was necessary to rotate Ranger

The first lunar soft landing was accomplished by Russia's Luna 9 on February 3, 1966, about 60 miles northeast of the crater Calaverius. Its pictures showed details down to a tenth of an inch five feet away. They indicated no loose dust layer, both rounded and angular rock fragments, numerous small craters, some with slope angles exceeding 40 degrees, and generally granular surface material. These results increased confidence that the Moon was not dangerously soft for a manned landing.

Surveyor I televised excellent pictures of the depth of the depression in the lunar soil made by its footpad when it soft-landed on June 2, 1966, four months after Luna 9. Calculations from these and similar images set at rest anxieties about the load-bearing adequacy of the Moon. Some scientists had theorized that astronauts could be engulfed in dangerously deep dust layers, but Surveyor's footpad pictures, as well as the digging done by the motorized scoop on board, indicated that the Moon would readily support the LM and its astronauts.

Like a tiny back hoe, the surface sampler fitted to some Surveyors could dig trenches in the lunar soil. Above, the smooth vertical wall left by the scoop indicated the cohesiveness of the fine lunar material. Variations in the amount of current drawn by the sampler motor gave indication of the digging effort needed. At left above, the sampler is shown coming to the rescue when the head of the alpha-scattering instrument failed to deploy on command. After two gentle downward nudges from the scoop, the instrument dropped to the surface.

"A dinosaur's skull" was the joking name that Surveyor I controllers used for this rock. Geologists on the team were more solemn: "A rock about 13 feet away, 12 by 18 inches, subangular in shape, with many facets slightly rounded. Lighter parts of the rock have sharper features, suggesting greater resistance to erosion."

Surveyor VI hopped under its own power to a second site about eight feet from its landing spot. This maneuver made it possible to study the effect of firing rocket engines that impinged on the lunar surface. Picture at left below shows a photometric chart attached to an omniantenna, which was clean after first landing. Afterward, the chart was coated with an adhering layer of fine soil blasted out of the lunar surface.

so that the cameras looked precisely along the flight path. This was not done on Ranger VII in order to avoid the risk of sending extra commands to the attitude-control system. I recall that on Ranger VIII JPL requested permission to make the final maneuver. NASA denied permission—we were still unwilling, after the long string of failures, to take the slightest additional risk. It was not until Ranger IX that JPL made the maneuver and achieved resolution approaching 1 foot in the last frame. This final Ranger, launched on March 21, 1965, was dedicated to lunar science rather than to reconnaissance of Apollo landing sites. It returned 5814 photographs of the crater Alphonsus, again showing craters within craters, and some rocks. Despite its dismal beginnings the Ranger program was thus concluded on a note of success. Proposed follow-on missions were cancelled in favor of upcoming Surveyor and Orbiter missions, whose development had been proceeding concurrently.

TESTING THE SURFACE

Surveyor, which had been formally approved in the spring of 1960, was originally conceived for the scientific investigation of the Moon's surface. As in the case of the Ranger, its use was redirected according to the needs of Apollo.

With the proposed addition of an orbiting version of Surveyor, later to become Lunar Orbiter, the unmanned lunar-exploration program in support of Apollo shaped up this way: Ranger would provide us with our first look at the surface; Surveyor would make spot checks of the mechanical properties of the surface; and Lunar Orbiter would supply data for mapping and landing-site selection. The approach was sound enough, but carrying it out led us into a jungle of development difficulties.

Few space projects short of Apollo itself embodied the technological audacity of Surveyor. Its Atlas-based launch vehicle was to make use of an entirely new upper stage, the Centaur, the world's first hydrogen-fueled rocket. It had been begun by the Department of Defense and later transferred to NASA. Surveyor itself was planned to land gently on the lunar surface, set down softly by throttlable retrorockets under control of its own radar system. It was to carry 350 pounds of complex scientific instruments. Responsibility for continuing the Centaur development was placed with the Marshall Space Flight Center, with General Dynamics the prime contractor. JPL took on the task of developing the Surveyor, and the Hughes Aircraft Company won the competition for building it. We soon found that it was a very rough road. Surveyor encountered a host of technical problems that caused severe schedule slips, cost growth, and weight growth. The Centaur fared little better. Its first test flight in 1962 was a failure. Its lunar payload dropped from the planned 2500 pounds to an estimated 1800 pounds or less—not sufficient for Surveyor. Its complex multistart capability was in trouble. Wernher von Braun, necessarily preoccupied with the development of Saturn, recommended cancelling Centaur and using a Saturn-Agena combination for Surveyor.

At this point we regrouped. Major organizational changes were made at JPL and Hughes to improve the development and testing phases of Surveyor. NASA management of Centaur was transferred to the Lewis Research Center under the leadership of Abe Silverstein, where it would no longer have to compete with Saturn for the attention it needed to succeed. Its initial capabilities were targeted to the minimum required for a Surveyor mission—2150 pounds on a lunar-intercept trajectory. This reduced weight complicated work on an already overweight Surveyor, and the scientific payload dropped to about 100 pounds.

It all came to trial on May 31, 1966, when Surveyor 1 was launched atop an Atlas-Centaur for the first U.S. attempt at a soft landing. On June 2, Surveyor 1 touched down with gentle perfection on a level plain in the Ocean of Storms, Oceanus Procellarum. A large covey of VIPs had gathered at the JPL control center to witness the event. One of them, Congressman Joseph E. Karth, whose Space Science and Applications Subcommittee watched over both Surveyor and Centaur, had been both a strong supporter and, at times, a tough critic of the program. The odds for success on this complex and audacious first mission were not high. I can still see his broad grin at the moment of touchdown, a grin which practically lighted up his corner of the darkened room. We sat up most of the night watching the first of the 11,240 pictures that Surveyor 1 was to transmit.

Four months prior to Surveyor's landing, on February 3, 1966, the Russian Luna 9 landed about 60 miles northeast of the crater Calaverius, and radioed back to Earth the first lunar-surface pictures. This was an eventful year in lunar exploration, for only two months after Surveyor I, the U.S. Lunar Orbiter I ushered in that successful and richly productive series of missions.

Surveyor found, as had Luna before it, a barren plain pitted with countless craters and strewn with rocks of all sizes and shapes. No deep layer of soft dust was found, and analysts estimated that the surface appeared to be firm enough for both spacecraft and men. The Surveyor camera, which was more advanced than Luna's, showed very fine detail. The first frame transmitted to Earth showed a footpad and its impression on the lunar surface, which we had preprogrammed just in case that was the only picture that could be received. At our first close glimpse of the disturbed lunar surface, the material seemed to behave like moist soil or wet sand, which, of course, it was not. Its appearance was due to the cohesive nature of small particles in a vacuum.

Surveyor II tumbled during a midcourse maneuver and was lost, but on April 19, 1967, Surveyor III made a bumpy landing inside a 650-foot crater in the eastern part of the Sea of Clouds. Its landing rockets had failed to cut off and it skittered down the inner slope of a crater before coming to rest. Unlike its predecessors, Surveyor III carried a remotely controlled device that could dig the surface. During the course of digging, experimenters dropped a shovelful of lunar material on a footpad to examine it more closely. When Surveyor III was visited by the Apollo 12 astronauts 30 months later in 1970, the little pile was totally undisturbed, as can be seen in the photograph reproduced at the beginning of Chapter 12.

The historic rendezvous of Apollo 12 with Surveyor III would never have been possible without the patient detective work of Ewen Whitaker of the University of Arizona. The difficulty was that the landing site of Surveyor was not precisely known. Using Surveyor pictures of the inside of the crater in which it had landed, Whitaker compared surface details with details visible in Orbiter photographs of the general area that had been taken before the Surveyor landing. He eventually found a 650-foot crater that matched, and concluded that that was where Surveyor must be. Thus the uncer-

The rolling highlands north of Tycho are portrayed with remarkable clarity in this mosaic assembled from among Surveyor VII's 21,038 photographs. To estimate scale, the boulder in the foreground is 2 feet across, the crater about 5 feet wide, and the far hills and ravines some 8 miles distant.

Surveyor VII's "garden" was a heavily worked-over area next to the spacecraft. Trenches were dug with the articulated scoop to give data on the mechanical properties of the surface. At left is the alpha-backscattering instrument that provided accurate measurements of the chemical composition of the surface.

tainty in Surveyor's location was reduced from several miles down to a single crater. By using Orbiter photographs as a guide, Apollo 12 was able to fly down a "cratered trail" to a landing only 600 feet away from Surveyor.

Surveyor IV failed just minutes before touchdown, but the last three Surveyors were successful. On September 10, 1967, Surveyor V landed on the steep inner slopes of a 30 by 40 foot crater on Mare Tranquillitatis. It carried a new instrument, an alpha-backscattering device developed by Anthony Turkevich of the University of Chicago. With this device he was able to make a fairly precise analysis of the chemical composition of the lunar-surface material, which he correctly identified as resembling terrestrial basalts. This conclusion was also supported by the manner in which lunar material adhered to several carefully calibrated magnets on Surveyor. Two days after landing, Surveyor V's engines were reignited briefly to see what effect they would have on the lunar surface. The small amounts of erosion indicated that this would pose no real problem for Apollo, though perhaps causing some loss of visibility just before touchdown.

Lunar Orbiter Photography

Lunar Orbiter was planned for use in conjunction with Surveyor; one spacecraft class was to sample the surface of the Moon, and the other was to map potential Apollo landing sites. Five Orbiters were flown, so successfully that they returned not only precision stereo-photography of all contemplated landing areas but also photographed virtually the entire Moon, including the far side. Included in the photographs returned were the landed Surveyor I, the impact crater caused by Ranger VIII, and many breathtaking images of high scientific value. Orbiter coverage is shown at the left; below is the equatorial Apollo landing zone with its precursor Ranger and Surveyor landing sites.

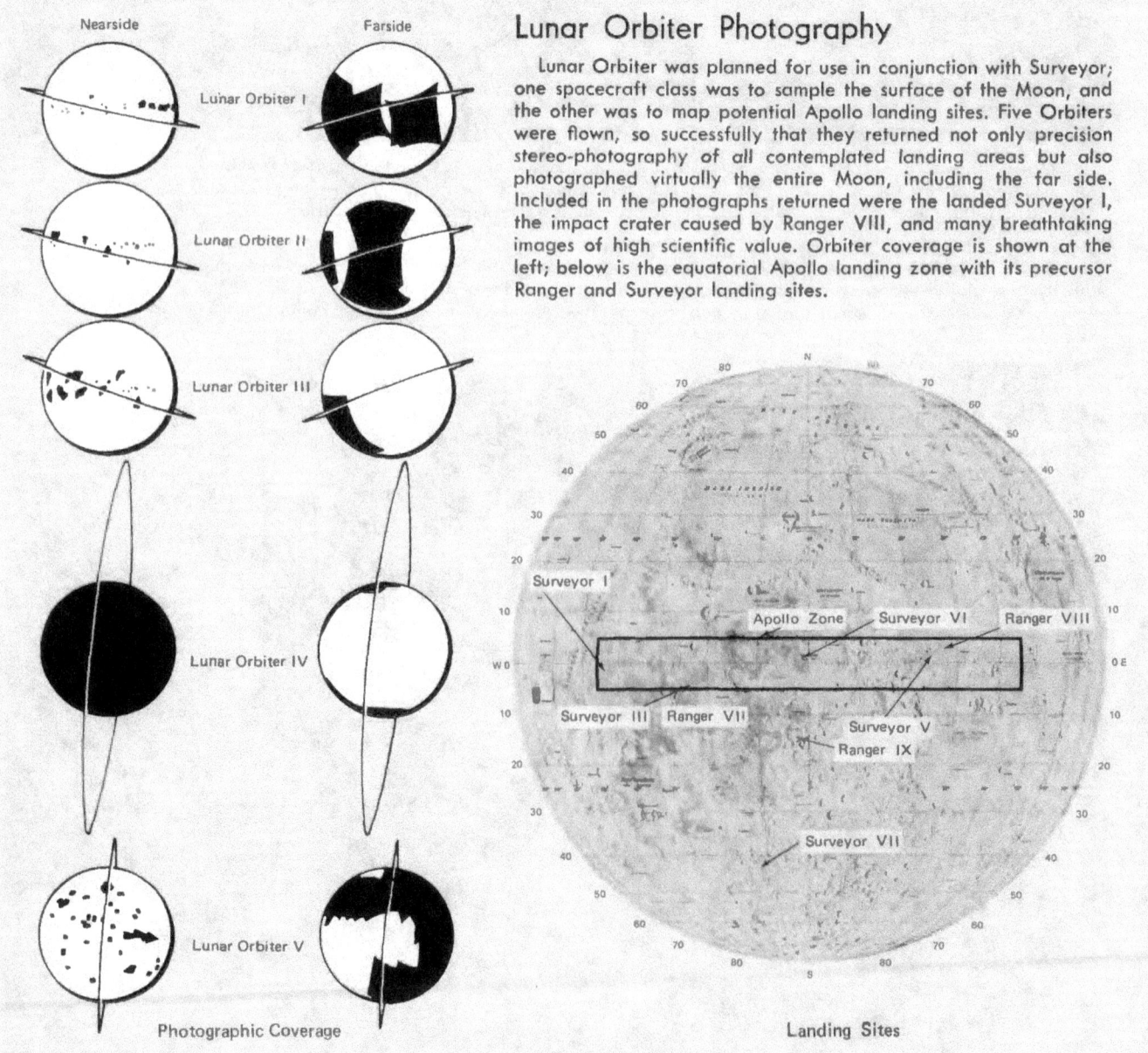

Photographic Coverage

Landing Sites

The two-eyed robot above is the spacecraft that mapped the Moon for Apollo planners. It was built by Boeing for the NASA Langley Research Center, and launched by an Atlas-Agena. Weighing 850 pounds, it drew electrical power from the four solar-cell arrays shown, which delivered a maximum of 450 watts. The rocket motor at top provided velocity changes for course corrections. Guidance was provided by inertial reference (three-axis gyros), celestial reference (Sun and Canopus sensors), and cold-gas jets to give attitude control. Because it would necessarily be out of touch with Earth during part of every orbit, it carried a computer-programmer that could accept and later carry out up to 16 hours of automatic sequenced operation.

It was in its photo system that Orbiter was most unconventional. Other spacecraft took TV images and sent them back to Earth as electrical signals. Orbiter took photographs, developed them on board, and then scanned them with a special photoelectric system— a method that, for all its complications and limitations, could produce images of exceptional quality. One Orbiter camera could resolve details as small as 3 feet from an altitude of 30 nautical miles. A sample complication exacted by this performance: because slow film had to be used (because of risk of radiation fogging), slow shutter speeds were also needed. This meant that, to prevent blurring from spacecraft motion, a velocity-height sensor had to insure that the film was moved a tiny, precise, and compensatory amount during the instant of exposure.

The five Orbiters accomplished more than photo reconnaissance for Apollo. Sensors on board indicated that radiation levels found near the Moon would pose no dangerous threat to astronauts. An unexpected benefit came from careful analysis of spacecraft orbits, which showed small perturbations suggesting that the Moon was not gravitationally uniform, but had buried concentrations of mass. By discovering and defining these "mascons," the Orbiters made possible highly accurate landings and the precision rendezvous that would characterize Apollo flights. Once their work was finished, the Orbiters were deliberately crashed on the Moon, so that their radio transmitters would never interfere with later craft.

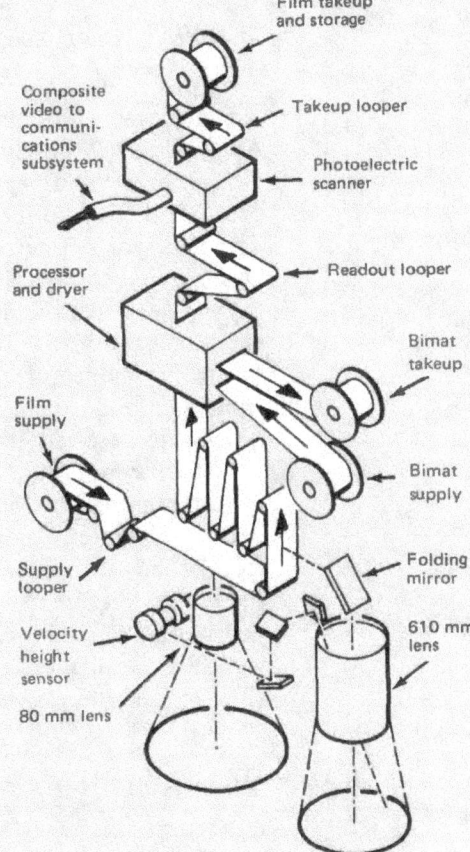

Surveyor VI checked out still another possible Apollo site in Sinus Medii. The rocket-effects experiment was repeated and this time the Surveyor was "flown" to a new location approximately 8 feet from the original landing point. Some of the soil thrown out by the rockets stuck to the photographic target on the antenna boom, as shown in the picture on page 88.

The last Surveyor was landed in a highland area just north of the crater Tycho on January 9, 1968. A panoramic picture of this ejecta field taken by Surveyor VII is shown on page 91 as well as a mosaic of its surface "gardening" area. I remember walking into the control room at JPL at the moment the experimenters were attempting to free the backscatter instrument, which had hung up during deployment. Commands were sent to the surface sampler to press down on it. The delicate operation was being monitored and guided with Surveyor's television camera. When I started asking questions, Dr. Ron Scott of Cal Tech crisply reminded me that at the moment they were "quite busy." I held my questions—and they got the stuck instrument down to the surface. It seemed almost unreal to be remotely repairing a spacecraft on the Moon some quarter of a million miles away.

Before the launch of Surveyor I, in the period when we faced cost overruns and deep technical concerns, NASA and JPL had pressed the Hughes Aircraft Company to accept a contract modification that would give up some profit already earned in favor of increased fee opportunities in the event of mission successes. They accepted, and this courageous decision paid off for both parties. NASA of course was delighted with five out of seven Surveyor successes.

MAPPING AND SITE SELECTION

Meanwhile the third member of the automated lunar exploration team had already completed its work. The fifth and last Lunar Orbiter had been launched on August 1, 1967, nearly half a year earlier. When JPL and Hughes began to experience difficulties with Surveyor development, and with the Centaur in deep trouble, NASA decided to back up the entire program with a different team and different hardware. The Surveyor Orbiter concept was scrapped, and NASA's Langley Research Center was directed to plan and carry out a new Lunar Orbiter program, based on the less risky Atlas-Agena D launch vehicle. Langley prepared the necessary specifications and Boeing won the job. Boeing's proposed design was beautifully straightforward except for one feature, the camera. Instead of being all-electronic as were prior space cameras, the Eastman Kodak camera for the Lunar Orbiter made use of 70-mm film developed on board the spacecraft and then optically scanned and telemetered to Earth. Low-speed film had to be used so as not to be fogged by space radiation. This in turn required the formidable added complexity of image-motion compensation during the instant of exposure. Theoretically, objects as small as three feet could be seen from 30 nautical miles above the surface. If all worked well, this system could provide the quality required for Apollo, but it was tricky, and it barely made it to the launch pad in time to avoid rescheduling.

The Orbiter missions were designed to photograph all possible Apollo landing sites, to measure meteoroid flux around the Moon, and to determine the lunar gravity

The youngest big crater on the Moon is Tycho, which is about 53 miles across and nearly 3 miles deep. These Orbiter V photographs reveal its intricate structure. (Area in the rectangle above is pictured in higher resolution at left.) A high central peak arises from the rough floor, and the crater wall has extensively slumped. The comparative scarcity of small craters within Tycho indicate its relatively recent origin. Flow features seen in both pictures could have been molten lava, volcanic debris, or fluidized impact-ejected material. Surveyor VII landed about 18 miles north of Tycho, in the area indicated by the white circle above. Enlargements of these pictures show an abundance of fissures and large fractured blocks, particularly near the uppermost wall scarp.

This breath-taking view was one of Lunar Orbiter II's most captivating photographic achievements. For many people who had only seen an Earth-based telescopic view looking down into the crater Copernicus, this oblique view suddenly transformed that static lunar feature into a dramatic landscape with rolling mountains, sweeping palisades, and tumbling landslides. The crater Copernicus is about 60 miles in diameter, 2 miles deep, with 3000-foot cliffs. Peaks near the center of the crater form a mountain range about 10 miles long and 2000 feet high. Lunar Orbiter II recorded this "picture of the year" on November 28, 1964, from 28.4 miles above the surface when it was about 150 miles due south of the crater.

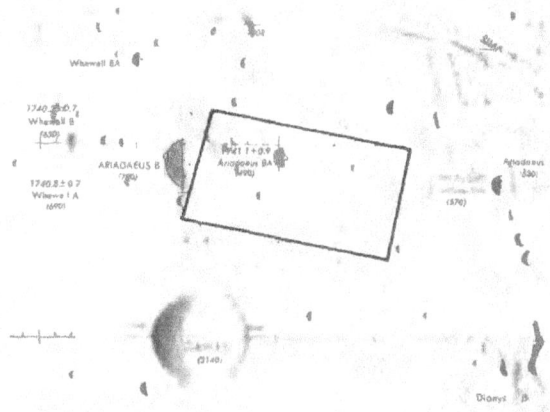

The best maps weren't good enough, even though they were based on years of telescopic photography from Earth. In early planning, the rectangle in the map at left was a possibility as a landing site. The handful of craters shown, it was innocently thought, should be easy enough to dodge during the last moments of a piloted landing. The site was an 11- by 20-mile rectangle located in the highlands west of Mare Tranquillitatis.

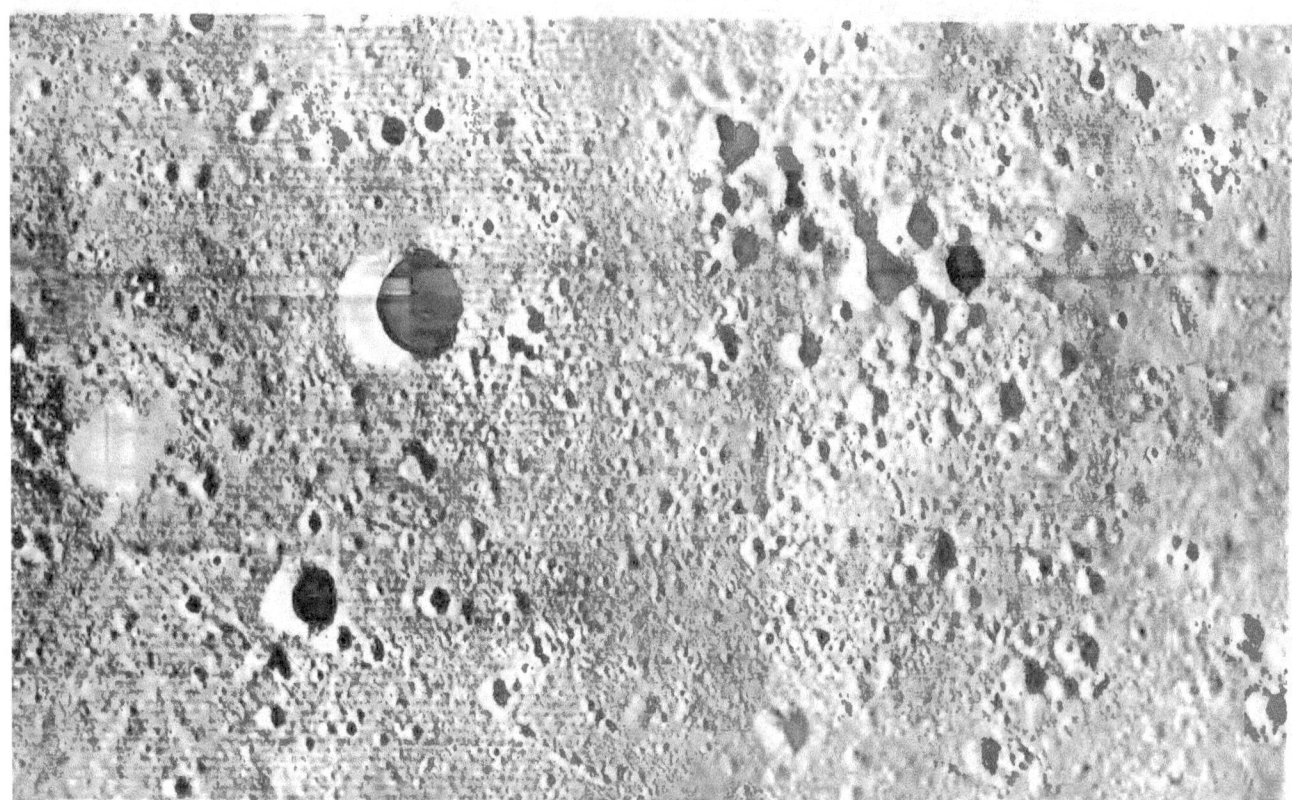

The truth about this site was revealed by the accurate eye of Lunar Orbiter II: it was far too rough to be attempted in an early manned landing. In fact, Orbiter pictures showed that parts of the Moon were as rough as a World War I battlefield, with craters within craters, and all parts of the surface tilled and pulverized by a rocky rain. No areas were found smooth enough to meet the original Apollo landing-site criteria, but a few approached it and the presence of a skilled pilot aboard the LM to perform last-minute corrections made landings possible. The high-quality imagery returned by the Orbiters also returned a harvest of new scientific information.

field precisely, from accurate tracking of the spacecraft. Orbiter did all these things—and more. As the primary objectives for Apollo program were essentially accomplished on completion of the third mission, the fourth and fifth missions were devoted largely to broader, scientific objectives—photography of the entire lunar nearside during Mission IV and photography of 36 areas of particular scientific interest on the near side during Mission V. In addition, 99 percent of the far side was photographed in more detail than Earth-based telescopes had previously photographed the front.

The first Lunar Orbiter spacecraft was launched on August 10, 1966, and photographed nine primary and seven secondary sites that were candidates for Apollo landings. The medium-resolution pictures were of good quality, but a malfunction in the synchronization of the shutter caused loss of the high-resolution frames. In addition, some views of the far side and oblique views of the Earth and Moon were also taken (see page 78). When we made the suggestion of taking this "Earthrise" picture, Boeing's project manager, Bob Helberg, reminded NASA that the spacecraft maneuver required constituted a risk that could jeopardize the company profit, which was tied to mission success. He then made the gutsy decision to go ahead anyway and we got this historic photograph.

The next two Lunar Orbiter missions were launched on November 6, 1966, and February 4, 1967. They provided excellent coverage of all 20 potential Apollo landing sites, additional coverage of the far side and other lunar features of scientific interest, and many oblique views of lunar terrain as it might be seen by an orbiting astronaut. One of these was a dramatic oblique photograph of the crater Copernicus, which NASA's Associate Administrator, Dr. Robert C. Seamans, unveiled at a professional society conference in Boston and which drew a standing ovation and designation as "picture of the year." Among the possible Apollo sites photographed by Orbiter III was the landing site of Surveyor I. Careful photographic detective work found the shining Surveyor and its dark shadow among the myriad craters.

The Apollo site surveys yielded surprises. Some sites that had looked promising in Earth-based photography were totally unacceptable. No sites were found to be as free of craters as had been originally specified for Apollo, so the Langley lunar landing facility was modified to give astronauts practice at crater dodging. Since the basic Apollo photographic requirements were essentially satisfied by the first three flights, the last two Orbiters launched on May 4 and August 1, 1967, were placed in high near-polar orbits from which they completed coverage of virtually the entire lunar surface.

The other Orbiter experiments were also productive. No unexpected levels of radiation or meteoroids were found to offer a threat to astronaut safety. Studies of the Orbiter motion, however, revealed relatively large gravitational variations due to buried mass concentrations—the phrase was soon telescoped to "mascons"—in the Moon's interior. This alerted Apollo planners to account properly for mascon perturbations when calculating precise Apollo trajectories.

With the completion of the Ranger, Surveyor, and Orbiter programs, the job of automated spacecraft in scouting the way for Apollo was done. Our confidence was high that few unpleasant surprises would wait our Apollo astronauts on the lunar surface. The standard now passed from automated machinery to hands of flesh and blood.

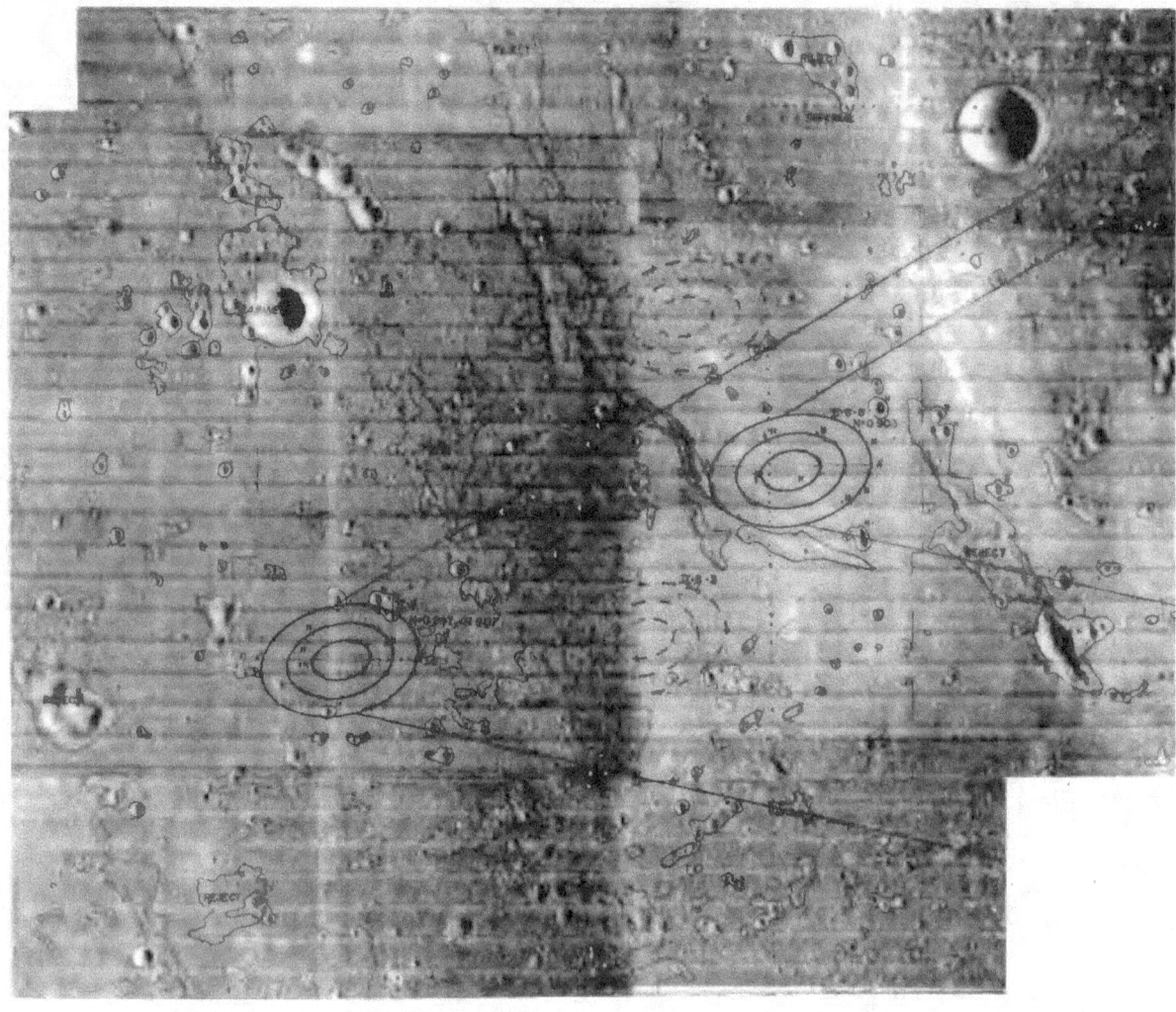

The Selection of Apollo Landing Sites

The search for places for astronauts to land began with telescopic maps and other observations from the Earth, and Ranger photos. The site-selection team considered landing constraints, potentials for scientific exploration, and options if a launch was delayed, which shifted chosen sites to the west. The team then designated a group of lunar areas as targets for Surveyors and Lunar Orbiters.

From the Orbiters' medium-resolution photos, mosaics were made and searched for geologic and topographic features that could make a landing risky: roughness, hills, escarpments, craters, boulders, and steep slopes.

Navigation errors could cause an Apollo landing module to miss a target point up to 1.5 miles north or south and 2.5 miles east or west.

So ellipses were drawn on the mosaics around possible target areas. Those ellipses represented 50, 90, and 100 percent dispersion possibilities. The surfaces within them were then examined to select the target points that appeared to be least hazardous.

Flight-path clearance problems were considered next. This is illustrated by drawing diverging lines eastward for 35 miles from the elliptical areas that otherwise looked best on the mosaics.

A typical marked mosaic is reproduced above. It is a view of a region in Mare Tranquillitatis, and the area within the set of ellipses at the far left was chosen as the target for the first manned landing.

Before it was selected, high-resolution Orbiter photos were used to examine details within the landing

ellipses. In those photos surface irregularities as small as 3 feet could be seen. One such mosaic is reproduced at the top of the facing page.

The black cross in a white circle on the upper picture marks the spot where the Apollo 11 astronauts' landing module descended. It was in an elliptical target area only 200 feet wide.

The lower picture is an oblique view of the same area. This Lunar Orbiter photograph illustrates more nearly the way it would look to an astronaut descending to land. The white lines indicate the elliptical target site and the approach boundaries. Processing flaws such as seen in this picture resulted occasionally from partial sticking of the moist bimat film development used aboard the Orbiter spacecraft.

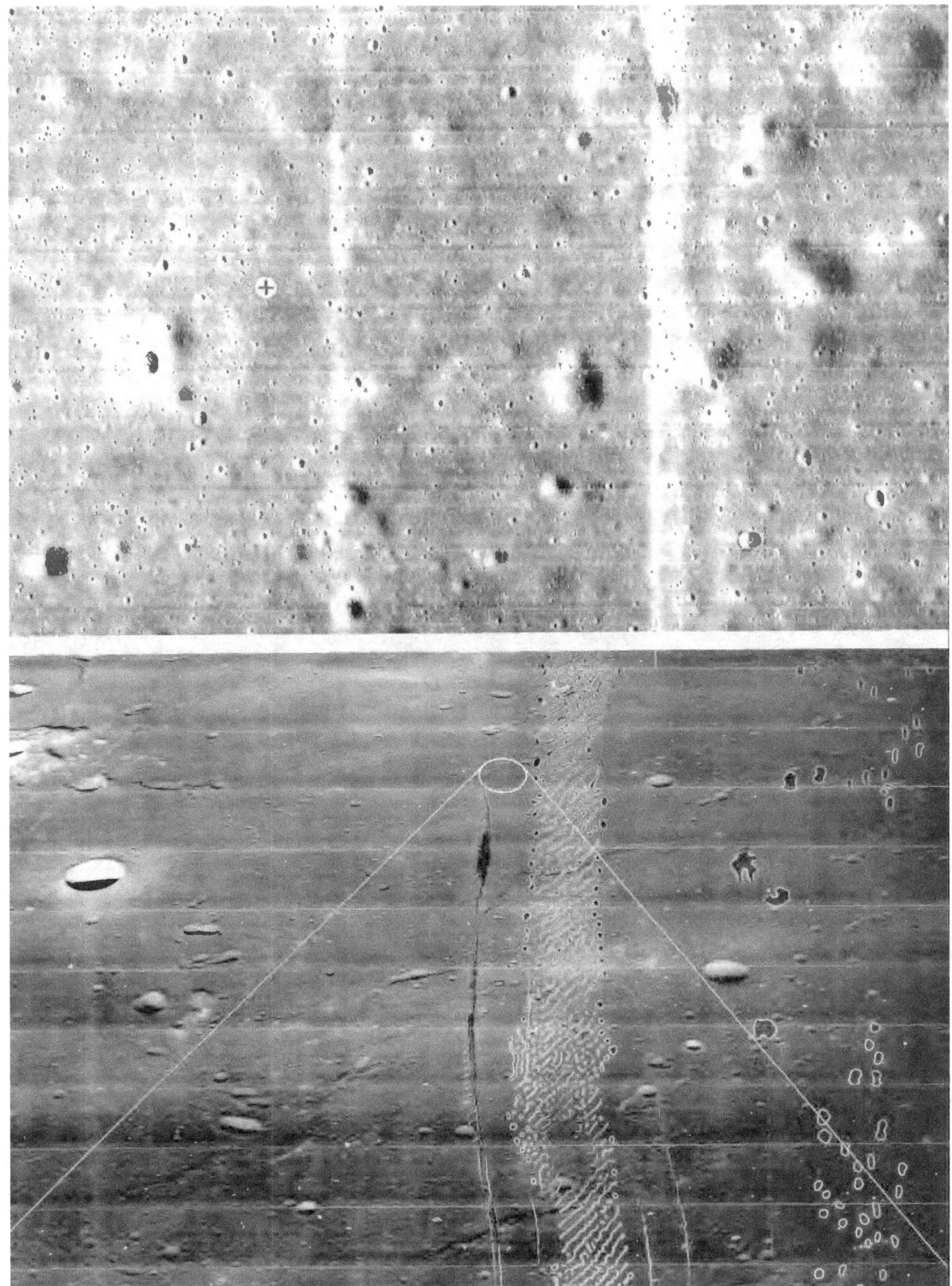

CHAPTER SIX

The Cape

By ROCCO A. PETRONE

At last everything was coming together—all those elements that had been committed piecemeal—the command module in Apollo 7, the first manned flight away from the bonds of Earth in Apollo 8, the flying of the lunar module with men on board in Apollo 9, and then Apollo 10, which went to the Moon and did everything short of landing. It all led up to that hot July morning in 1969, when Apollo's moment of truth was irreversibly upon us.

Uncounted things had to be done before we reached that moment. Before the operational phase even began, we had to pass through the conceptualization and construction phase, remembering something Jim Webb once said, "The road to the Moon will be paved by bricks and steel and concrete here on Earth." For Apollo we had to build Complex 39 at a cost of half a billion dollars, that is, we had to finish Stage Zero, before we could proceed to Stage One, the flying of the birds.

Complex 39 was to Moon exploration what Palos was to Columbus: the takeoff point. Man had never attempted any such thing before, and I wonder when he will again. Everything was outsize—among the impressive statistics about the core of Complex 39, the Vehicle Assembly Building, were its capacity (nearly twice as big as the Pentagon) and its height (525 feet, thirty feet short of the Washington Monument's).

During 39's construction phase we were, of course, flying Saturn Is—ten of them —and in 1966 three Saturn IBs, all from Pads 34 and 37. Everything—launches and construction—had to mesh; it was like building a thousand different homes for a thousand demanding people. It was on the last of these IBs that I made my rookie appearance as director of launch operations, August 25, 1966. The mission was a 1 hour 23 minute suborbital flight to test the command and service module subsystems and the heatshield. CM 011 was recovered near Wake Island in good condition; its shield had withstood the heat of reentry at 19,900 mph.

During the construction phase of Complex 39 (1961–66) I was Apollo Program Manager at the Cape. The first thing we had to do was decide where to build the

Pad A of Launch Complex 39, shortly before a launch. The Mobile Service Structure is parked back by the crawlerway, the crawler separated, and only the Mobile Launcher is on the pad with the vehicle.

moonport. My boss, Dr. Kurt Debus, and Maj. Gen. Leighton Davis, USAF, were directed to find a place from which to launch huge vehicles like the projected Nova or the Saturn V—Cape Canaveral's 17,000 acres weren't nearly large enough. In this study we considered sites in Hawaii, the California coast, Cumberland Island off Georgia, Mayaguana Island in the Bahamas, Padre Island off the coast of Texas, and several others. Eventually we concluded that the most advantageous site was Merritt Island, right next to the Air Force's Cape Canaveral facilities, which had been launching missiles since 1950 and NASA vehicles since 1958. Our report was completed July 31, 1961, and we spent all night printing it, after which Dr. Debus and I flew up to Washington and briefed Mr. Webb and Dr. Seamans. So 84,000 acres of sand and scrub were acquired for NASA by the government, plus 56,000 additional acres of submerged lands, at a total cost of $71,872,000.

We lost no time in raising the curtain on Stage Zero. No one who was involved can ever forget the driving urgency that attended Apollo. Nor the dedication of those who worked on it, including the construction crews, who by 1965 numbered 7000 persons at the Cape.

At one time we had considered preparing the space vehicle horizontally, and then erecting it vertically on the pad, but this was simply out of the question for a 360-foot bird. So we had to erect the Saturn V stage by stage, which meant that, because of rain and wind, we had to have an enclosed building. Even a 10- to 15-knot wind would have given us trouble while we were erecting outdoors, and higher winds could prove disastrous. Thus the Vehicle Assembly Building became an enclosed structure. Should the high bays be strung out four in a row, or built back-to-back? We decided on the latter format because only two big cranes (250-ton bridge cranes) would be needed instead of four, and because the box-like structure would better withstand hurricane winds of 125 mph. The possibility of hurricanes also dictated that we have two crawlers, one to carry the Mobile Service Structure away from the pad, and one to bring the Apollo-Saturn V and its Mobile Launcher to shelter in the VAB. The height of the building was dictated by the hook height, and we started planning for 465 feet; the final height was 525. Remember that when we started planning the VAB in 1961 we weren't sure what size bird would roost in this big nest. We also had to begin design work before we knew whether the trip to the Moon would involve an Earth-orbit or lunar-orbit rendezvous.

I think of the VAB not as a building but as an intricate machine that assembled the vehicle in its final phases. People were surprised to learn that the various stages had never seen each other until they were introduced in the VAB drydock. The first stage had been built by Boeing at Michoud outside New Orleans, the second stage by North American at Redondo Beach, Calif., the third by McDonnell Douglas at Huntington Beach, Calif., and the Instrument Unit by IBM at Huntsville, Ala. The spacecraft that went on top of this stack were also introduced for the first time in the VAB, the CSM from the North American plant at Downey, Calif., and the LM from Grumman in Bethpage, Long Island.

Could these pieces, arriving from all over the country, play together? Every wire in every plug had to join exactly the right wire, with no electrical interference or

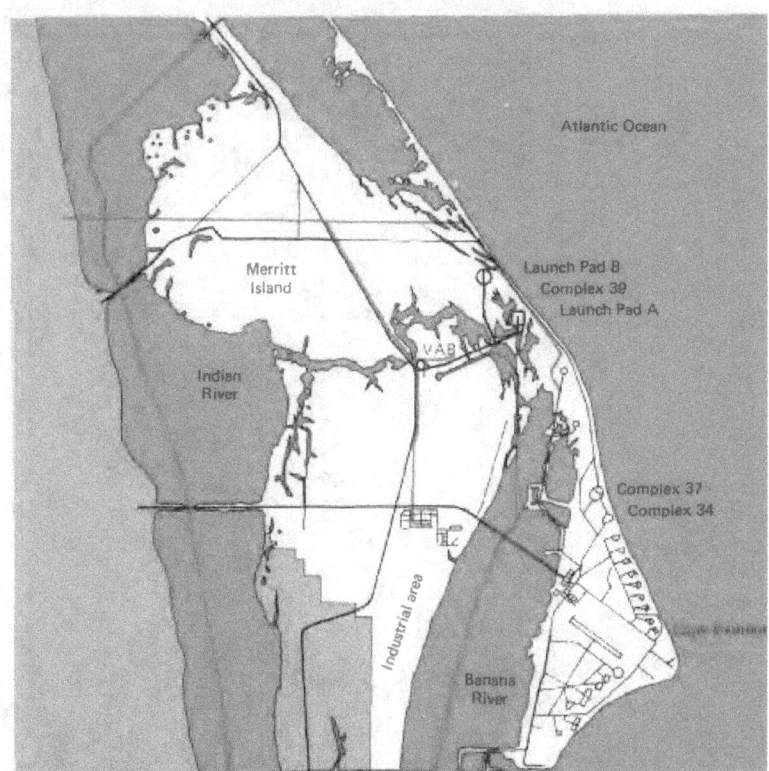

A launch pad looked different in the early days at the Cape. Here Redstone 4 is in final stages of preparation in August 1954. The plume of vapor at the base is oxygen boiling off, a hazard being ignored by the men at work; but the presence of an ambulance suggests awareness that all might not necessarily go well. Central sections of the deployment track are concreted, to keep exhaust gases from tearing up the pad. This launch was successful.

John F. Kennedy Space Center, on Merritt Island across the Banana River from Cape Canaveral, was acquired because the Air Force's launch complex at Canaveral did not provide the area needed for the Apollo program.

The Vehicle Assembly Building—an "intricate machine" to put rockets and spacecraft together—took shape early in 1965. In the foreground is the turning basin for barges to bring in giant rocket stages. At right three big Mobile Launchers are also abuilding.

change of signal strength, and a command signal had to work from the Instrument Unit through the third and second stages into the first. After the various stages had been put together they had to be checked out as an entity, and once this had been done you naturally don't want to break the electrical connections again. When the Instrument Unit orders, "Go right," you don't want an engine three stages down to go left. The pieces not only had to match each other, but also the ground equipment. Lines for the flow of liquid hydrogen from the ground had to match the stages, and so did others for the flow of liquid oxygen and still others for gaseous nitrogen and helium. Television monitors had to be designed and installed so that we could know what was happening in hundreds of places at any given time—which meant all the time.

The skill required of the technicians working within Complex 39—and throughout the Kennedy Space Center—had to be fine-honed. One of the legends had it that the crane operator who set the 88,000-pound second stage on top of the first stage had to qualify for the job by lowering a similar weight until it touched a raw egg without cracking the shell. Of the Cape's 26,500 workers—the peak number in 1968—a high percentage were men and women who possessed such skills. One can speculate whether such a crew is likely to be assembled in one place again.

The pads of Complex 39 were abuilding at the same time the VAB rose on its 4225 16-inch steel pipe pilings that had been driven more than 160 feet below the sandy surface into bedrock. Pad A and Pad B were twins, each occupying about 160 acres; we had also planned a Pad C, which explains why the crawlerway from A to B has an elbow-like crook in it—the elbow would have led to C. (We even had a contingency plan for a Pad D in case launchings became more and more frequent.) The pads were built 8700 feet apart so that an explosion could not wreck more than one of them. They were located three and one-half miles from the VAB and the Launch Control Center. In the early days we'd had to control a launch from a site close to the pad, to avoid electrical problems induced by a longer run of electric cabling, which in turn meant that we worked from a blockhouse heavily protected against fire and blast. Now, by the time the Saturns were ready, digital data technology had advanced to the point where firing rooms could be in a structure miles away.

How to get the Apollo-Saturn V from its birthplace in the VAB to the pad from which it would fly for the first and only time in its life? Early in the program we considered moving it on its three and one-half mile journey by water. The barge concept was deep in our thinking: the first and second stages had to come to the Cape from Louisiana and California, respectively, by man's oldest form of transportation, since they were too large to go through tunnels or under bridges. On this short trip why not also float the Saturn V and its Mobile Launcher standing upright on a barge? We got the Navy to run tests at the David Taylor Model Basin in Washington, which showed that the hydrodynamic requirements of such a topheavy barge would be too demanding. We looked into a rail system, into pneumatic-tire transporters, and ground effects machines but all were impractical or too expensive.

Then somebody in our shop came up with the idea of using giant tracked machines like those used in strip mining. What evolved was the unique crawler or,

In this super-barn, launch vehicles and spacecraft were delicately built up, interconnected in myriad ways, and then exhaustively checked out for any mismatch.

Lowering the second (S–II) stage of a Saturn V to mate with the first stage, which is already standing on the launcher. Here the camera is about 200 feet up in a high bay of the Vehicle Assembly Building.

Looking down at the space vehicle as it leaves the VAB on the base platform of the Mobile Launcher. The nine swing arms that run from the launcher's tower to the vehicle's various levels are in extended position.

The first Saturn V space vehicle and its Mobile Launcher on their way out to Launch Pad A atop a transporter. The pad is 3½ miles east of the VAB. The Mobile Service Structure is parked at the loop where the crawlerway to Pad B diverges to the north (right). The barge canal to the VAB runs parallel to the main crawlerway.

The last Moonbound Apollo-Saturn rolls slowly out of the huge VAB into the morning Sun. Each door of the four VAB bays opens 456 feet high and 75 feet wide. The VAB encloses a volume of 129,482,000 cubic feet. But the tale that clouds form inside and rain falls is only folklore.

to find the guards and tell *them* to come back, and that took another hour before we could launch. At other times the count would be held up because diesel engines powering a tractor didn't work, or a key to the tractor had been lost.

But we learned as we traveled this long road. The really big event was the launch of the first Saturn V, November 9, 1967, which we designated Apollo 4. To me that was the real mark; its success meant we were really going to make the Moon landing. To bring together the massive hardware and the complex ground equipment exactly when we wanted to was an achievement. You don't look at a thing like the Saturn V without a lot of humility in your heart. Consider the three stages, the Instrument Unit, the vast amount of automation, the many computers on the ground and in the stack, the swing arms, the hold-down arms, the propellant loading system—all these were intricate and potentially troublesome elements that had to be brought up to speed, and at a vast launch complex that was being used for the first time.

We kept having problems during the Countdown Demonstration Test in September and October; it was scheduled for three and one-half days, but lasted 23. We'd go so far, and we'd find a leak. We had ground equipment problems, then procedural problems; batteries failed, and pressure gauges developed faults. But these problems melded our team, as in a cauldron. Under pressure our people came of age, in the firing room and on the pad. When the five engines of Saturn 501 fired up at 7 a.m. we had confidence, which proved to be justified.

By the time of Apollo 11, the number of printed pages, including interface control documents, that were required to check out a space vehicle actually surpassed 30,000. We had to make so many copies that a boxcar would have been required to hold the documents necessary to launch a Saturn V. The more contractors involved, the greater the need for formality. No more holding up launches for fishing skiffs; no more offhand decisions such as we sometimes had to make in the early days. The schedule was always upon us in Apollo. We had to work concurrently on different launch flows. When Apollo was at its peak we had three firing rooms working simultaneously in the launch control center, with three crews of 500 apiece manning the consoles. That took a lot of manpower, and the person I depended on most heavily to keep the operation moving was Paul Donnelly, an unflappable veteran of Mercury and Gemini days.

A LAUNCH EVERY OTHER MONTH

The centers—that's what we called intervals—between launches were two months. That is, Apollo 9 went in March, Apollo 10 in May, and Apollo 11 in July. But each vehicle took five months from the time its components arrived in the VAB until its launch. Thus the overlapping.

Look at the situation in early March 1969. Apollo 9 on Pad A was ready for launch—delayed three days because the astronauts caught colds—and the team of 500 —engineers and technicians—was working twelve- or thirteen-hour days. Apollo 10 was ready to be rolled out of the 456-foot hangar doors of the VAB for two months of intensive checkout on Pad B. But the components of Apollo 11 had already arrived and were undergoing tests in the VAB and in the vacuum chambers of the Manned

Mining technology for a Moon launch. Utilizing power shovel design concepts, two 131-foot long, 3000-ton, track-mounted crawlers were built to transport the MSS and an assembled Apollo-Saturn V from the VAB to the pad, and back should hurricane weather threaten a launch. Shipped in sections from Marion, Ohio, the 114-foot wide, X-framed crawlers were assembled on site at the Cape. Four large diesel engines coupled to six electrical generators power crawler's motive, leveling, jacking, and steering systems. Operator's cab, lower right, is matched diagonally by another.

The Mobile Service Structure coming down from the launch pad. About 11 hours before launch, a crawler carries the MSS back from the pad to its parking area. Although the crawler is descending a 5-degree slope, the hydraulic cylinders, shown here fully extended, keep its platform level.

A driver eases the crawler beneath a Mobile Launcher; once in place it lifts up both launcher and space vehicle. The driver, who wears a seat belt for his 1-mph trip, is like the helmsman of a ship; the total crawler crew, in control and engine rooms inside, is about 15 men.

Dress rehearsal. During a Countdown Demonstration Test, as for an actual Apollo launch, the Mobile Service Structure was removed from the pad at about T-minus-11 hours, leaving the fueled space vehicle in place. The MSS owes its awkward shape to the many platforms built out over the base of the Mobile Launcher to service the rocket. The designers then balanced the MSS by placing the elevators and their machinery at the other side of the structure.

Balancing its load on its head, a crawler ascends the grade to the launch pad. The unfueled space vehicle and its Mobile Launcher together weigh 6000 tons, and the crawler weighs another 3000 tons. Two tracks at each corner drive the crawler. The individual cleats of the tracks weigh a ton apiece.

Getting set for the flames of an Apollo launch. The thrust chambers of the first stage's five engines extend into the 45-foot-square hole in the Mobile Launcher platform. Until liftoff, the flames will impinge downward onto a flame deflector that diverts the blast lengthwise in the flame trench. Here, a flame deflector coated with a black ceramic is in place below the opening, while a yellow (uncoated) spare deflector rests on its track in the background. It takes a tremendous flow of water (28,000 gallons per minute) to cool the flame deflector and trench. The pumps, which start 8 seconds before ignition, can deliver that flow for 30 seconds, and then a reduced flow for an indefinite period. Another 17,000 gpm of water curtains the Mobile Launcher tower from the rising flames.

The white room is the work platform, 400 feet in the air, through which a flight crew is loaded into the spacecraft. The crew arrives at the pad by van, ascends the launcher tower by elevator, and then crosses a swing arm to reach the white room.

Spacecraft Operations Building. Apollo 11 would roll out at 12:30 p.m., May 20, one month and 26 days before it lifted off for the Moon.

The pressure on these people was pretty severe. At a launch a person just sat there glued to his console, watching the needles for any sudden changes, knowing that he would be committing this big vehicle, with men aboard: a $400 million commitment. And it wasn't the money only, or the men. The entire world was watching for the success of the United States.

We had both to prepare the bird and to make sure our people could detect and understand any anomaly. I used to walk through the console panel area right up to about the last 45 minutes before liftoff. I'd be checking on alertness, especially among men who had been working long hours. Were they fatigued? Were they concentrating on the dials? Was there any unnecessary chit-chat going on?

When we had long holds during the Countdown Demonstration Tests I had to judge how far we could go, whether we were pushing too hard, whether we had to call a delay and wait until the next day. The team had to be as well rehearsed as any ballet, or any football team. You do not get the commitment for launch without a lot of hard days and weeks and months of practice.

TESTING THE TEAM

About eighty percent of the people on these teams worked for the contractors; the rest were NASA employees. All of them had to go through examinations. We'd call a man in—say a swing-arm console operator—before a board of three or four examiners and we'd have his part of the mission simulated on a console. He would have five minutes to get set, before making split-second decisions. We'd say, "Okay, here's your console, and here's your condition." The examiners would move a red-green slide, or put a yellow light on. The operator would look at the simulation on his console, and say, "Okay, that's green, and it means the pressure is okay; that's red and it means the pressure is too low." Dozens of other simulators would test his proficiency. We had to make sure. We had to be able to say, "We understand the problems, we've done the detective work; we've found the solution and we've tested it, and we have confidence everything will work."

In our testing we had a building block approach, very logical, very methodical; you built each test on the last test, and the whole sequence expanded in the process. Everything culminated in the two main tests, Flight Readiness and Countdown Demonstration. Flight Readiness would take us through the total flight, including an abbreviated trip to the Moon, with all the valves working, all the sequences following according to the logic we had worked out for them. It was a total test of the electrical system and the software.

The Flight Readiness Test was dry (that is, without propellants) but the Countdown Demonstration Test (93 hours) was loaded with propellants, including several thousand tons of cryogenics in the three stages and tons of RP–1 fuel in the first stage. This one we took right up to the point of 14 seconds before ignition. We had four or five different ways to stop the countdown sequence at 14 seconds, and I would customarily look at Ike Rigell and say, "How many stops have we got?" The test had to be stopped at T minus 14 seconds because if it went down to 9 we would activate the ignition sequence. So everybody wore a sort of tense smile when it came to 14 seconds. We never had an accidental ignition, which would have meant chaos. (We did not have the astronauts in the CM during this part of the CDDT.)

Then we would unload the cryogenic propellants and dry out the tanks, which took five or six hours, a little longer than it took to load. Next day we would pick up the count at about three hours and run through the schedule up to simulated lift-off, now with the astronauts on board. It was important for the flight crew to go through this final exercise: to suit up in the Manned Spacecraft Operations Building, get in the vans, ride out to the pad, load into the CM, and check the flight systems.

I have often been asked why it took hundreds of men to launch the astronauts to the Moon, whereas just two of them on the Moon can launch themselves back to Moon orbit. Well, the two of them were there on the Moon in the LM's ascent stage. They had everything they needed: their fuel was loaded; they had water; their cooling system was working and so was their oxygen supply. Their radar was tracking and their communications to Earth were functioning, and long before launch we had checked to see that they had no electrical interference. These systems were working because of the preparations and check-out efforts of hundreds of people on the ground before the spacecraft was committed to launch.

WHEN THE RED LIGHT LIT UP

It was remarkable that every manned Apollo launch lifted off exactly on schedule, up to the last one. (Apollo 14's forty-minute delay was due to weather.) But Apollo 17, the only night launch, was delayed 2 hours and 40 minutes, until 12:33 a.m., because of the failure of an automatic countdown sequencer in the ground equipment. The way we had the launch set up was that the last three-minute period in effect was a series of automatic commands, all done by a timer. If you didn't get through a certain gate in the automatic sequencer the next command would not be given. This protected us against a faulty liftoff.

This is what the term "terminal sequence" meant, which took a great deal of check-out time in the months preceding launch. When we got down to 30 seconds before lift-off, the indication of pressurization for one of the propellants in the S-IVB stage hadn't registered, so the sequencer stopped the count. The red light on the overhead indicator in the firing room lit up. The engineer monitoring that read-out on the strip chart told us the S-IVB was not pressurized. The ladder in the sequence wasn't met, so we got a cut-off at 30 seconds.

The team went through a back-out act, as they had practiced, the arming command was withdrawn, the on-board batteries were taken off line, the radio-frequency transmitters turned off, and within three or four minutes the space vehicle had been returned to a status where we could safely hold. Everything was done very coolly, very gingerly. Gene Cernan, the commander of the flight, said later that he kept his hand—very tightly—on the abort switch, "because you never know." But once again the launch escape tower went unused.

The problem turned out to be a faulty diode in the terminal sequencer. Among

The newly completed Launch Complex 39 attracted many VIPs. Here Petrone briefs President Johnson and Chancellor Ludwig Erhard of the Federal Republic of Germany in 1966 on the characteristics of the Mobile Launcher.

Checkout and assembly of the Apollo 17 lunar module in a clean room of the Manned Spacecraft Operations Building. In the foreground, the Lunar Roving Vehicle is undergoing its final checkout (with Astronauts Schmitt and Cernan aboard) prior to being packaged and stowed into the descent stage of the lunar module.

TV screens and display panels ablaze, Firing Room 2 of the Launch Control Center, adjacent to the VAB, is the hub of activity for the start of Apollo 6's unmanned Earth-orbit mission. Large wall screens show the Saturn V in readiness.

President **John F. Kennedy** is briefed on the developing plans for Launch Complex 39, designed to carry out his call for men on the Moon before 1970.

Apollo 11 lifting off the pad on July 16, 1969, the culmination of years of intense activity at the Cape. In this wide-angle view from the press site, all eyes squint in the direction of the hot morning Sun and the distant Launch Pad 39A, where the exhaust gases of the rocket's first stage have been split by the flame deflector into two distinct columns of flame and smoke.

the hundreds of commands given in sequence, one was not forthcoming, so everything stopped—which is one of the marvels of Apollo. At this late date in the program nobody batted an eye, including Walter J. Kapryan, the able engineer who had succeeded me as launch director when I went to Washington as Apollo Program Director three years earlier.

No story about the Cape—Canaveral, then Kennedy, then Canaveral again—would be complete without a mention of the visitors. By the time the Apollo program ended in 1972 we had attracted more than 6 million of them, and that doesn't count people who lined the roads and watched the lift-offs (there were a million of those, it was estimated, for Apollo 11). We had VIP visitors in a steady stream—Presidents of the United States, members of the Supreme Court, members of Congress, almost any prominent person you could mention.

We had leaders from many countries—I recall the Shah of Iran, King Hussein of Jordan (he was a jet pilot), the King of Afghanistan, King Baudoin of Belgium, Haile Selassie of Ethiopia, Chancellor Erhard of Germany, and President Radhakrishnan of India, who had been a professor of philosophy at Oxford. Other visitors included foreign ministers and cabinet members from many countries. We talked to most of them in informal sessions, explaining as best we could the mysteries of spaceflight.

The one visitor who impressed me most came in November 1963, and we briefed him on a Saturn-Apollo unmanned mission due to fly in January—which would be the first Saturn I to carry two stages (with a total of 14 engines, still the record for launch vehicles). He promised to come back for the launch if he possibly could. But he never made it because he was assassinated in Dallas six days later.

CHAPTER SEVEN

"This Is Mission Control"

By CHRISTOPHER C. KRAFT, JR.

The Mission Operations Control Room was the focal point for all the activities of the Mission Control Center. At any time during an Apollo mission, one of the four flight control teams would be manning these consoles. Each controller in this room was supported in his operation by other

The last Apollo flight to the Moon has been called "the end of the beginning." It represented more than just the end of a program to me. It brought to a close a phase of my career. Apollo had become intimately interwoven in the fabric of the waking hours of my life and often caused the remaining hours to be fewer than they should have been. My first involvement with the program had occurred at Langley Field, Va., 11 years before the Apollo 17 flight. During those formative years of the lunar program I was faced with the challenge of flying Mercury, and of necessity my commitment to Apollo could not assume the proportions it would in later years. Only twenty-

people and facilities, both in the staff support areas within the building and in the world outside the Center. Since the completion of the Apollo missions, this room and the support areas have been reconfigured for other manned space programs like Skylab and Apollo-Soyuz.

three days before my first Apollo meeting at North American Aviation, we were flying *Friendship 7* and John Glenn on the country's first manned orbital flight.

In those naive early days I had no idea I would be charged with the responsibility not only for flight operations but for managing the computer software programs that would be used for landing two astronauts on the Moon and returning them to Earth. Although in 1962 we had decided we were going to the Moon, we had yet to figure out how we were going to get there and return, let alone determine the equipment, facilities, and personnel we would need. Many difficult hours were yet to be spent in conference rooms, visiting contractor plants and test sites, and waiting at airports. I had yet to experience the frightening experience of disarming an angry young man with a gun on one of many flights to Cape Kennedy. The future held both periods of despair and frustration and those exciting and satisfying moments when we flew that were to make it all seem worthwhile. Now, with Apollo 17, it was coming to an end. I found it difficult to accept the finality of that landing on December 19, 1972, near the USS *Ticonderoga* in the Pacific Ocean. The challenge would never again be quite the same. Apollo was like an intoxicating wine and certainly the last of the vintage.

VIGILANCE AND JUDGMENT

"The accomplishments of this last Apollo mission and the successes of the previous Apollo flights were the result of the dedicated efforts and the sacrifices of thousands of individuals." I have difficulty recollecting how many times I stood on the platform at Ellington Air Force Base welcoming the returning flight crews and heard those words repeated. But they are nevertheless quite true. The people in Houston were with their astronauts each step of the way. The interchange between Mike Collins, serving as the CapCom (capsule communicator), and Bill Anders as Apollo 8 orbited the Moon clearly demonstrated this feeling. Mike called Apollo 8, saying "Milt says we are in a period of relaxed vigilance." Bill came back with "Very good. We relax; you be vigilant." They came to rely on the controllers, as they well knew their very lives depended on their vigilance and judgment. Mike later put it well in his book, *Carrying the Fire.* He writes of the Gemini 10 reentry and their reliance on "Super Retro" John Llewellyn. Mike says that they knew if they made a mistake John would be so angry that he would stick up his strong Welsh arm and yank them out of the sky. John's dominant personality is illustrated by the time he was coming on duty for his shift in the Control Center and, finding his parking space taken, he simply parked on the walk next to the door rather than waste time looking. Like his compatriots, John was thoroughly dedicated. His type is at its best when fighting wars or flying missions.

Many individuals were involved in the building and testing of the spacecraft and its systems, but once given the spacecraft and the necessary facilities and equipment, the Apollo Operations Team was charged with the awesome responsibility for the accomplishment of the mission. This team was composed of hundreds of individuals—government and contractor personnel, as well as representatives of the Department of Defense and of foreign nations such as Australia and Spain. Each team member had been carefully selected and subjected to countless hours of training and simulations; each had also participated in Mercury, Gemini, or Apollo testing and flight operations. Time and time again, these young men had to rely on their technical knowledge to assess the unexpected and determine the right course of action. The "luck" the Operations Team had in overcoming adversity is exemplified by the words of University of Texas football coach Darrell Royal: "Luck is what happens when preparation meets with opportunity." The luck of the Operations Team was the result of thorough and careful planning and training and the development of both people and procedures.

Thinking back over the events of the past years, I realize the Operations Team was always prepared when the opportunity presented itself. I'll certainly always remember their performance on Apollo 11. It takes an awful lot of events all going right to get you to the Moon, let alone return. It was our first attempt at the landing and we had somehow, incredibly, reached the point where we were starting the descent for the landing. Thus far, all had gone astonishingly well. The first phase of the firing of the lunar module engine went well as the descent started; and then, approximately five minutes after ignition, the first of a series of computer alarms was received via telemetry in the Mission Control Center and was also displayed to the crew onboard the lunar module *Eagle*. I was responsible for the software in that computer, the logic that made it all work. You can imagine the thoughts racing through my mind: Had we come all this way for naught? What was wrong? The flight controller responsible for assessing the problem, 27-year-old Steve Bales, was faced with an immediate decision: Should we continue the descent or initiate an abort? An abort meant there would be no landing for Apollo 11: we would have to try again. When Flight Director Gene Kranz pressed him for his answer, young Mr. Bales' response was the loudest and most emphatic "go" I have ever heard.

But it wasn't over yet. The lunar module was under automatic control as it approached the surface. Neil realized that the automatic descent would terminate in a boulder field surrounding a large rim crater. He took over control of the spacecraft and steered the *Eagle* toward a smooth landing site. The low-level fuel light for the engine came on, indicating about enough fuel for only 116 seconds of firing time on the engine. With 45 seconds of fuel left, *Eagle* set down with a jolt and we were there.

A LIGHTNING STRIKE

I could recall any one of hundreds of incidents that have occurred over the years as we flew Apollo. Launch has always been an uneasy time for me, and I always looked forward to successful separation from the booster. When one adds to this an apprehension caused by bad weather over the Cape, I become even more concerned. It turned out that all of the elements were present for Apollo 12. The launch was made into a threatening gray sky with ominous cumulus clouds. Pete Conrad's words 43 seconds after liftoff, electrified everyone in the Control Center: "We had a whole bunch of buses drop out," followed by "Where are we going?" and "I just lost the platform." The spacecraft had been struck by lightning. Warning lights were illuminated, and the spacecraft guidance system lost its attitude reference.

The spacecraft was still climbing outbound, accelerating on its way to orbit. There was not much time to decide what should be done. The crew was given a "go" for staging and separation from the first stage of the Saturn V launch vehicle. Within

Mission Control candid photography
by Andrew R. Patnesky

Flight Director Gene Kranz watches his console display tensely as the Apollo 11 lunar module *Eagle* slowly settles down with its descent engine fuel supply all but exhausted. Gerry Griffin, a Flight Director during other phases of the mission, looks on in complete absorption.

As the Apollo 11 lunar module begins the descent toward its historic touchdown, off-duty Operations Team members watch unobtrusively from a few extra chairs in the Mission Operations Control Room.

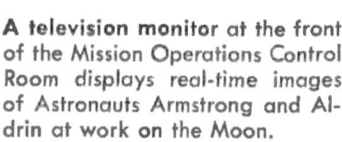

A television monitor at the front of the Mission Operations Control Room displays real-time images of Astronauts Armstrong and Aldrin at work on the Moon.

The lightning bolt that struck Apollo 12 aloft also hit the crane and platform of the mobile launcher.

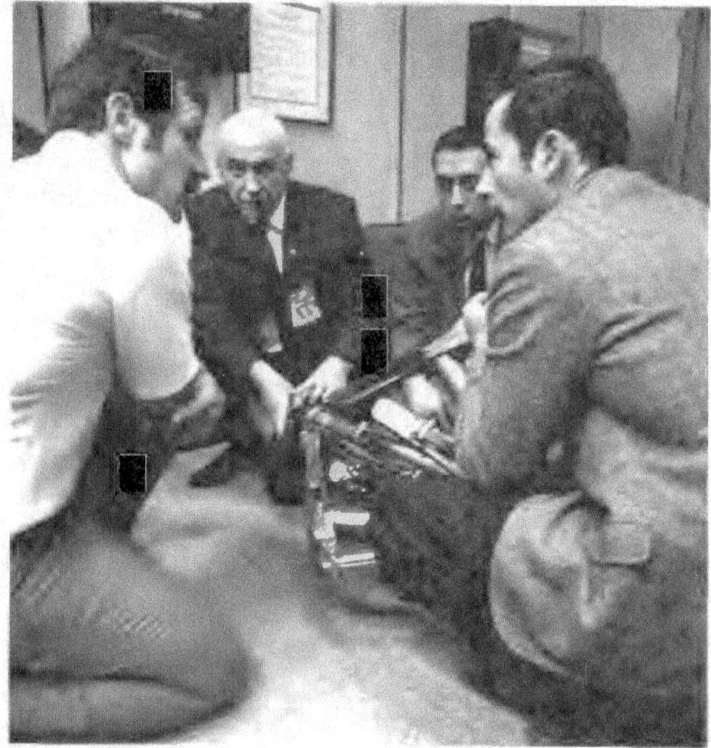

During the **Apollo 13 crisis** the Mission Control directors discussed possible landing recovery options. Because of the unique configuration (the LM still attached to the CM) new procedures leading to reentry were developed. Ten phone lines were open between Mission Control and experts at the Grumman plant. Engineers in Downey, Calif., where *Odyssey* was built, ran emergency problems through computers. And at MIT a team worked through the night on the guidance system and prepared new trajectories. Perseverance and ingenuity were rewarded with a safe landing in the Pacific less than 4 miles from the USS *Iwo Jima*.

When the **Apollo 14 crew was unable**, after repeated attempts, to dock with the lunar module, the Operations Team was faced with the prospect of having to abort the mission. In order to work out new procedures, Mission Control hastily located a docking probe and drogue. Flight Controller John Llewellyn (left) discusses possible solutions with Bob Gilruth, George Abbey, and John Young. The crew docked successfully with the new procedure, and had no trouble docking again.

seconds, John Aaron, the CSM electrical and environmental systems engineer, found what had happened. Pete was asked to switch to the secondary data system so that telemetry would show the status of the electrical system. The crew was then asked to reset the fuel cells, which came back on line, and Apollo 12 continued on its way into orbit. Additional checks were made of the spacecraft electrical system and a guidance reference was reestablished. Apollo 12 went on to the Moon.

A chapter of this book is devoted to Apollo 13. As I moved up in the organization, I reluctantly relinquished the job of flight director. But there were many well qualified young men to assume this responsibility. My faith in their abilities was confirmed by their actions during this epic flight. Following the successful return of the Apollo 13 crew, the performance of the Operations Team was recognized with the presentation of the Medal of Freedom by the President of the United States to Sig Sjoberg, my colleague through all the tribulations of Mercury, Gemini, and Apollo.

Docking was another major hurdle that had to be overcome if we were to make it to the Moon. Normally, it went well but I always breathed easier when it was behind us. There had been no major docking problems in the program until Apollo 14. After five unsuccessful attempts by Al Shepard and his crew, we still had not made the initial docking with the lunar module. Previously we'd always had a docking probe and drogue available in the Control Center, as well as experts on the system, but now there were frantic calls for assistance and the absent docking system had to be hurriedly located to help understand what might be going on thousands of miles out in space. Procedures were worked out and another attempt proved successful.

REPROGRAMMING IN FLIGHT

The next Apollo 14 problem occurred just prior to the final descent for landing at Fra Mauro. An abort command was received by the lunar module's guidance computer. Had the abort command been initiated, it would have separated the ascent stage from the descent stage and terminated the landing. The descent had to be delayed; and, as Al Shepard and Ed Mitchell orbited the Moon, the ground valiantly tried to determine the cause of the problem. It was isolated to one set of contacts of the abort switch on the instrument panel. Recycling the switch or tapping on the instrument panel removed the signal from the computer. A computer program was developed and verified within two hours by the Operations Team and inserted manually into the computer, allowing the computer to disregard the abort command. The unexpected came again within minutes. As the crew started the descent to the Moon, the altitude and velocity lights of the computer display indicated that the landing radar data were not valid. This information provided essential updates to the computer. Flight Controller Dick Thorson made a call to recycle the landing radar circuit breaker. The crew complied. The lights were extinguished and the necessary computer entry update was made at an altitude of about 21,000 feet. Apollo 14 and Al Shepard's and Ed Mitchell's climb almost to the top of Cone Crater are now history.

There were occasions when the problems that came up did not require an instant decision but rather resulted in long hours in conference in Mission Control. For example, on Apollo 15, the flight of *Endeavour* and *Falcon,* as the spacecraft traveled from the Earth to the Moon, the service propulsion system developed a problem. This is the system that is required to place the spacecraft in orbit around the Moon and on its trajectory back to Earth. Needless to say, this was a critical system. A light had illuminated showing that the engine was firing while it obviously was off. This had to be caused by a short in the ignition circuitry. Had this circuit been armed while the short was present, the service propulsion engine would have fired. The Operations Team, working with Don Arabian, a legend in his own time, and Gary Johnson, an excellent young electrical engineer, isolated the short to one of two systems. A test firing was initiated by the crew to verify that the short existed on the ground side of one of two sets of valves. Procedures were then developed by the ground, working with the flight crew, and the mission continued.

LONG-DISTANCE SOLUTIONS

Apollo 16 had its unique problems and one was a major one of the instantaneous and serious variety. Just after separation of the CSM from the LM, prior to initiating final descent for the landing, a maneuver was to be performed by the command and service module *Casper* to circularize its orbit around the Moon. Preparations for the burn went well until a check was made of the secondary yaw gimbals. These gimbals controlled the direction of thrust in yaw plane for the service propulsion system, a system that was essential to insuring that the astronauts could get out of lunar orbit. The gimbals appeared normal until the motor was started and then they exhibited rapidly diverging oscillations. The two spacecraft were asked to rendezvous; and Jim McDivitt, the Apollo Spacecraft Program Manager, met with Bob Gilruth and me to tell us that it appeared to him that the mission would have to be terminated. Another meeting in an hour was scheduled to review the bidding. By the time we had the second meeting, the Operations Team, through extensive testing and simulations, determined that the oscillations would have damped and the secondary servo system was safe to use. John Young and Charlie Duke proceeded with the landing, as I reflected on the phenomenal capabilities of a group of young engineers who had solved a problem of a spacecraft 240,000 miles away from Earth.

Apollo 17, the final mission to the Moon, clearly demonstrated the maturity of the Operations Team. For the first time, a manned launch was made at night. A landing was made in the valley of Taurus-Littrow, the most difficult of any of the Apollo landing sites. The spacecraft performed in an outstanding fashion, and there were no major problems. Minor ones that did occur were handled without difficulty.

The problems encountered were all overcome due to the careful premission preparation, rigorous testing, planning, training, and hours and hours spent simulating critical phases of the mission with the flight crew. These simulations prepared the controllers and the crew to respond to both normal and abnormal situations. Their record speaks for itself on the adequacy of the training. This was not brought together overnight, and in 1962 we were a long way from Taurus-Littrow.

The basic flight-control concepts used for Apollo were developed by a small group of people on the Mercury Operations Team. In 1958, under the leadership of Robert R. Gilruth, the Space Task Group had been given the fantastic responsibility of placing a man in orbit around the Earth. Those few young men who assumed this task did not have any previous experience on which to rely. It had never been done before. What they did have was the willingness to tackle any job, and a technical capability that they had attained through an apprenticeship in what I consider to have been the Nation's finest technical organization, the National Advisory Committee for Aeronautics. Other members of the Mercury Operations Team had experience with aircraft development and flight testing with the Air Force and Navy or with major aircraft companies, both within this country and in particular with AVRO of Canada. That country's cancellation of the CF–105 with its attendant effect on the AVRO program proved to be a blessing to the United States space program. Many fine engineers came to work as members of the Space Task Group at Langley: Jim Chamberlin, John Hodges, Tecwyn Roberts, Dennis Fielder, and Rod Rose, to name a few. The operational concepts that were developed by this cadre on Mercury were improved as experience was gained on each flight. As the Operations Team assumed the responsibility for flying Gemini, the concepts were further developed, expanded, and improved. There were many essential steps that had to be taken to get to the Moon. For the Operations Team, Gemini was one.

Only a small group of people were involved in Mercury operations. When the team was given the responsibility for flying Gemini, and with the Mercury flights continuing, the organization had to be expanded. A conscious effort was made to bring young people into the organization. With an abundance of recent college graduates, the team took on a young character. The additions brought with them the aggressiveness, initiative, and ingenuity that one finds in the young engineer. They did not all come from major colleges; there were graduates of Southwestern State College in Oklahoma, Willamette University in Oregon, San Diego State College, Texas Wesleyan College, and Northeastern University in Boston, to name a few. A large contingent of officers was also made available by the U.S. Air Force and this group provided excellent support. I came to rely on these young people and I can honestly say they never let me down.

AN ADVANCED COMPUTER COMPLEX

As the team was being built, the facilities and equipment were also being defined, developed, constructed, and brought on line. The Mercury flights were directed from a control center at Cape Canaveral, Fla. In 1962, the Space Task Group moved to Houston to form the Manned Spacecraft Center. The construction of the Mission Control Center in Houston, designed to accomplish the lunar missions, was started in 1962. Thirty-six months later it was to be used to control Jim McDivitt's and Ed White's flight in the Gemini IV spacecraft. Its full capability was not used for Gemini, however, as much work still had to be accomplished. One of the most advanced computer complexes in the world had to be integrated with a global tracking network. Tracking and telemetry data had to be relayed from stations in Australia, Spain, the Canary Islands, Guam, Ascension Island, California, Bermuda, Hawaii, Tananarive, and Corpus Christi. Tracking ships were built to provide additional communication coverage in ocean areas. Special Apollo Range and Instrumenta-

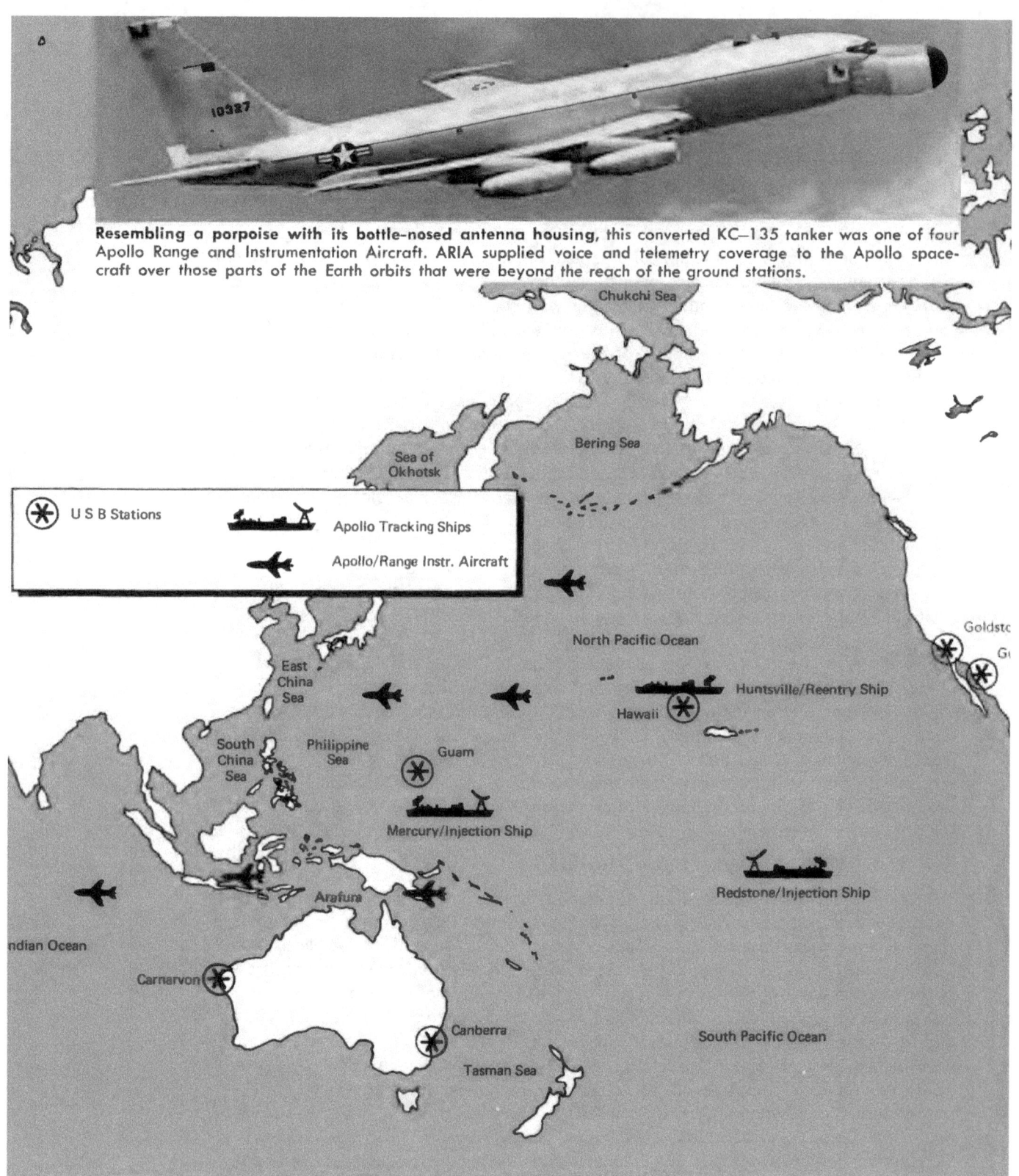

Resembling a porpoise with its bottle-nosed antenna housing, this converted KC–135 tanker was one of four Apollo Range and Instrumentation Aircraft. ARIA supplied voice and telemetry coverage to the Apollo spacecraft over those parts of the Earth orbits that were beyond the reach of the ground stations.

U S B Stations

Apollo Tracking Ships

Apollo/Range Instr. Aircraft

Chukchi Sea

Bering Sea

Sea of Okhotsk

North Pacific Ocean

Goldsto

East China Sea

Huntsville/Reentry Ship

Hawaii

South China Sea

Philippine Sea

Guam

Mercury/Injection Ship

Redstone/Injection Ship

Arafura

ndian Ocean

Carnarvon

Canberra

South Pacific Ocean

Tasman Sea

The Manned Space Flight Network (MSFN) is built around a set of land stations whose antennas supply all spacecraft tracking and communication functions via a single radio telecommunication link. Very large antennas at three of the stations provide continuous tracking at lunar distances. The land stations were supplemented by tracking ships to cover such critical phases of the Apollo missions as insertion into Earth and translunar orbits, and Earth reentry. In addition, instrumented aircraft filled the gaps in communication coverage during the Earth orbital phase.

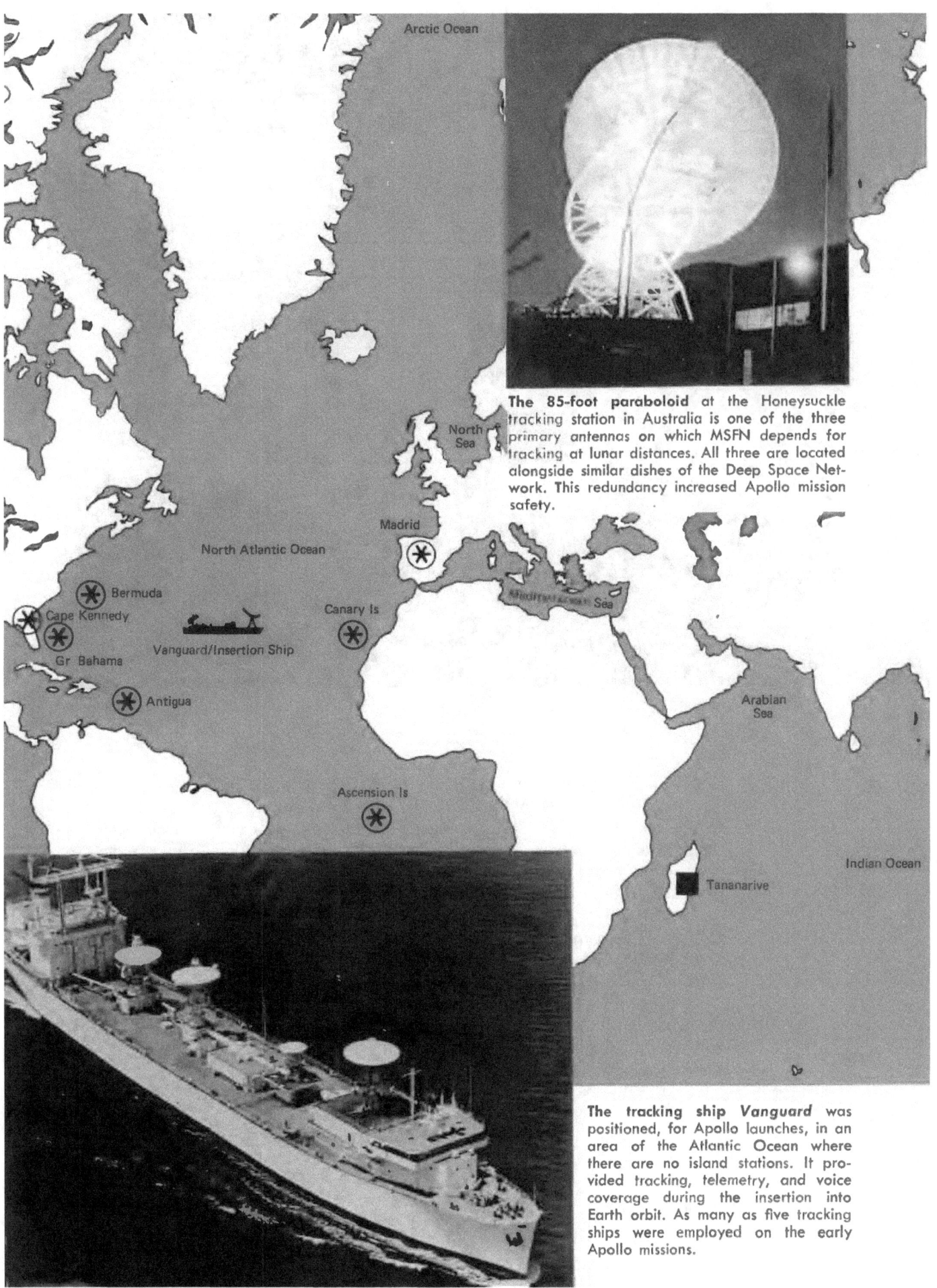

The 85-foot paraboloid at the Honeysuckle tracking station in Australia is one of the three primary antennas on which MSFN depends for tracking at lunar distances. All three are located alongside similar dishes of the Deep Space Network. This redundancy increased Apollo mission safety.

Arctic Ocean

North Sea

Madrid

North Atlantic Ocean

Bermuda

Cape Kennedy

Vanguard/Insertion Ship

Canary Is

Gr Bahama

Mediterranean Sea

Antigua

Arabian Sea

Ascension Is

Tananarive

Indian Ocean

The tracking ship *Vanguard* was positioned, for Apollo launches, in an area of the Atlantic Ocean where there are no island stations. It provided tracking, telemetry, and voice coverage during the insertion into Earth orbit. As many as five tracking ships were employed on the early Apollo missions.

In Mercury days, with a one-man spacecraft in Earth orbit, this control center was sufficient.

tion Aircraft (modified Boeing KC-135 jets) were deployed around the world.

All this was being done concurrently with the evolution of operational concepts. During the Mercury and Gemini flight programs, teams of flight controllers at the remote tracking stations were responsible for certain operational duties somewhat independent of the main Control Center. The advantages of having one centralized operations team became more apparent, and for Apollo, two high-speed 2.4-kilobit-per-second data lines connected each remote site to the Mission Control Center in Houston. This permitted the centralization of the flight control team in Houston. Provisions were also made to tie into the Control Center, through a communications network, the best engineering talent available at contractor and government facilities.

As engineers from the Goddard Space Flight Center were intently determining the requirements for this ground communications network, building and installing equipment, and laboriously testing and verifying the network's capabilities, engineers in Houston, led by a young Air Force officer, Pete Clements, and a fine young engineer, Lynwood Dunseith, were feverishly working to integrate the computer com-

By **Apollo**, with three men in two spacecraft at lunar distances, Mission Control had grown.

An array of specialists manned the consoles during an Apollo mission

Key numbers above identify the locations of flight controllers. 1 was the Booster Systems Engineer, responsible for the three Saturn stages. 2 was the Retrofire Officer, keeping continuous track of abort and return-to-Earth options. 3 was the Flight Dynamics Officer, in charge of monitoring trajectories and planning major spacecraft maneuvers; he also managed onboard propulsion systems. 4 was the Guidance Officer, who watched over the CSM and LM computers and the abort guidance system. In the second row, 5 was the Flight Surgeon, keeping an eye on the condition of the flight crew. At 6 was the Spacecraft Communicator, an astronaut and member of the support crew, who sent up the Flight Director's instructions. (He was usually called CapCom, for Capsule Communicator, from Mercury days.) 7 concerned CSM and LM systems, including guidance and navigation hardware; and electrical, environmental, and communications systems. After Apollo 11, all communications systems were consolidated as a separate task. On the next row in the middle was 8, the Flight Director, the team leader. 9 was the Operations and Procedures Officer, who kept the team—in and out of the Center—working together in an integrated way. 10 was the Network Controller, who coordinated the worldwide communications links. 11 was the Flight Activities Officer, who kept track of flight crew activities in relationship to the mission's time line. 12 was the Public Affairs Officer who served as the radio and TV voice of Mission Control. 13 was the Director of Flight Operations; 14 the Mission Director from NASA Headquarters; and 15 the Department of Defense representative. During activity on the lunar surface an Experiments Officer manned the console at 1 to direct scientific activities and relay word from the science team.

Only minutes before this picture was taken, Jack Swigert had made the call, "Houston, we've had a problem." Left to right, Christopher C. Kraft, Jr., Deputy Director of the Manned Spacecraft Center; James A. McDivitt, Apollo Spacecraft Program Manager; and Robert R. Gilruth, Director of the Manned Spacecraft Center.

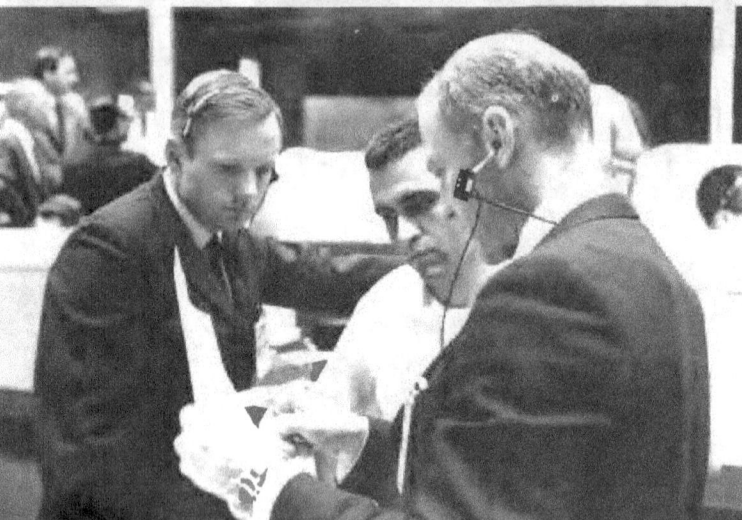

Astronauts assigned to an upcoming mission took particular interest in following the current flight from the Mission Control Center. In this picture, taken during the Apollo 10 mission, Neil Armstrong (left) and Buzz Aldrin (right) discuss the lunar orbit activities in progress with astronaut-scientist Jack Schmitt.

A long moment of quiet satisfaction in the Mission Operations Control Room during the Apollo 11 mission, as George Low and Robert Gilruth look past their consoles toward a television monitor where they can watch Astronauts Armstrong and Aldrin walking on the Moon.

plex and Control Center displays with the network. The critical parameters and limits that had to be monitored in flight needed to be defined; the necessary sensors for measuring the parameters needed to be incorporated in the design of the spacecraft; and rules for utilizing the measurements needed to be developed. But it was not only a question of ensuring that the right measurements were made. Spacecraft and subsystem design also had to have the redundancy and the flexibility needed to overcome failures and contingencies as they arose. And time was relentlessly marching on. Testing of the spacecraft revealed new problems, and new techniques and procedures often had to be developed to avoid potential difficulties in flight. Programs had to be developed for operating the spacecraft and Control Center computers, and the programs had to be verified, tested, and incorporated in the computers. The end of the decade moved closer each day. The complexity of the spacecraft and launch vehicle was exceeded only by the complexity of a worldwide ground-control system.

Then came January 27, 1967, and the AS–204 fire, a day I'll never forget. I was at the console in Houston monitoring the test at the Cape, together with a group of flight controllers. We thought we had considered every eventuality, and now we were struck down by an event that did not occur in space but happened during a ground test. There were no excuses that could be offered but, out of the despair of the fire, there came a rededication to the successful accomplishment of the goal and an intensified effort on the part of every individual.

The Operations Team had many functions not associated with testing and checking out the spacecraft and controlling the mission. These functions were nevertheless essential to success. One was recovery operations. Recovery techniques for the spacecraft and the crew had to be worked out in conjunction with the Department of Defense and the U.S. Navy. Bob Thompson organized and led this effort during Mercury and Gemini. The organizational team he established provided the same excellent recovery support for Apollo as it had for Mercury and Gemini.

MANEUVER TARGETING

The team also developed the techniques for flying the spacecraft and controlling its trajectory. It had the primary responsibility for developing the programs or logic used in the computers onboard the lunar module and the command and service module as well as those in the Control Center. Except for rendezvous maneuvers, the Control Center was the only source of maneuver targeting; that is, determining the exact magnitude, direction, and the time for executing each flight maneuver. Bill Tindall, a truly outstanding engineer, contributed significantly to this effort. Operations were planned in detail before a flight. Plans were based both on everything working properly and on the "what if" situations that might occur. The "what if" situations could not be carried to the point of actually reducing reliability by introducing confusion or complexity into the system. This was quite often a fine line to walk. Techniques also had to be developed for monitoring all essential systems during critical mission phases. The procedures, the techniques, the personnel, and an organization all had to be defined and developed, a task of no small magnitude. Each landing demonstrated how well the task was performed. Apollo 12 was a classic example, with an incredible pinpoint landing some 600 feet from the Surveyor spacecraft that had previously landed on the Ocean of Storms.

To conduct operations for the flights, four complete flight-control teams were organized and used for all Apollo missions. Each team was headed by a flight director: Gene Kranz, Cliff Charlesworth, Glynn Lunney, Gerry Griffin, Pete Frank, Milt Windler, Neil Hutchinson, Phil Shaffer, and Chuck Lewis were all assigned this responsibility during various phases of the program. To simplify the overall training program, each team was assigned different events or activities. The individuals on each team could thus devote their full attention and energy to developing proficiency in accomplishing a few things, as opposed to having to cover an impossible spectrum.

The team was responsible for developing the mission plans to demonstrate the capability of the spacecraft, the systems, and the team to land a crew on the Moon. A series of unmanned developmental flights was planned with well defined objectives to be demonstrated on each flight. Apollo 7, the first manned flight, occurred in October 1968, and the first flight to the Moon, Apollo 8, occurred two months later. Even while Apollo 7 was flying, the Operations Team was performing simulations and training for the Apollo 8 mission. As Apollo 8 was flying, training and simulations were being conducted for Apollo 9, the first Earth-orbital flight of the LM and CSM in March 1969. The next step, Apollo 10—a dress rehearsal for the first landing—was taken in May 1969. On this flight, on the far side of the Moon, a fuel cell was lost and taken off line. The team had trained for this contingency and reacted accordingly.

TRAINING CREWS AND CONTROLLERS

Astronaut training and development of the flight plans and crew procedures were directed by Deke Slayton. He accomplished these tasks in an outstanding manner. The training that Deke provided the crews, as well as the training provided the flight controllers, gave them the capability to react to the unexpected. Quite often it resulted in unique training devices and equipment. The zero-gravity environment was simulated by using a modified KC-135 aircraft that flew parabolas, thus creating 20 to 30 seconds of weightlessness. A neutral buoyancy water tank was also used to simulate the weightless environment. The unique Lunar Landing Training Vehicle (LLTV) was developed to train the astronauts in controlling the lunar module during the final phase of its descent and landing. The test flights of the LLTV, for example, saw the successful emergency ejection of three pilots—Joe Algranti and Stu Present, research pilots, and Neil Armstrong, the commander of the first lunar landing mission—because of vehicle failures. Bob Gilruth and I both believed that flying this craft was more hazardous than flying the actual lunar module.

Simulators also had to be developed to provide training for the flight crew in the operation of the spacecraft. These simulators were tied in with the Mission Control Center so that an integrated training could be accomplished with the flight controllers. These simulations allowed the flight crew to train realistically for all phases of the mission, including the landing itself. Unique display techniques were used with actual models of each landing site. The models allowed the crew to gain familiarity with the terrain and recognizable landmarks. Detailed lunar maps that were based primarily

Standing at the rim of the Rio Grande gorge near Taos, N. Mex., Apollo 15 Astronauts Jim Irwin and Dave Scott see a landscape remarkably like the one they visited at the Hadley Rille landing site on the Moon. Each astronaut team participated in a series of geology field trips to acquaint them with the kinds of field observation that would be most useful to lunar scientists, the types of rock specimen they should particularly try to sample, and the special problems in working with their equipment on the general terrain they would encounter.

Astronauts and their instructor take notes on their field observations during a geology training trip into the Grand Canyon, in Arizona. Although the rocks that are exposed at the Grand Canyon do not resemble lunar rocks in any way, the trip here was an important step in familiarizing the astronauts with the basics of geology, so that they could function well as observers and collectors.

Apollo 13 Astronauts Fred Haise and Jim Lovell observe features of a lava flow near Hilo, Hawaii, during a geology field training trip. They used such items of lunar equipment as the handtool carrier behind them and the Hasselblad cameras mounted on their chest packs. As fate would have it, this pair did not have the chance to use their training.

Although it was past 2 a.m., a crowd of more than 2000 people was on hand at Ellington Air Force Base to welcome the members of the Apollo 8 crew back home. Astronauts Frank Borman, James Lovell, and William Anders had just flown to Houston from the Pacific recovery area by way of Hawaii. The three crewmen of the first manned lunar-orbit mission are standing at the microphones in the center of the picture.

Apollo 8 crewman Frank Borman gets a warm greeting from Robert Gilruth, the Manned Spacecraft Center Director, upon his arrival at Ellington Air Force Base, just outside Houston. Looking on is Edwin Borman, the astronaut's 15-year-old son. William Anders and his family are in the background.

on data provided by the NASA unmanned lunar orbit program were prepared by the Air Force Information and Charting Service and by the U.S. Geological Survey.

As the system matured after Apollo 11, greater emphasis was placed on scientific training and on ensuring that the astronauts were prepared to perform scientific experiments when they arrived on the Moon. Prominent scientists both from within the government and from universities throughout the country offered their time and talent to ensure that the crew and the Operations Team were adequately trained to perform the demanding scientific tasks. During the mission, they also participated as members of the Operations Team. Apollo brought a new aspect to spaceflight as man on the surface of the Moon worked in conjunction with a science team on Earth that capitalized on his observations, judgments, and abilities. They assessed his comments and evaluations, and modified the science planning and objectives in real time. This was not accomplished, however, without moments of frustration and anguish during the early flights, when the acceptability of the spacecraft and its systems was yet to be proved. During the later lunar missions, the crew and the Operations Team were working with proven procedures and a proven spacecraft, and the capabilities of the science organization were effectively integrated in the performance of the missions.

As a means of saying thanks, on March 5, 1973, this group of scientists held a dinner for a number of the program and operations personnel they had worked with over the years. The events of that night clearly showed how well this relationship had developed. As late as 1969, there were very few that would have been brave enough to predict such a dinner would have ever occurred.

CONTRIBUTIONS OF FLIGHT SURGEONS

Chuck Berry and Dick Johnston and their medical personnel also played an important role as members of the Operations Team. Working with engineering personnel, they developed the monitoring techniques used to observe the critical medical parameters of man in flight. The flight surgeons' judgment and ability to assess the astronauts' well-being in flight as well as their confidence in the crew's readiness to undertake each of the missions were very necessary to achieving success. In the beginning, there were some who doubted man's capability to even exist, let alone work, in the environment of space. Chuck Berry had no such doubts and worked hard to alleviate such concerns. I do not believe that we could have gotten to the Moon without the contributions of the flight surgeons.

The Apollo Operations Team was a unique group brought together to accomplish a successful landing on the Moon and return to Earth. I do not believe that the dedication and the capabilities of these people have ever previously been duplicated, and I doubt that such a group will ever be brought together again. A great amount of preparation preceded the actual flying of an Apollo mission. The spacecraft had to be designed, built, and tested, but the group that actually flew the mission was faced with an awesome responsibility. President Truman had a sign on his desk in the White House stating that "The buck stops here." This comment could well be applied to the Apollo Operations Team. For these young men and women, the Apollo missions were their finest hour—the truly great adventure of their lives as well as of mine.

Footprints on the plain at Hadley, beneath the unearthly Apennines, were made by men who had walked the long path of astronaut selection and training. To be one of the dozen men who have so far walked the Moon was to have survived close screening for physical and mental excellence, and to have emerged successfully from long, intensive, and often competitive training.

CHAPTER EIGHT

Men for the Moon

By ROBERT SHERROD

On a June day in 1965, following their spectacular Gemini 4 flight, James McDivitt and Edward White flew up to Washington from Houston with their wives and children. The helicopter bearing them from Andrews Air Force Base, Md., had no sooner settled on the White House lawn than Lady Bird Johnson said she wanted them all to spend the night; babysitters would be provided. The two astronauts heard the President call them "Christopher Columbuses of the twentieth century," and he pronounced the United States now caught up with the Russians.

The two astronauts had a parade. They lunched with Vice President Humphrey and congressional leaders, and in the evening they went to the State Department for a reception. Before a packed assemblage of foreign diplomats they showed a 20-minute movie of their flight, which included the first American walk in space by Ed White.

In strode Lyndon B. Johnson himself, who told McDivitt and White, " I want you to join our delegation in Paris." Furthermore, the President wanted them to go now, as soon as they and their wives could pack. He was seething because the Russians had humbled the Americans at the Paris Air Show, where Yuri Gagarin was standing by his spacecraft, shaking hands with everybody and passing out Vostok pins. The French press noted that the lackluster American pavilion was shunned by the crowds.

Patricia McDivitt and Patricia White wailed in unison, "But we have nothing to wear!" Never mind, said LBJ, Lady Bird and Lynda Bird and Luci have plenty of clothes. The ladies retired to the White House bedrooms, and the two Pats were duly outfitted. Long after midnight the Presidential plane took off, bearing as additional passengers Hubert Humphrey, James Webb, and Charles Mathews, the Gemini program manager.

The astronauts made it only in time for the last day and a half of the eleven-day show, but they gave the Russians some real competition. Wherever they appeared, the American *jumeaux de space* were followed by masses of Frenchmen. "A partial recovery for the United States" was the Paris newspapers' verdict.

Apollo 1   Edward H. White II, senior pilot
Virgil I. Grissom, command pilot
Roger B. Chaffee, pilot

Apollo 7   R. Walter Cunningham, lunar module pilot
Walter M. Schirra, Jr., commander
Donn F. Eisele, command module pilot

Apollo 10   Eugene A. Cernan, lunar module pilot
John W. Young, command module pilot
Thomas P. Stafford, commander

Apollo 11   Neil A. Armstrong, commander
Michael Collins, command module pilot
Edwin E. Aldrin, Jr., lunar module pilot

Apollo 14   Stuart A. Roosa, command module pilot
Alan B. Shepard, Jr., commander
Edgar D. Mitchell, lunar module pilot

Apollo 15   David R. Scott, commander
Alfred M. Worden, command module pilot
James B. Irwin, lunar module pilot

Apollo 8   James A. Lovell, command module pilot
William A. Anders, lunar module pilot
Frank Borman, commander

Apollo 9   James A. McDivitt, commander
David R. Scott, command module pilot
Russell L. Schweickart, lunar module pilot

Apollo 12   Charles Conrad, Jr., commander
Richard F. Gordon, command module pilot
Alan L. Bean, lunar module pilot

Apollo 13   Fred W. Haise, Jr., lunar module pilot
James A. Lovell, commander
John L. Swigert, Jr., command module pilot

Apollo 16   Thomas K. Mattingly II, command module pilot
John W. Young, commander
Charles M. Duke, Jr., lunar module pilot

Apollo 17   Harrison H. Schmitt, lunar module pilot
Ronald E. Evans, command module pilot
Eugene A. Cernan, commander

Such were the glory days, when to be an astronaut was to be in heaven—if one could endure the slavery that was the obverse of the coin.

Altogether there were seven groups of astronauts, a total of 73, of whom 43 flew before the long night settled on manned space flight after the Apollo-Soyuz mission in July 1975. Twenty-nine of these filled the 33 slots in Apollo, with which we are principally concerned here.

What did it take to become an astronaut? Dr. Robert Voas, a psychologist who was the Mercury astronauts' training director, detailed the required characteristics as he saw them: intelligence without genius, knowledge without inflexibility, a high degree of skill without overtraining, fear but not cowardice, bravery without foolhardiness, self-confidence without egotism, physical fitness without being muscle-bound, a preference for participatory over spectator sports, frankness without blabbermouthing, enjoyment of life without excess, humor without disproportion, fast reflexes without panic in a crisis. These ideals were fulfilled to a high degree.

THE STRENUOUS SELECTION PROCESS

In December 1958, plans had been made to post civil service notices inviting applications for astronaut service, GS–12 to GS–15, salary $8,330 to $12,770. President Eisenhower thought this ridiculous, and decided that the rolls of military test pilots would furnish all the astronauts necessary. "It was one of the best decisions he ever made," said Robert Gilruth sixteen years later. "It ruled out the matadors, mountain climbers, scuba divers, and race drivers and gave us stable guys who had already been screened for security." From the records of 508 test pilots, 110 were found to meet the minimum standards (including the height and age limitations, 5 feet 11 inches and 40 years).

After further examination, the 110 were narrowed to 69, then to 32, who were put through strenuous physical tests: How much heat could the man stand? How much noise? How many balloons could he blow up before he collapsed? How long could he keep his feet in ice water? How long could he run on a treadmill?

Worst of all, the astronauts thought, were the 25 psychological tests that entailed minute and painful self-examination ("Write 20 answers to the question: 'Who am I?'") From the 18 survivors, seven were chosen in April 1959, and they would remain the Nation's only astronauts for three and one-half years. Their IQs ranged from 130 to 145, with a mean of 136. Even before they had accomplished anything they became instant heroes to small boys and other hero-worshipers around the world.

Among those who flunked the first round were James Lovell and Charles Conrad, who were picked up in the Second Nine in 1962 and went on to make four spaceflights apiece—a record they shared only with John Young and Tom Stafford. The Second Nine proved even more stable than their predecessors. Excepting Ed White, killed in the spacecraft fire of January 1967, and Elliott See, who died in a plane crash, all commanded Apollo flights.

The Second Nine were test pilots, too, but two of them were civilians: Neil Armstrong, who had flown the X–15 for NASA, and See a General Electric flier. By the time the third group of fourteen was selected in 1963 the test-pilot requirement had been dropped—as had most of the outlandish physical tests—but the educational level had risen to an average of 5.6 years of college, even though the IQ average fell a couple of points below the first two groups. Four of the fourteen would die in accidents without making a spaceflight.

The fourth group of six men wasn't even required to be pilots because they were scientists (doctors of geology, medicine, physics, and electrical engineering) who, after selection, had to take an extra year to learn to fly. Because three missions were cut from the program, only one, Harrison Schmitt, was to fly in Apollo. Three others flew in Skylab.

The fifth group was the biggest of all, nineteen pilots, of whom twelve would fly in Apollo, three in Skylab, and one in Apollo-Soyuz. The sixth group, eleven more scientists, scored the highest mean IQ, 141, but they came too late to fly and six resigned before 1975. The final group of seven transferred from the Air Force's defunct Manned Orbital Laboratory in 1969 and their hope had to rest on the resumption of flight with the Space Shuttle about 1979.

One thing all astronauts had in common: hard work. Each astronaut was assigned one or more specialties, which he had to learn with dizzying completeness. Neil Armstrong, for example, was assigned trainers and simulators, John Young environmental control systems and personal and survival equipment, Frank Borman boosters. In the third group Buzz Aldrin, who had earned a doctorate with a dissertation on orbital rendezvous, was a natural for mission planning; Bill Anders, who had a master's degree in nuclear engineering, drew the environmental control system; Mike Collins had the spacesuit and extravehicular activity. The astronauts worked with and learned from scientists and engineers, and suggested many ideas from a crewman's viewpoint.

PREPARING FOR ALL EVENTUALITIES

Training was the name of the game, and they trained until it seemed the labors of Hercules were child's play—how to make a tent out of your parachute in case you came down in a desert; how to kill and eat a snake in the jungles of Panama; how to negotiate volcanic lava in Hawaii. An Air Force C–135 flew endless parabolas so the astronauts could have repeated half-minute doses of weightlessness. They wore weights in huge water tanks in Houston and in Huntsville to get a feel of movement in zero and one-sixth gravity.

Hair-raising was the device called the Lunar Landing Training Vehicle, a sort of flying bedstead, which had a downward-pointing jet engine, gimbal-mounted and computer-controlled to eliminate five-sixths of gravity. In addition, it had attitude-controlling thrusters to simulate the way the LM would act before touchdown on the Moon. If the trainer ran out of fuel at altitude, or if it malfunctioned, the Apollo commander—for he was the only one who had to fly the thing—had to eject, which meant he was catapulted several hundred feet into the air before his parachute opened. That was exactly what happened to Neil Armstrong a few months before his Apollo 11 mission, when his bedstead started to tilt awry a hundred feet above the ground. Armstrong shot into the air, then floated to safety; the machine crashed and burned.

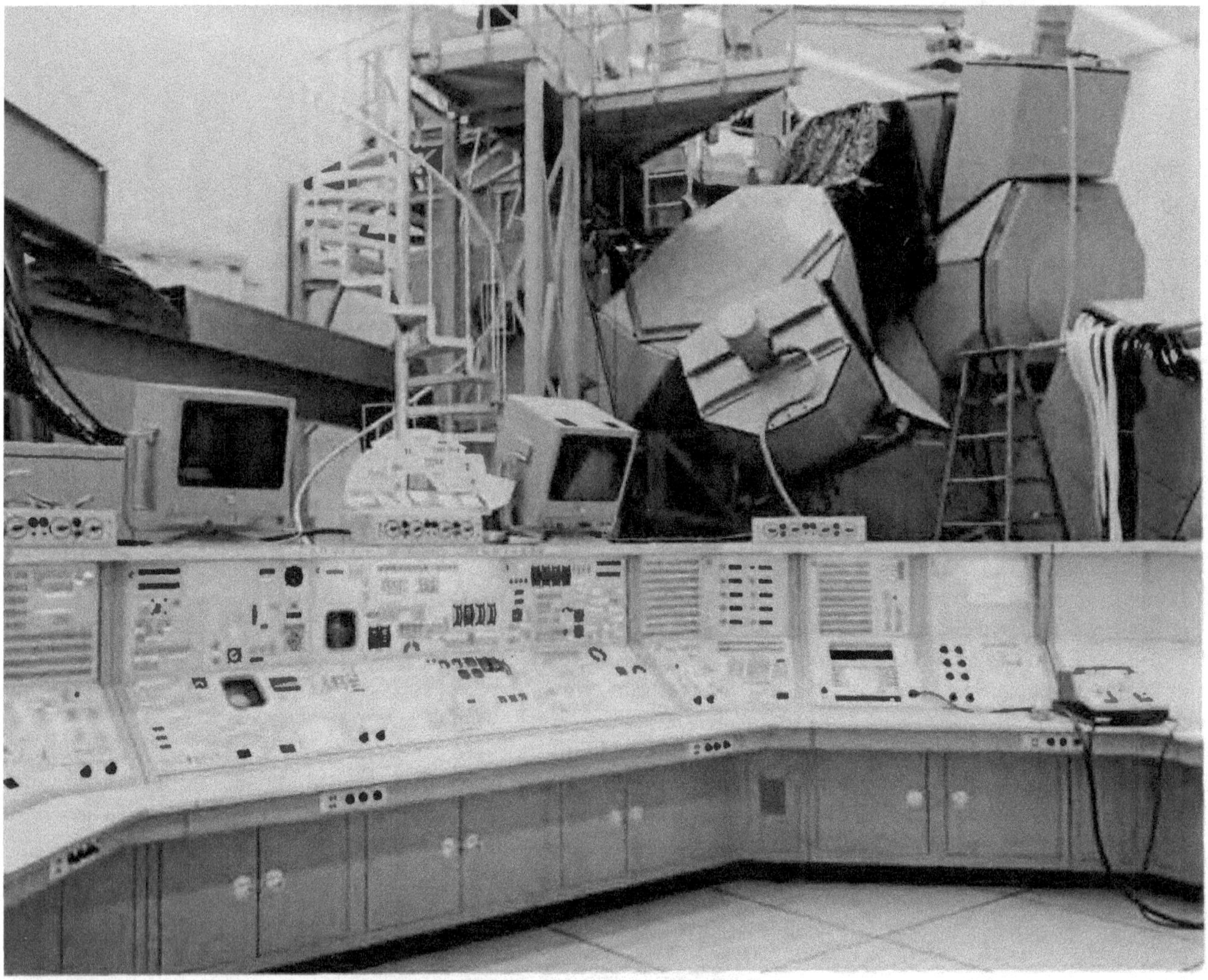

"The great train wreck" was John Young's description of the contraption beyond the console. At the top of the stairs was a compartment that exactly duplicated a command module control area, with all switches and equipment. Astronauts spent countless hours lying on their backs in the CM simulator in Houston. Panel lights came on and off, gauges registered consumables, and navigational data were displayed. Movie screens replaced the spacecraft windows and reflected whatever the computer was thinking as a result of the combined input from the console outside and astronaut responses. Here the astronauts practiced spacecraft rendezvous, star alignment, and stabilizing a tumbling spacecraft. The thousands of hours of training in this collection of curiously angled cubicles paid off. Many of the problems that showed up in flight had already been considered and it was then merely a matter of keying in the proper responses. At left, Charles Conrad and Alan Bean in the LM simulator at Cape Kennedy prepare to cope with any possible malfunctions that the controllers at the console outside could think up to test their familiarity with the spacecraft and its systems.

Neil Armstrong contemplates the distance between the footpad and the lowest rung: would he be able to get back up? (The bottom of the ladder had to end high to allow for shock-absorber compression of the LM leg.) He decided he could do it. Ascent proved no problem in reduced lunar gravity.

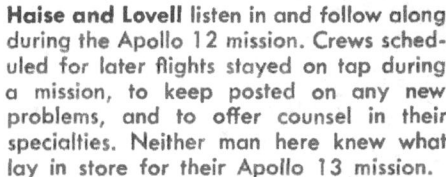

Haise and Lovell listen in and follow along during the Apollo 12 mission. Crews scheduled for later flights stayed on tap during a mission, to keep posted on any new problems, and to offer counsel in their specialties. Neither man here knew what lay in store for their Apollo 13 mission.

Preparing for the unknown was a challenge. How much work could be done by a man within a pressurized (and hence stiff-jointed) spacesuit? What effect would the lesser lunar gravity have on his efforts? This truck-borne hoist, adjusted to take out five-sixths of his weight, gave preliminary indications. It also previewed the loping and kangaroo-hopping gaits that would occur on the Moon. A different way to simulate lunar gravity was also tried out; see the rig on page 162.

Dozens of training aids sharpened astronaut skills, but the most indispensable were flight simulators, contraptions built around copies of CM and LM control areas and complete to every last switch and warning light. Astronauts on prime status for the next mission would climb in to flip switches and work controls. The simulator would be linked to a computer programmed to give them practice too. What made it exciting was that training supervisors could also get in the loop to introduce sneaky malfunctions, full-bore emergencies, or imminent catastrophes to check on how fast and well the crews and their controllers would cope. Surrounding the mockup spacecraft were huge boxes for automatic movie and TV display of what astronauts would see in flight: Earth, Moon, stars, another spacecraft coming in for docking. When John Young first encountered a simulator he exclaimed, "the great train wreck!" Hour after weary hour the spacemen had to solve whatever problems the training crews thought up and fed into the computer. The Apollo 11 crew calculated they spent 2000 hours in simulators between their selection in January and their flight in July 1969.

Some of this bone-cracking training was done in Houston but much of it at the Cape, in Downey, Calif. (the CM), or Bethpage, Long Island (the LM). When the astronauts were not training they were flying in their two-seater T–38 planes from one place to the other, or doing aerobatics to sharpen their edge, or simply to unwind. Their long absences proved a plague on their home lives, and there was hardly a man among them who did not consider quitting the program at one time or another "to spend some time with my family."

THEIR ORIGINS AND CHARACTERISTICS

The 29 Apollo astronauts tended to originate in the American heartland, in such places as Weatherford, Okla.; Columbus, Ohio; Jackson, Mich.; and St. Francis, Kans. Four states gave birth to three astronauts each: Texas, Ohio, New Jersey, and Illinois. Only one astronaut was born in New England, and none in New York City; only two in the Deep South and two on the West Coast. Birthplaces in the twentieth century can be deceptive, however: Frank Borman was born in a steel-mill town, Gary, Ind., but because he was a sickly child (sinus problems and mastoiditis) his family moved to Arizona; John Young first saw light in San Francisco, but his accent was authentic grits and red-eye because he grew up in Orlando and rural Georgia and attended Georgia Tech (where he stood second in engineering). Two were born overseas of U.S. families, Anders in Hong Kong and Collins in Rome.

The service schools educated nearly half the Apollo astronauts: the U.S. Naval Academy eight, West Point five. As might be expected of superachievers they were good students, the median class-standing pegged at the top sixteen percent. Only two other colleges produced more than one Apollonian: the University of Colorado and Purdue with two apiece. Three of the twenty-nine earned doctorates: Aldrin and Mitchell at MIT, Schmitt, the only one with no military experience, at Harvard. (Pete Conrad was the lone astronaut to do his undergraduate work at an Ivy League institution, Princeton.)

The average Apollo astronaut was medium-sized, slightly under 5 ft 10 in., about 160 pounds. He was 38.6 years old when he made his flight or, in the case of

Coming down easy on an unknown surface with limited fuel would take great piloting skill. This giant gantry at the Langley Research Center was used by research pilots to aid LM design, and to explore piloting techniques having the least risk of damage or upset. This multiple-exposure shot shows a landing with little forward movement at touchdown.

the four who flew two Apollo missions, his first one. The blue-eyed outnumbered the brown, 16 to 8, and five had eyes described as hazel or green. Nineteen had (or had had) brown hair, seven blond, one black, and there were two redheads, Schweickart and Roosa. Six of them were well over on the bald side, Stafford almost at the peeled-egg stage.

The twenty-seven who were married (Swigert and Schmitt were not) produced an average of 2.8 children. Generally the astronauts—twenty-three Protestants and six Catholics—adhered more closely to formal religion than their contemporaries; a high proportion of them served as elders, stewards, deacons, or vestrymen. Presbyterian Aldrin administered holy communion to himself inside *Eagle* after it landed on the Moon, and when Frank Borman was orbiting the Moon he apologized to his fellow members of St. Christopher's Episcopal Church because his absence made it impossible for him to serve as a lay reader on Christmas Eve. The prayer he did say reached a somewhat larger audience and caused an atheist to sue NASA, unsuccessfully, on a separation-of-state-and-church issue.

The astronauts were big on sports and flashy cars, but weak in the classics. Mike Collins, who combined the fastest game of handball (nobody could even come close to beating him) with a taste for literature, used to say that the space program needed more English majors to get away from the engineerese in the exhaustive checklists. He proved his versatility by writing a literate, perceptive, and witty book called *Carrying the Fire*.

ADVANTAGES OF BEING THE ELDEST SON

Let us proceed to the anomalies, space jargon for the surprises. Who would have imagined that seven of these superboys-next-door would turn out to be left-handed (over twice what the percentages would predict), particularly since it is more convenient for test pilots to be right-handed? Was it another symptom of the determination to beat the odds? Similarly, nine of the twenty-nine didn't make it the first time they applied for admission (two of them had to apply three times). Was it the "I'll show 'em" obsession that prompted them to try again—and win?

Then there is the eldest-son syndrome. As long ago as 1874 a psychologist noted that the first child in a family tended to excel. He got a headstart: his doting parents taught him to read early, to study hard, to take on responsibility for himself. A survey of the twenty-nine confirms this thesis to an astonishing degree. Twenty-two were first-born (six of them in the only-child category). Five others had older sisters but were eldest sons. Only two, Stuart Roosa and Mike Collins, had older brothers, and in the case of Collins there is a qualification: he hardly knew his brother, who was thirteen years older, and Major General and Mrs. Collins gave Mike the only-son treatment the second time around.

To be dubbed astronaut, by way of the thrilling telephone call from Deke Slayton or, in some cases, his deputy Al Shepard, was to achieve knighthood. This occasion the chosen one never forgot, like the day he got married or the day JFK was shot.

Soon he learned that this was only the beginning. An astronaut didn't join the

The Flying Bedstead, officially the Lunar Landing Training Vehicle, made everyone a bit tense. (Inset shows Al Shepard beginning a run.) Its jet engine and thrusters gave an excellent feel for landing the LM, but it was cranky, unforgiving, and uncontrollable if allowed to tilt too

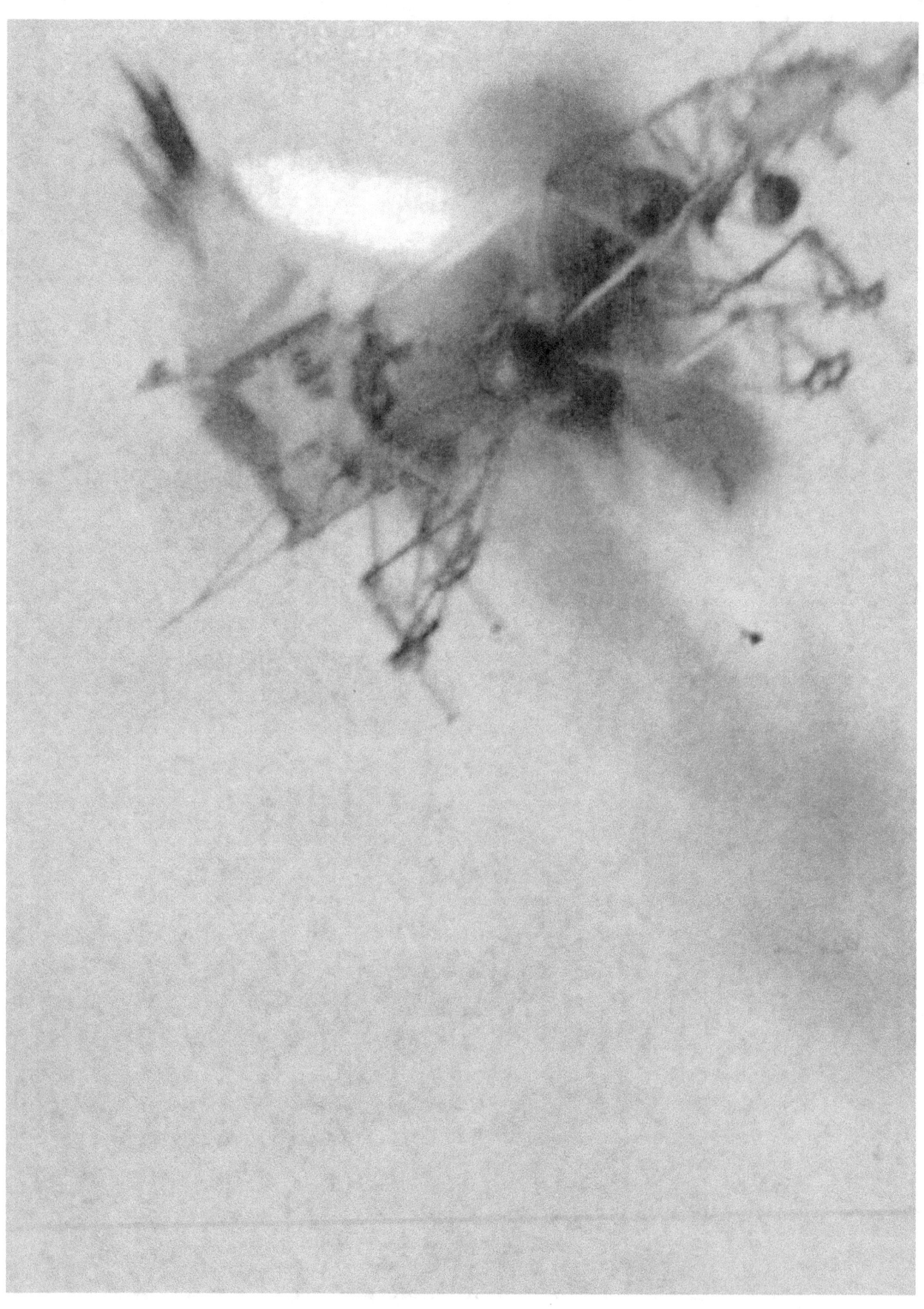

peerage until he got assigned to a flight, initially to a back-up crew and then later prime. "I would have flown alone, or with a kangaroo—I just wanted to fly," Mike Collins wrote.

Men assigned to a crew not only were in the public eye. They and their families achieved status with their peers; they went to the head of the line when it came to time in the simulators, or in access to the T–38 planes.

The men with missions got invited to the White House; Congress welcomed them. Even as late as Apollo 13, whose crew appeared before the Senate Space Committee, Senator Curtis credited the astronauts with the ability "to increase the attendance of this committee more than anything I know of." Senator Symington observed that Lovell & Co. represented "all the best in this country," and Senator Margaret Chase Smith of Maine said that "this is certainly one of the most momentous occasions of my career."

Sometimes fate played a hand, as in the case of Buzz Aldrin. Deep in despair over his failure to be assigned to a prime Gemini crew—he was back-up on Gemini 10, and there would be no Gemini 13—Aldrin was suddenly catapulted to the last mission because his next-door neighbor Charles Bassett was killed in a plane crash. From his superb walk in space on his Gemini mission Aldrin went on to back up Apollo 8, succeeded to prime on Apollo 11, and assured himself a front seat in the history books.

Spaceflight was the *sine qua non*, but ironically the man who sat atop this bastion of power was an unflown astronaut. Donald K. Slayton, a dairy farmer's son from Sparta, Wis., got his wings at age 19 and flew 56 B–25 missions in the Mediterranean theater before he was 21, followed by seven missions in the Pacific. After World War II he went to the University of Minnesota (aeronautical engineering) and became an Air Force test pilot.

A blue-eyed, steel-nerved flier, he was a natural for the second orbital Mercury mission, after John Glenn's. But doctors discovered he had a slightly irregular heartbeat, which raised an issue that assumed vast proportions—his six colleagues even appealed to President Kennedy to overrule the faction of doctors who wanted to ground Deke. Kennedy assigned this hot potato to his Vice President, who invited the astronauts, along with Robert Gilruth, to a weekend at the LBJ Ranch to thrash out this complaint and others. Deke remained grounded but his comrades elected him their leader, thereby conferring on him (with Gilruth's concurrence) immense power that would control the destinies of all astronauts for the next decade.

WHICH MEN FOR WHICH FLIGHT CREWS?

How did Deke select the men to fly? Mostly by seniority. An astronaut stood in line until his turn came, though the order of assignment within his own group was important. "They are a durn good bunch of guys, real fine troops, a bunch of chargers," said Slayton in 1972. "It's not the kind of organization where you have to keep pointing people in the right direction and kicking them to get them to go. Everybody would like to fly every mission, but that's impossible, of course. They all understand that, even if it makes some of them unhappy." Slayton was rarely accused of unfairness, a remarkable achievement considering the stakes involved; his own even-

If the bedstead went flooey, it was time to leave. Here research pilot Stuart Present ejects, to parachute to safety. In all, two research pilots and Neil Armstrong had to bail out. But when NASA brass suggested dropping the risky trainer, astronauts who'd already made Moon landings vetoed it, insisting it accurately forecast LM handling.

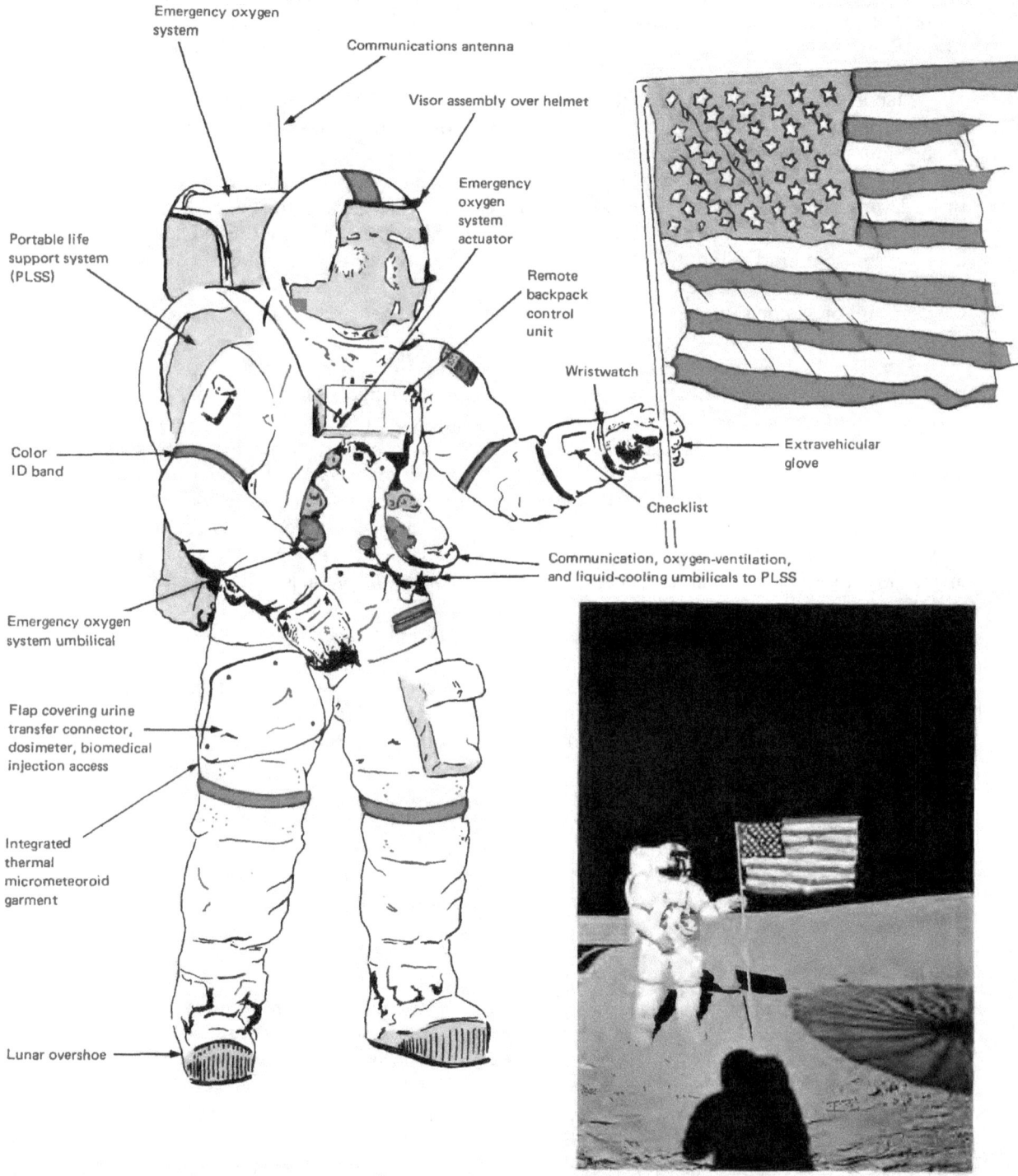

Emergency oxygen system

Communications antenna

Visor assembly over helmet

Emergency oxygen system actuator

Portable life support system (PLSS)

Remote backpack control unit

Wristwatch

Color ID band

Extravehicular glove

Checklist

Communication, oxygen-ventilation, and liquid-cooling umbilicals to PLSS

Emergency oxygen system umbilical

Flap covering urine transfer connector, dosimeter, biomedical injection access

Integrated thermal micrometeoroid garment

Lunar overshoe

The lunar environment is hostile to man. It lacks oxygen and water; it is too hot in the day's sunshine (250° F) and too cold in the night's darkness (—250° F). Hazardous micrometeoroids and space radiation dart about. With no atmosphere and therefore no atmospheric pressure, fluids in an exposed human body will boil on the Moon. How can man survive, much less operate, in such an unfriendly environment? Answer: by taking his own friendly environment with him in the form of a spacesuit.

The spacesuit, or Extravehicular Mobility Unit (EMU), shown on opposite page, is actually three suits in one. These three may roughly be described as: a union suit, a diver's suit, and a suit of armor. Each is custom-tailored to an astronaut's dimensions. Precise models of astronaut's hands, for example, (shown below right), were used to prepare fitted gloves.

The union suit is climbed into first. Made up of a network of flexible tubes embedded in a mesh fabric, this water-cooled underwear is linked to the vital backpack portable life support system (PLSS) where the water and oxygen are stored and metered for precise circulation. The cooling is necessary because body heat cannot dissipate adequately.

Over this liquid cooling garment the astronaut wears the pressure garment assembly, a kind of super-sophisticated diver's suit. Constructed of rubbers and fabrics, and with cleverly contrived joints, this assembly retains the oxygen atmosphere without leakage while facilitating movement. It too is linked into the PLSS, where carbon dioxide and other contaminants are removed from the oxygen stream. A chest-mounted remote control unit permits the astronaut to adjust oxygen flow and cooling temperature to match his preferences. The PLSS will sustain activity for up to 8 hours before its oxygen, water, and battery power must be recharged. A separate oxygen system, atop the PLSS, is available for emergencies; up to 75 minutes' worth of oxygen is available, depending on the draw-off rate.

Protecting the integrity of the pressure garment is the outer suit of armor, a 13-layer composite more like a coat of mail than the rigid clanking uniform of castle-haunting ghosts. The various layers of the outer garment protect the knight of Apollo from the slings and arrows of outer space: micrometeoroids, ultraviolet rays, and other radiation. Visors and shades similarly protect the eyes.

All suited up the astronaut is not unlike the small boy who has been so bundled up that he can hardly move, much less frolic in the snow. Actually, given all the constraints and requirements, spacesuit designers were highly successful in affording the astronaut extensive mobility in his EMU. Nevertheless, like the knights of old, squires are needed to assist suited astronauts in

mounting their steeds and once aboard their spacecraft the astronauts help each other in donning and doffing their outfits. Going to the bathroom, as one would imagine, would be nigh impossible in such garb. So designers built in a sheath urinal and collector bag. Aside from specially absorbent underwear, there was no provision for bowel movements during EVA. (Within the spacecraft, these were coped with by sealable disposable bags.)

A compressed food bar for the astronaut to nibble on is positioned inside the helmet, as is a straw-like tube for sipping a beverage from a neck-ring suspended bag. Close to the mouth also are voice-actuated microphone pick-ups, which are an integral part of a skull cap worn under the helmet. A backpack-housed communications system enables the astronaut out on the airless Moon to converse with his colleagues in the spacecraft and in Mission Control. The latter also receive, through the PLSS, telemetered biomedical data picked up from sensors on the astronaut.

Astronauts actually had a wardrobe of suits, each appropriately designed for use in training, flight, surface and free-space EVA, and as backup. Expensive, sartorial splendor was not one of the suit's design criteria; survival plus the abilities to communicate, move about, and deal with equipment were.

On mock lunar terrain, wearing restrictive pressure suits, Schmitt and Cernan practiced collecting geological samples. They were drilled in formal sampling procedure: locate, radio description of size and color, photograph in place by the gnomon, and then collect in numbered plastic bags.

It's a long way down. The astronauts had to train for the possibility that during countdown their launch vehicle could turn into a bomb. The rig shown here was practice for a ride for life from the 320 ft level of the Mobile Launcher. The astronauts would, if necessary, enter the cab and zoom down the guide wire into an underground, padded and insulated room, safe from explosion. Astronaut Roosa prepares to climb down.

Getting in and out of the hatch: a necessary element in training. Before launch, an astronaut had to fit into a very cramped space without rearranging preset console switches. During return from the Moon, it was also necessary for one man to get out and return through the CM hatch to retrieve the film and data cassettes from the service module.

Fitted with Earth wheels and performance characteristics to match those expected on the Moon, a version of the Moon buggy was driven on rough terrain until its handling traits were second nature to the astronauts. They learned its steering feel, braking ability, and grew familiar with its guidance and navigation calculator. Here Astronauts Scott and Irwin practice for Apollo 15.

tual assignment at age 48 to the Apollo-Soyuz mission was universally popular.

Not all astronauts were equal, of course. Before the first manned Apollo flights six crews had been formed, commanded by Schirra, Borman, McDivitt, Stafford, Armstrong, and Conrad, who had shaken down as the natural leaders with appropriate seniority. Before he flew Apollo 7 Schirra announced he was quitting afterward, and that left five. Which one would land first on the Moon? It depended on the luck of the draw. "All the crews were essentially equal," said Slayton, "and we had confidence that any one of them could have done that first job."

Either Borman or McDivitt seemed likeliest to be first on the Moon; after they flew their early Gemini missions they were sent straight to Apollo instead of being recycled into later Gemini flights. But in 1968 two things happened to derail this prospect: McDivitt, scheduled to lift off on the first Saturn V (Apollo 8), declined the opportunity because it would not carry the LM, on which he had practiced so long. Borman, scheduled for Apollo 9, was "highly enthusiastic" about Apollo 8, LM or no LM, but in deference to wife and children he decided that this would be his last flight.

Borman's back-up was Armstrong; McDivitt's was Conrad. In each case, the back-up shifted with the prime, so Armstrong in the normal rotation became Apollo 11, and Conrad lost his chance to be first man on the Moon by becoming Apollo 12. There was also the possibility that Apollo 10 (Stafford) might be the first Moon lander—George Mueller initially saw no point in going to the Moon a second time without touching down. But for this one the LM wasn't completely adapted for the task (it weighed too much) and the program management decided they were not ready for the big step. When the batting order was aligned in the summer of 1968, nobody could have forecast how the assignments would sort out.

CHOOSING THE FIRST MAN ON THE MOON

Once it was fairly certain that Apollo 11 was it, newspaper reporters and some NASA officials predicted that Aldrin would be the first man to step on the Moon. The logic was that in Gemini the man in the right-hand seat had done the EVA, and the early time line drawn up in MSC's lower engineering echelons showed him dismounting first. But the LM's hatch opened on the opposite side. For Aldrin to get out first it would have been necessary for one bulky-suited, back-packed astronaut to climb over another. When that movement was tried, it damaged the LM mockup. "Secondly, just on a pure protocol basis," said Slayton, "I figured the commander ought to be the first guy out . . . I changed it as soon as I found they had the time line that showed that. Bob Gilruth approved my decision." Did Armstrong pull his rank, as was widely assumed? Absolutely not, said Slayton. "I was never asked my opinion," said Armstrong. "It was fine with me if it was to be Neil," Aldrin wrote, half-convincingly.

Five days before he sent McDivitt and White on that 1965 midnight ride to Paris, Lyndon B. Johnson had thrown a monkey wrench into the Pentagon's machinery by jubilantly announcing that he was promoting those astronauts, both Air Force officers, from major to lieutenant colonel (he didn't bother to find out that both had only recently made major). The astronauts were naturally delighted.

In justice to Maj. Virgil Grissom USAF and Lt. Comdr. John Young USN, who

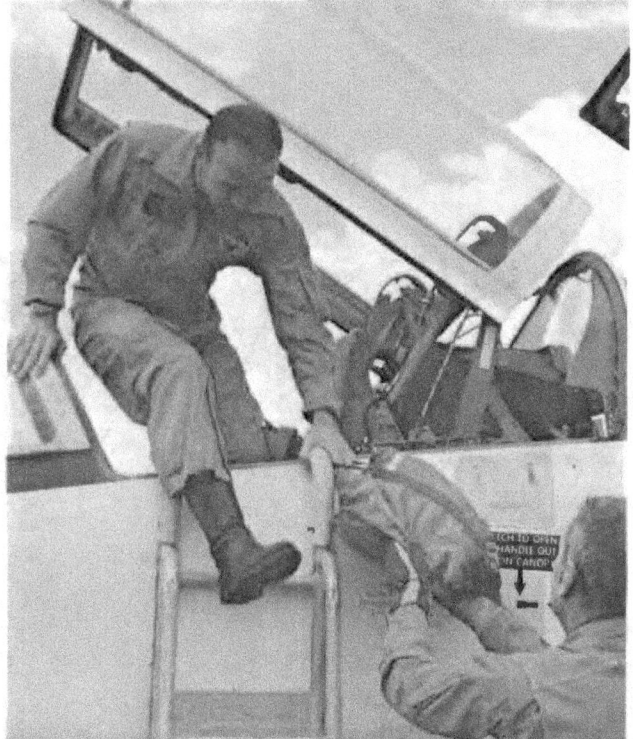

A constant companion to an astronaut during his training was the graceful twin-engined T–38, a two-seat jet that was fine for aerobatics. T–38's were handy for the incessant travel—to California, New York, the Cape, and way stations—that was called for by the policy of involving astronauts in spacecraft development. And to men who had in the main been expert test pilots, the agile T–38 was both a means of keeping sharp and a resource offering privacy and pleasure.

Mike Collins, left, lands after an exhilarating session of aerobatics. The T–38 was useful not just as a means of keeping piloting skills fine-honed but also to keep up g-load tolerances and inner-ear response to weightlessness. Plenty of flight hours before launch seemed to reduce the tendency toward nausea during initial exposure to weightlessness during spaceflight.

A parabolic flight path in a jet transport could create up to 30 seconds of zero gravity, enough to practice exit through a spacecraft hatch (above). Two earth-bound simulations of reduced or zero gravity are shown at right and below. Wearing pressure suits carefully weighted to neutral buoyancy, astronauts in a big water tank learn the techniques needed to work effectively in space. Below, ingenious slings are supported by wires running to a trolley high above. The angled panels on which the man walks or runs are offset just enough from directly under the trolley to simulate the sixth of Earth gravity that prevails on the Moon.

had flown Gemini 3 three months earlier, the President accelerated promotions for them, too, again without saying anything to NASA or the Defense Department. He also went back and picked up some unpromoted Mercury astronauts. Admiral W. Fred Boone, NASA's liaison officer to the Pentagon, noting "some dissatisfaction both among the astronaut community and in the Pentagon," undertook a study. Wrote Boone: "We agreed it would be preferable that meritorious promotions be awarded in accordance with established policy rather than on a 'spur of the moment' basis."

The upshot was a policy, approved by the President, providing that each military astronaut be promoted after his first successful flight, but not beyond colonel USAF or captain USN. Civilians would be rewarded by step increase in civil service grade. Only one promotion to any individual.

That policy came unstuck on Apollo 12, flown by three Navy commanders, Pete Conrad, Dick Gordon, and Alan Bean. Conrad and Bean, having been upped from lieutenant commander to commander after their Gemini 11 flight, were ineligible for another promotion. Rookie Bean was. But should Bean be promoted over the heads of his seniors? Hang the policy, said President Nixon, promoting all three.

Of all the amenities accruing to astronauts, the hard cash came from "the *Life* contract." Between 1959 and 1963 *Life* magazine paid the Seven Original astronauts a total of $500,000 for "personal" stories—concerning themselves and their families— as opposed to "official" accounts of their astronautical duties. This arrangement increased the astronauts' military income by about 200 percent. It also simplified NASA public relations, since the famous young men's bylines would be concentrated in one place and the contract called for NASA approval of whatever they said.

There were drawbacks. The rest of the press took a sour view of what it considered public property being put up for exclusive sale (the dividing line between "personal" and "official" was wafer-thin). Since the same ghostwriters put their stories in print, the astronauts (and their families) all seemed shaped in the same mold, utterly homogenized for the greater glory of home, motherhood, and the space program. "To read it was to believe we were the most simon-pure guys there had ever been," wrote Buzz Aldrin in *Return to Earth*. "The contract almost guaranteed peaches and cream, full-color spreads glittering with harmless inanities," was the way Mike Collins's book had it.

WHAT HAPPENS TO EX-ASTRONAUTS?

The exclusive-story gambit almost ended when the Kennedy Administration took over, and Kennedy's press secretary actually announced there would be no more contracts after Mercury ended. But John Glenn went sailing with the President one summer day in 1962 and enumerated the costs and risks that came with fame. The President relented and more contracts were signed after the Second Nine entered, this time with not only *Life* but also Field Enterprises. But as more astronauts were selected, the pie sliced thinner until finally each astronaut was receiving only $3000 per year for his literary output. One last surge came with *Life*'s European syndication of the stories of Apollo 8 through 11 in 1969, which brought about $16,000 for each of sixty astronauts and widows.

Desert survival training was part of the regular program of what-ifs. If any flight had ended with an emergency landing in a desert, sun-protective dress and tents could have been fashioned from spacecraft parachutes. The astronauts were taught the best tricks for survival in the desert. Left to right, seated: Borman, Lovell, Young, Conrad, McDivitt, White. Standing: training officer Zedehar, Stafford, Slayton, Armstrong, and See.

Saying a few words to a sea of friendly faces was the lot of the Apollo 11 astronauts, whose world tour aboard Air Force One took them to a dizzying 24 countries in 45 days.

Children of Kinshasa dance a special welcome for the men from the Moon. Tact, diplomacy, an iron constitution, and a knack for public speaking were what the astronauts needed on tours.

A principal advantage of the contracts was the insurance, $50,000 worth from both *Life* and Field for each astronaut, and the widows of the accident victims were left with nest eggs. Congress might have been able to provide extra income and extra insurance, but the Vietnam War got in the way, and who was to say a man dying in space was more deserving than one who stepped on a land mine in a jungle path?

Unfortunately, this easy money led, directly or indirectly, to money acquired less scrupulously when the Apollo 15 astronauts sold 400 unauthorized covers to a German dealer in exchange for $8000 each (the dealer got $150,000). The three returned the money, and were subsequently reprimanded. It also turned out that each of fifteen astronauts had sold 500 copies of his signature on blocks of stamps for $5 apiece, without saying anything to bosses Slayton and Shepard about it (five of the fifteen gave the proceeds to charity). Deke and Al were incensed, but threw up their hands. If a man has a claim to owning anything, it is his own signature. Nevertheless NASA put a stop to this business and also placed heavy restrictions on what astronauts could and could not carry into space.

Homogeneous the astronauts never were. Frank Borman learned to fly a plane before he was old enough to get a driver's license, and so did Neil Armstrong, but at the same age Dick Gordon was considering the priesthood, Mike Collins was more interested in "girls, football, and chess" than in planes, and Jim Irwin had never flown until he rode a commercial aircraft to begin flight training.

John Glenn went into politics and, after several disappointments, was elected U.S. Senator from Ohio on the Democratic ticket in 1974. Alan Shepard's $125,000 from the *Life* and Field contracts became the egg that hatched a fortune in real estate. Borman, success-prone as always, spent a semester at Harvard Business School, went to work for Eastern Airlines, and became its president. For several years Donn Eisele served as director of the Peace Corps in Thailand, which was as different from Dick Gordon's executive job with the New Orleans Saints football team as was Mike Collins's directorship of the Smithsonian Institution's Air and Space Museum. Armstrong became a professor of engineering at the University of Cincinnati. Ed Mitchell founded an organization devoted to extrasensory perception, and Jim Irwin became a fundamentalist evangelist. Jim McDivitt became an executive of Consumers Power Company in his home town, Jackson, Mich., and Jim Lovell stayed put with the Bay-Houston Towing Company.

Once in awhile some them still turned up on television or radio: Schirra plugging the railroads, Aldrin Volkswagens, Armstrong and Carpenter banks, and Lovell insurance. Collins was offered $50,000 to advertise a beer but he turned it down, "although I like beer." Lest he appear too upright, Collins did confess that he once made an unpaid commercial for U.S. Savings Bonds, although he had never seen one in his life.

If the astronauts sometimes dwelt in an aura of public misconception, they nonetheless performed dazzling feats with skill and finesse. You may search the pages of history in vain for deeds to match theirs, and many years will pass before similar feats occur again. All hail, then, to these daring young men who married technique to valor and in barely a decade transformed the impossible into the commonplace.

Lifeless and slowly tumbling, the S—IB stage that put Apollo 7 in orbit gave Astronauts Schirra, Eisele, and Cunningham man's first ride atop a load of liquid hydrogen. Now the spent 59-foot stage served as a passive target for practice in rendezvous, with one run starting from a distance of 80 miles.

CHAPTER NINE

The Shakedown Cruises

By SAMUEL C. PHILLIPS

Consider the mood of America as it approached the end of 1968, by any accounting one of the unhappiest years of the twentieth century. It was a year of riots, burning cities, sickening assassinations, universities forced to close their doors. In Southeast Asia the twelve-month toll of American dead rose 50 percent, to 15,000, and the cost of the war topped $25 billion. By mid-December the country's despair was reflected in the Associated Press's nationwide poll of editors, who chose as the two top stories of 1968 the slayings of Robert Kennedy and Martin Luther King; *Time* magazine picked a generic symbol, "The Dissenter," as its Man of the Year. The poll and the Man were scheduled for year-end publication.

This condition was changed dramatically during the waning days of the year, figuratively with two out in the last half of the ninth, and that is what this chapter is about.

Nineteen sixty-seven, which began as a bad year for the space program, had had its own sensational upturn with the first flight, unmanned, of the giant Saturn, on November 9—a landmark on the path to the Moon. Then 1968 started fairly well with Apollo 5, the first flight of the lunar module on January 22, which proved the structural integrity and operating characteristics of the Moon lander, despite overly conservative computer programming that caused the descent propulsion system to shut off too soon.

The second unmanned Saturn V mission, numbered Apollo 6, on April 4, 1968, was less successful. For a time it raised fears that the Moon-landing schedule had suffered another major setback. Three serious flaws turned up. Two minutes and five seconds after launch the 363-foot Saturn V underwent a lengthwise oscillation, like the motion of a pogo stick. Pogo oscillation subjects the entire space-vehicle stack to stresses and strains that, under certain circumstances, can grow to a magnitude sufficient to damage or even destroy the vehicle. This motion, caused by a synchronization of engine thrust pulsations with natural vibration frequencies of the vehicle structure, tends to be self-amplifying as the structural oscillations disturb the flow of propellants and thus magnify the thrust pulsations.

The second anomaly, loss of structural panels from the lunar module adapter (the structural section that would house the lunar module), was originally thought to have been caused by the pogo oscillations. Engineers at Houston were able to establish, however, that a faulty manufacturing process was to blame. This was quickly corrected.

The third problem encountered on this flight was more serious. After the first stage had finished its work, the second stage was to take over and put Apollo 6 into orbit; then the third stage, the S–IVB, would take the CSM up to 13,800 miles, from where reentry from the Moon would be simulated, retesting the command module heat shield at the 25,000 mph that Apollo 4 had achieved five months earlier.

The second stage's five J–2 engines ignited as scheduled. About two-thirds of the way through their scheduled burn, no. 2 engine lost thrust and a detection system shut it down. No. 3 engine followed suit. With two-fifths of the second stage's million-pound thrust gone, Apollo 6, with the help of the single J–2 engine of the third stage, still achieved orbit, though in an egg-shaped path. But when the attempt was made to fire the third stage's engine a second time—as would be necessary to send astronauts into translunar trajectory—the single J–2 failed to ignite. The mission was saved when the service module's propulsion-system engine—20,000 pounds of thrust as against the third stage's 225,000—took over and sent the CSM up to the desired 13,800-mile altitude from which it reentered the atmosphere and landed in the Pacific.

DETECTIVE WORK ON THE TELEMETRY

What had happened? Ferreting out clues from mission records and the reams of data recorded from telemetry was a fascinating story of technical detective work. More than a thousand engineers and technicians at NASA Centers, contractor plants, and several universities were involved in establishing causes and designing and testing fixes.

The solution for pogo was to modify the pre-valves of the second-stage engines so that they could be charged with helium gas. This provided shock-absorbing accumulators that damped out the thrust oscillations.

Finding the culprit that cut off the J–2 engines involved long theorizing and hundreds of tests that finally pinpointed a six-foot tube, half an inch in diameter, carrying liquid hydrogen to the starter cup of the engine. This line had been fitted with two small bellows for absorbing vibration. It worked fine on ground tests because ice forming on the bellows provided a damping effect. But in the dryness of space—eventually simulated in a vacuum chamber—no ice formed because there was no air from which to draw moisture, and there the lines vibrated, cracked, and broke. The fix: replace the bellows with bends in the tubes to take up the motion.

With careful engineering analysis and extensive testing we satisfied ourselves that we understood the problems that plagued Apollo 6 and that the resulting changes were more than adequate to commit the third Saturn V to manned flight.

At this point we planned that the next Saturn V would be the D mission, launching Apollo 8 in December with Astronauts McDivitt, Scott, and Schweickart. Their main

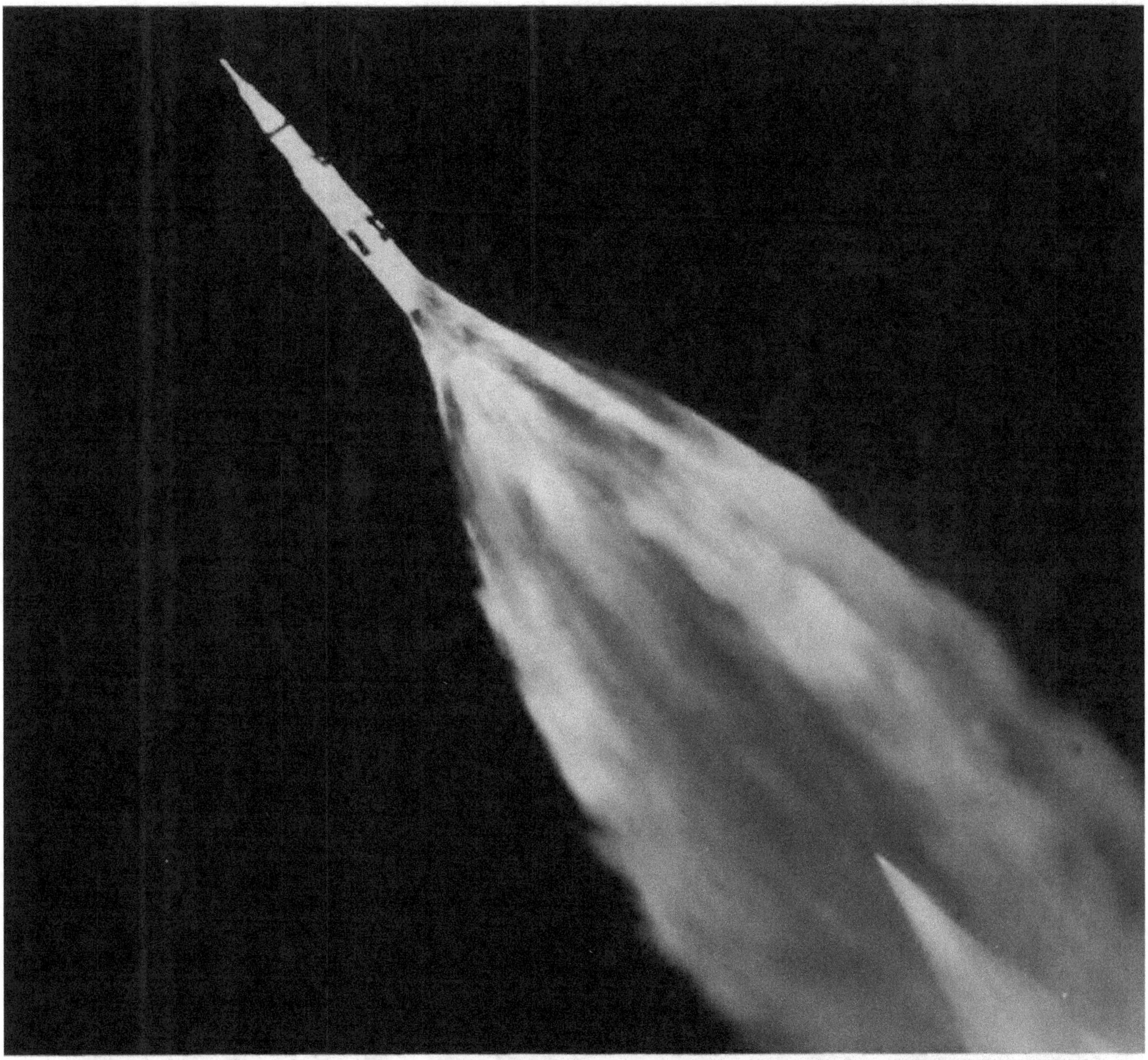

A fiery exhaust plume trails Apollo 6 during the first stage of launch. Second-stage burn was marred by premature shutdown of three of the five J-2 engines, causing the craft to enter an elliptical rather than a circular orbit. Many months of technical detective work identified the bellows in one liquid hydrogen fuel line (bottom right) as the culprit. In flight the bellows had flexed excessively, cracked, and leaked fuel. A redesigned fuel line substituted specially placed bends for the bellows.

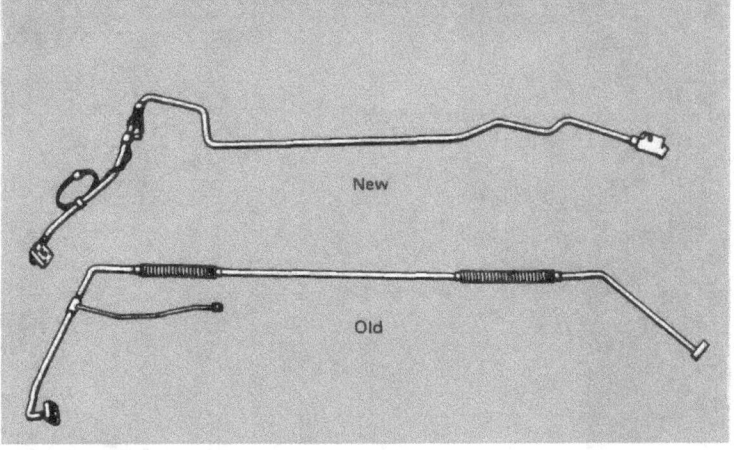

New

Old

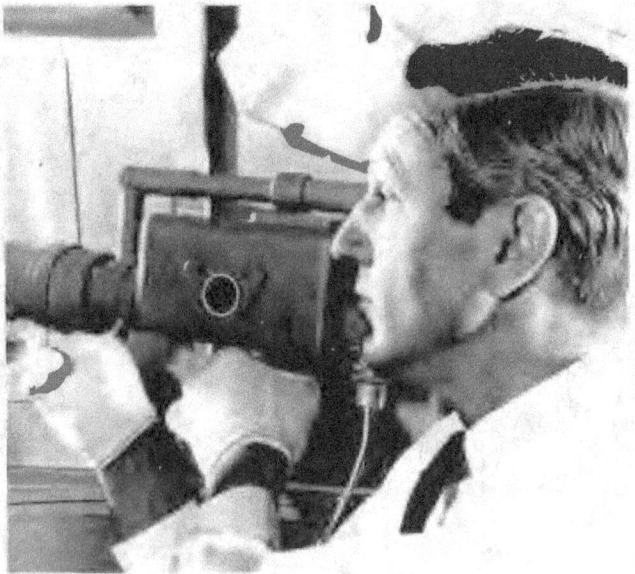

Snug-fitting cocoon housing the Apollo 7 service module is carefully extracted by workmen from the *Super Guppy,* the specially built cargo transport plane. The SM will be mated with its CM and then fully tested to confirm compatibility.

Mini TV camera is demonstrated by engineer. On Apollo 7 it produced the first live television broadcast from space, a seven minute segment with Astronauts Schirra, Eisele, and Cunningham displaying hand-printed signs, and head colds. A similar camera was used on Apollo 8.

A prelaunch conference at Cape Kennedy shows no sign of prelaunch tension. Apollo 7 Commander Wally Schirra raises his coffee cup, with Cunningham at his right. Eisele is at the extreme right, in sweater. Al Shepard looks over General Phillips' shoulder. At Phillips' left is George Low; across the table is Rocco Petrone.

objective would be to test the spider-like lunar module then abuilding at the Grumman plant on Long Island. Early in 1968 we had set the objective of flying the D mission before the year was out. It was a reasonable target at the time, considering progress across the program, and would put us in an excellent position to complete the preparatory missions and have more than one shot at the landing in 1969.

Meanwhile, the command and service modules, after almost two years of reworking at the North American Aviation plant in Downey, Calif., would have their crucial flight test on Apollo 7, the C mission, after launch in October 1968 by the smaller Saturn IB. On board this first manned Apollo mission would be Astronauts Schirra, Eisele, and Cunningham.

By midsummer it was apparent that Apollo 7 would fly in October, but that the lunar module for the D mission would not be ready for a December flight. Electromagnetic interference problems were plaguing checkout tests, and it was obvious that engineering changes and further time-consuming tests were needed. After a comprehensive review in early August, my unhappy estimate was that the D mission would not be ready until March 1969.

AN EARLY TRIP AROUND THE MOON

George Low, the spacecraft program manager, then put forward a daring idea: fly the CSM on the Saturn V in December, with a dummy instead of the real LM, all the way to the Moon. We would then make maximum progress for the program, while we took the time necessary to work out rigorously the LM problems. Low had discussed the feasibility of such a mission with Gilruth, Kraft, and Slayton at the Manned Spacecraft Center; I was at the Kennedy Space Center in Florida when he called to voice his idea. The upshot was a meeting that afternoon of the Apollo management team. The Marshall Space Flight Center at Huntsville, Ala., was a central point, considering where we all were at the moment. George Hage, my deputy, and I joined with Debus and Petrone of KSC for the flight to Huntsville. Von Braun, Rees, James, and Richard of MSFC were there. Gilruth, Low, Kraft, and Slayton flew from Houston.

We discussed designing a flexible mission so that, depending on many factors, including results of the Apollo 7 flight, we could commit Apollo 8 to an Earth-orbit flight, or a flight to a few hundred or several thousand miles away from Earth, or to a lunar flyby, or to spending several hours in lunar orbit. The three-hour conference didn't turn up any "show-stoppers." Quite the opposite; while there were many details to be reexamined, it indeed looked as if we could do it. The gloom that had permeated our previous program review was replaced by excitement. We agreed to meet in Washington five days later. If more complete investigation uncovered no massive roadblocks, I would fly to Vienna for an exegesis to my boss, Dr. George E. Mueller, Associate Administrator for Manned Space Flight, and to the NASA Administrator, James Webb, who were attending a United Nations meeting on the Peaceful Uses of Outer Space. (Going to Vienna was at first considered necessary, lest other communications tip off the Russians, believed to be planning a Moon spectacular of their own. Eventually it was decided that my appearance in Vienna would trip more alarms than overseas telephone conversations.)

Many problems remained. The high-gain antenna was an uncertain quantity; but Kraft agreed that the mission could be flown safely with the omniantennas, even though television might be lost. What should we carry in lieu of the LM? The answer was the LM test article that had been through the dynamics program at Marshall. Deke Slayton wanted to leave McDivitt and his crew assigned to the first LM mission; so the next crew, Borman, Lovell, and Anders, scheduled for the E mission, were brought forward for the newly defined Apollo 8 mission. McDivitt's mission retained its D designation, and Borman's was labeled "C-Prime."

Upon returning to Washington, I presented the plan to Thomas O. Paine, Acting Administrator in Webb's absence. Paine reminded me that the program had fallen behind, pogo had occurred on the last flight, three engines had failed, and we had not yet flown a manned Apollo mission; yet "now you want to up the ante. Do you really want to do this, Sam?" My answer was, "Yes, sir, as a flexible mission, provided our detailed examination in days to come doesn't turn up any show-stoppers." Said Paine, "We'll have a hell of a time selling it to Mueller and Webb."

He was right. A telephone conversation with Mueller in Vienna found him skeptical and cool. Mr. Webb was clearly shaken by the abrupt proposal and by the consequences of possible failure.

On August 15 Paine and I sent Webb and Mueller a seven-page cable with sug-

Graduated missions led confidently to a landing on the Moon

We designed seven types of missions to test the suitability and safety of all equipment in all mission phases. These were designated by letters A through G:

A. Unmanned flights of launch vehicles and the CSM, to demonstrate the adequacy of their design and to certify safety for men. Five of these flights were flown between February 1966 and April 1968; Apollo 6 was the last.

B. Unmanned flight of the LM, to demonstrate the adequacy of its design and to certify its safety for men. The flight of Apollo 5 in January 1968 accomplished this.

C. Manned flight to demonstrate performance and operability of the CSM. Apollo 7, which flew an eleven-day mission in low Earth orbit in October 1968, was a C mission. Apollo 8, which flew the CSM into lunar orbit in December 1968 was also a C mission, but designated as C-Prime, to distinguish it from the prior flight.

D. Manned flight of the complete lunar landing mission vehicle in low Earth orbit to demonstrate operability of all the equipment and (insofar as could be done in Earth orbit) to perform the maneuvers involved in the ultimate mission. Apollo 9, which flew in March 1969, satisfied this requirement.

E. Manned flight of the complete lunar-landing mission vehicle in Earth orbit to great distances from Earth. When the time came to commit this mission to flight, we decided that we had already accomplished its objectives and that it was not required. But because this mission was in the program, we had made detailed plans for it, and in fact pulled much of the planning, preparation, and training forward to use in the Apollo 8 lunar-orbit mission.

F. This was a complete mission except for the final descent to and landing on the lunar surface. Apollo 10, flown by Stafford, in May 1969, was an F mission. The need for this mission was hotly debated. Here we would be, 50,000 feet above the Moon, having accepted much of the risk inherent in landing. The temptation to go the rest of the way was great; but this mission demonstrated the soundness of the strategy of "biting off chunks." The training and confidence of readiness that the Apollo 10 mission gave the entire organization was of inestimable value.

G. The initial lunar-landing mission. This, of course, was accomplished by Armstrong, Aldrin, and Collins in July 1969. ——S. C. P.

Lifting off for the first time with men aboard, the Saturn IB and its hydrogen-fueled upper stage carry the Apollo 7 command and service module toward Earth orbit. This was the first trial of the intensively reengineered CSM, and to the relief of NASA it performed beautifully, staying in orbit for 10.8 days, longer than a Moon landing mission would require.

An exciting Earthrise greets the Apollo 8 crew as they return from the far side of the Moon. This was the first time men had ever directly seen Earthrise or the far side, though photos had been taken earlier. Potential landing sites were photographed from the 70-mile-high orbit.

gested wording of a press release saying lunar orbit was being retained as an option in the December flight, the decision to depend on the success of Apollo 7 in October. Webb replied from Vienna via State Department code, accepting the crew switches and schedule changes. But he proposed saying only that "studies will be carried out and plans prepared so as to provide reasonable flexibility in establishing final mission objectives" after Apollo 7.

Paine interpreted Webb's instructions "liberally," and authorized me to say in an August 19 press conference, upon announcing the McDivitt-Borman switch, that a circumlunar flight or lunar orbit were possible options. I am told that I diminished the possibility so thoroughly, by saying repeatedly "the basic mission is Earth orbit," that the press at first mostly missed the point.

October 11 at Cape Kennedy was hot but the heat was tempered by a pleasant breeze when Apollo 7 lifted off in a two-tongued blaze of orange-colored flame at 11:02:45. The Saturn IB, in its first trial with men aboard, provided a perfect launch and its first stage dropped off 2 minutes 25 seconds later. The S–IVB second stage took over, giving astronauts their first ride atop a load of liquid hydrogen, and at 5 minutes 54 seconds into the mission, Walter Schirra, the commander, reported, "She is riding like a dream." About five minutes later an elliptical orbit had been achieved, 140 by 183 miles above the Earth. The S–IVB stayed with the CSM for about one and one-half orbits, then separated. Schirra fired the CSM's small rockets to pull 50 feet ahead of the S–IVB, then turned the spacecraft around to simulate docking, as would be necessary to extract an LM for a Moon landing. Next day, when the CSM and the S–IVB were about 80 miles apart, Schirra and his mates sought out the lifeless, tumbling 59-foot craft in a rendezvous simulation and approached within 70 feet.

A SUPERB SPACECRAFT

During the 163 orbits of Apollo 7 the ghost of Apollo 204 was effectively exorcised as the new Block II spacecraft and its millions of parts performed superbly. Durability was shown for 10.8 days—longer than a journey to the Moon and back. A momentary shudder went through Mission Control when both AC buses dropped out of the spacecraft's electrical system, coincident with automatic cycles of the cryogenic oxygen tank fans and heaters; but manual resetting of the AC bus breakers restored normal service. Three of the five spacecraft windows fogged because of improperly cured sealant compound (a condition that could not be fixed until Apollo 9). Chargers for the batteries needed for reentry (after fuel cells departed with the SM) returned 50 to 75 percent less energy than expected. Most serious was the overheating of fuel cells, which might have failed when the spacecraft was too far from Earth to return on batteries, even if fully charged. But each of these anomalies was satisfactorily checked out before Apollo 8 flew.

The CSM's service propulsion system, which had to fire the CSM into and out of Moon orbit, worked perfectly during eight burns lasting from half a second to 67.6 seconds. Apollo's flotation bags had their first try-out when the spacecraft, a "lousy boat," splashed down south of Bermuda and turned upside down; when inflated, the brightly colored bags flipped it aright.

Gag card is held before TV camera by Apollo 7 Commander Wally Schirra during third day of the first manned Apollo mission. CM pilot Donn Eisele looks on. TV coverage using the small, hand-held camera was to have begun on the second day but minor tasks used more than the expected time. Another sign displayed during the nation-wide broadcast greeted viewers from "the lovely Apollo room high atop everything."

"Stable two," an engineering euphemism for upside-down, was one of the ways that the command module could float, and this was the way that Apollo 7 splashed down. The astronauts hung from their restraining belts for a few minutes until three righting bags were inflated to flip the spacecraft. The photo sequence above and at right, not the actual Apollo 7 landing, shows a training session, one of many constantly held to drill recovery teams and astronauts. Not used in this exercise was the flotation collar, normally fixed around the command module, that provided insurance against swamping from water taken aboard through the open hatch.

In retrospect it seems inconceivable, but serious debate ensued in NASA councils on whether television should be broadcast from Apollo missions, and the decision to carry the little 4½-pound camera was not made until just before this October flight. Although these early pictures were crude, I think it was informative for the public to see astronauts floating weightlessly in their roomy spacecraft, snatching floating objects, and eating the first hot food consumed in space. Like the television pictures, the food improved in later missions.

Apollo 7's achievement led to a rapid review of Apollo 8's options. The Apollo 7 astronauts went through six days of debriefing for the benefit of Apollo 8, and on October 28 the Manned Space Flight Management Council chaired by Mueller met at MSC, investigating every phase of the forthcoming mission. Next day came a lengthy systems review of Apollo 8's Spacecraft 103. Paine made the go/no-go review of lunar orbit on November 11 at NASA Headquarters in Washington. By this time nearly all the skeptics had become converts.

At the end of this climactic meeting Mueller put a recommendation for lunar orbit into writing, and Paine approved it. He telephoned the decision to the White House, and the message was laid on President Johnson's desk while he was conferring with Richard M. Nixon, elected his successor six days earlier.

LIFTING FROM A SEA OF FLAME

In the pink dawn of December 21 a quarter million persons lined the approaches to Cape Kennedy, many of them having camped overnight. At 7:51, amid a noise that sounded from three miles away like a million-ton truck rumbling over a corrugated road, the first manned Saturn V, an alabaster column as big as a naval destroyer, lifted slowly, ever so slowly, from the sea of flame that engulfed Pad 39–A. The upward pace quickened as the first stage's 531,000 gallons of kerosene and liquid oxygen were thirstily consumed, and in 2 minutes 34 seconds the big drink was finished, whereupon the second stage's five J–2 engines lit up. S–II's 359,000 gallons of liquid hydrogen and liquid oxygen boosted the S–IVB and CSM for 6 minutes 10 seconds to an altitude of 108 miles. After the depleted S–II fell away, the S–IVB, this time the third stage, fired for 2 minutes 40 seconds to achieve Earth orbit. Except for slight pogo during the second-stage burn, Commander Borman reported all was smoothness.

During the second orbit, at 2 hours 27 minutes, CapCom Mike Collins sang out "You are go for TLI" (translunar injection), and 23 minutes after that Lovell calmly said, "Ignition." The S–IVB had restarted with a long burn over Hawaii that lasted 5 minutes 19 seconds and boosted speed to the 24,200 mph necessary to escape the bonds of Earth. "You are on your way," said Chris Kraft, from the last row of consoles in Mission Control, "you are *really* on your way." The anticlimactic observation of the day came when Lovell said, "Tell Conrad he lost his record." (During Gemini 11 Pete Conrad and Dick Gordon had set an altitude record of 850 miles.) After the burn the S–IVB separated and was sent on its way to orbit the Sun.

In Mission Control early in the morning of December 24 the big center screen, which had carried an illuminated Mercator projection of the Earth for the past three

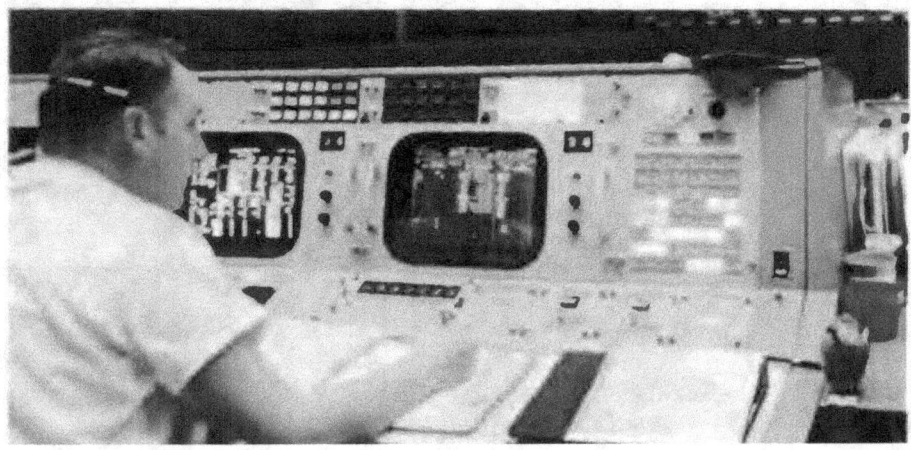

Intently watching the dials above them, the crew of Apollo 8, Anders, Lovell, Borman, left to right, rehearse for their lunar orbit mission inside a simulator at the Kennedy Space Center. Simulation was a central feature of the training given flight crews and mission controllers.

Like a whirling dervish, the path of Apollo 8 about the Earth and Moon spanned seven days and well over half a million miles. Ten lunar revolutions, at distances as close as 60 miles, were made. Translunar trip takes about 20 percent longer than the return trip because going out one has to overcome the stronger gravity of the Earth but can capitalize on it coming back. Not shown is the solar orbit trajectory taken by the burned S–IVB stage.

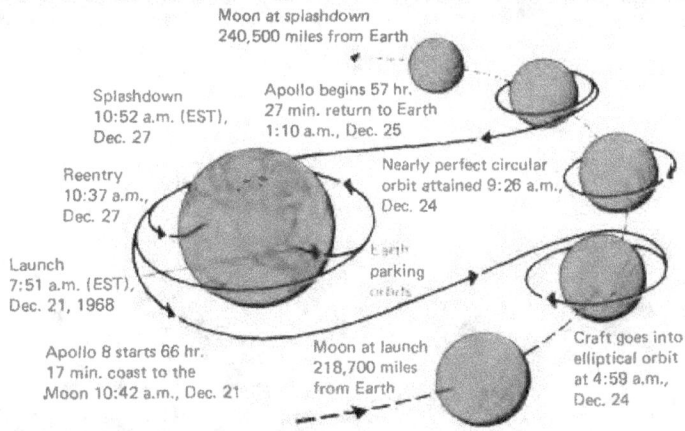

Moon at splashdown
240,500 miles from Earth

Splashdown
10:52 a.m. (EST),
Dec. 27

Apollo begins 57 hr.
27 min. return to Earth
1:10 a.m., Dec. 25

Reentry
10:37 a.m.,
Dec. 27

Nearly perfect circular
orbit attained 9:26 a.m.,
Dec. 24

Launch
7:51 a.m. (EST),
Dec. 21, 1968

Earth
parking
orbits

Apollo 8 starts 66 hr.
17 min. coast to the
Moon 10:42 a.m., Dec. 21

Moon at launch
218,700 miles
from Earth

Craft goes into
elliptical orbit
at 4:59 a.m.,
Dec. 24

Brightly lit panel lights and screens confront Green Team Flight Director for Apollo 8, Cliff Charlesworth, at his console in the Mission Operations Control Room in Houston. The radio signal between here and the Moon took three seconds roundtrip.

Walt Cunningham, lunar module pilot on Apollo 7 (which carried no LM), makes notes while a spare film magazine floats weightlessly a few inches above his pen. To "park" something in space, it had to be left with zero motion.

Slashing across the floor of the crater Goclenius, which is about 40 miles in diameter, are strange trenches called rilles. One rille extends over the entire crater floor, across the central peak, and continues up over the rim and out along the surrounding mare. This is only part of an Apollo 8 telephoto negative.

Terraced inner walls lead from the rim of the crater Langrenus down to the smooth crater floor, broken by some central peaks. Langrenus is about 85 miles in diameter and its smooth, worn walls suggest that it is fairly old. The photo was taken from an altitude of some 150 miles.

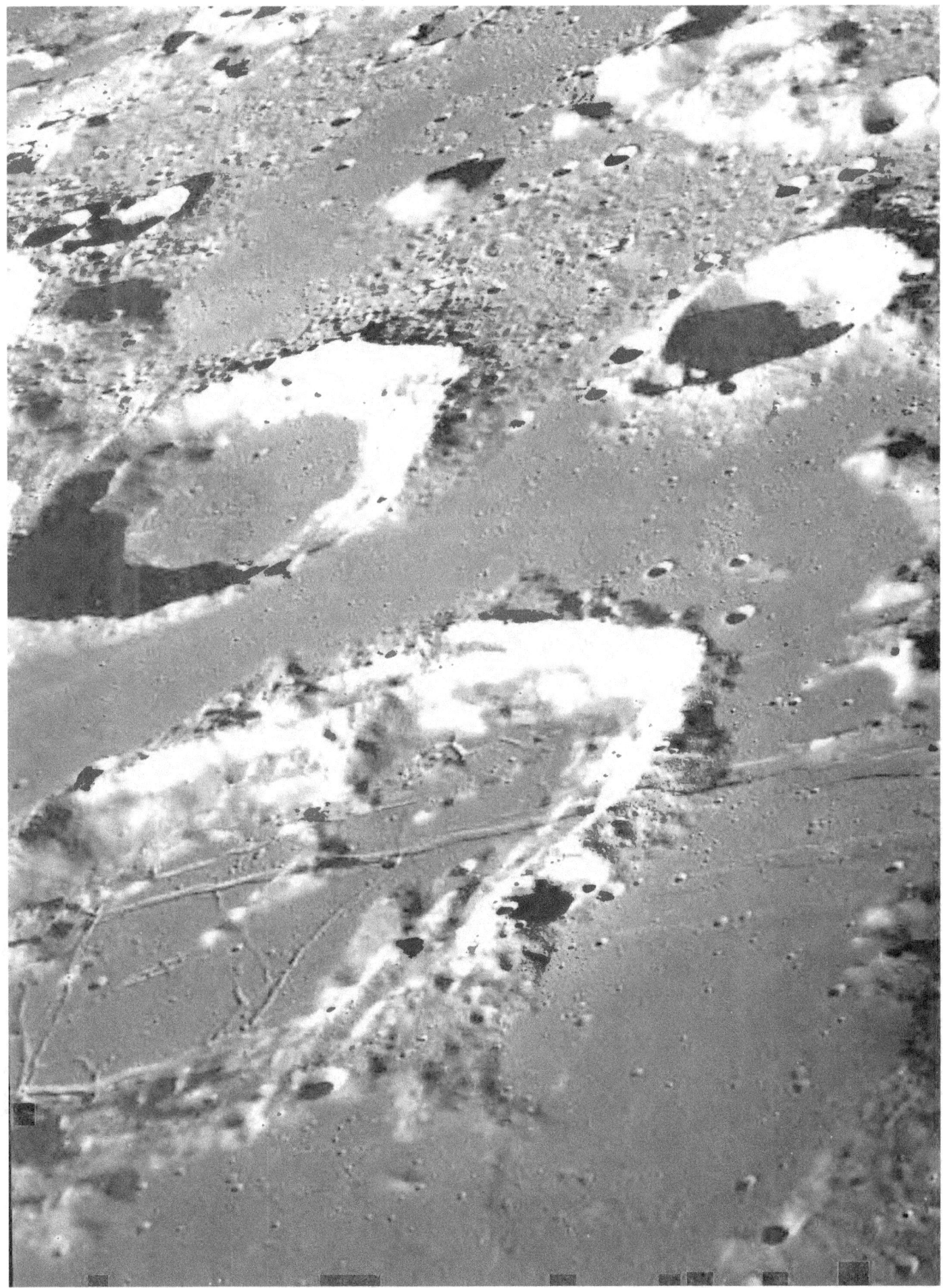

and one-half years—a moving blip always indicated the spacecraft's position—underwent a dramatic change. The Earth disappeared, and upon the screen was flashed a scarred and pockmarked map with such labels as Mare Tranquillitatis, Mare Crisium, and many craters with such names as Tsiolkovsky, Grimaldi, and Gilbert. The effect was electrifying, symbolic evidence that man had reached the vicinity of the Moon.

CapCom Gerry Carr spoke to the three astronauts more than 200,000 miles away, "Ten seconds to go. You are GO all the way." Lovell replied, "We'll see you on the other side," and Apollo 8 disappeared behind the Moon, the first time in history men had been occulted. For 34 minutes there would be no way of knowing what happened. During that time the 247-second LOI (lunar orbit insertion) burn would take place that would slow down the spacecraft from 5758 to 3643 mph to enable it to latch on to the Moon's field of gravity and go into orbit. If the SPS engine failed, Apollo 8 would whip around the Moon and head back for Earth on a free-return trajectory (a la Apollo 13); during one critical half minute if the engine conked out the spacecraft would be sent crashing into the Moon.

ORBITING THE MOON CHRISTMAS EVE

"Longest four minutes I ever spent," said Lovell during the burn, in a comment recorded but not broadcast in real time. At 69 hours 15 minutes Apollo 8 went into lunar orbit, whereupon Anders said, "Congratulations, gentlemen, you are at zero-zero." Said Borman, "It's not time for congratulations yet. Dig out the flight plan."

Unaware of this conversation, Mission Control buzzed with nervous chatter. Carr began seeking a signal to indicate that the astronauts were indeed in orbit: "Apollo 8, Apollo 8, Apollo 8." Then the voice of Jim Lovell came through calmly, "Go ahead, Houston."

Mission Control's viewing-room spectators broke into cheers and loud applause. Apollo 8 was in a 168.5 by 60 mile orbit on this day before Christmas.

"What does the old Moon look like from 60 miles?" asked CapCom.

"Essentially gray; no color," said Lovell, "like plaster of paris or a sort of grayish beach sand." The craters all seemed to be rounded off; some of them had cones within them; others had rays. Anders added: "We are coming up on the craters Colombo and Gutenberg. Very good detail visible. We can see the long, parallel faults of Goclenius and they run through the mare material right into the highland material."

During the second egg-shaped orbit the astronauts produced the Moon on black-and-white television (color would not come until Apollo 10). It proved to be a desolate place indeed, a plate of gray steel spattered by a million bullets. "It certainly would not appear to be an inviting place to live or work," Borman said later.

On the third revolution the SPS engine fired nine seconds to put the spacecraft into a circular orbit, 60.7 by 59.7 miles, where it would stay for sixteen hours longer (each orbit lasted two hours, as against one and one-half hours for Earth).

At 8:40 p.m. the astronauts were on television again. First, they showed the half Earth across a stark lunar landscape. Then, from the other unfogged window, they tracked the bleak surface of the Moon. "The vast loneliness is awe-inspiring and it makes you realize just what you have back there on Earth," said Lovell. The pictures aroused great wonder, with an estimated half billion people vicariously exploring what no man had ever seen before.

"For all the people on Earth," said Anders, "the crew of Apollo 8 has a message we would like to send you." He paused a moment and then began reading:

In the beginning God created the Heaven and the Earth.

After four verses of *Genesis,* Lovell took up the reading:

And God called the light Day, and the darkness he called Night.

At the end of the eighth verse Borman picked up the familiar words:

And God said, Let the waters under the Heavens be gathered together unto one place, and let the dry land appear; and it was so. And God called the dry land Earth; and the gathering together of the waters He called seas; and God saw that it was good.

The commander added: "And from the crew of Apollo 8, we close with good night, good luck, a Merry Christmas, and God bless all of you—all of you on the good Earth." It was a time of rare emotion. The mixture of the season, the immortal words, the ancient Moon, and the new technology made for an extraordinarily effective setting.

A LUNAR CHRISTMAS

"At some point in the history of the world," editorialized *The Washington Post,* "someone may have read the first ten verses of the *Book of Genesis* under conditions that gave them greater meaning than they had on Christmas Eve. But it seems unlikely . . . This Christmas will always be remembered as the lunar one."

The New York Times, which called Apollo 8 "the most fantastic voyage of all times," said on December 26: "There was more than narrow religious significance in the emotional high point of their fantastic odyssey."

As Apollo 8 began its tenth and last orbit, CapCom Ken Mattingly told the astronauts: "We have reviewed all your systems. You have a GO to TEI" (trans-earth injection). This time the crew really was in thrall to the SPS engine. It had to ignite in this most apprehensive moment of the mission, else Apollo 8 would be left in lunar orbit, its passengers' lives measured by the length of their oxygen supply. Ignite it did, in a 303-second burn that would effect touchdown in just under 58 hours. Apollo 8 reentered at 25,000 mph and splashed down south of Hawaii two days after Christmas.

The stupendous effect of Apollo 8 was strengthened by color photographs published after the return. Not only was the technology of going to the Moon brilliantly proven; men began to view the Earth as "small and blue and beautiful in that eternal silence," as Archibald MacLeish put it, and to realize as never before that their planet was worth working to save. The concept that Earth was itself a kind of spacecraft needing attention to its habitability spread more widely than ever.

During the last week of 1968 the Associated Press repolled its 1278 newspaper editors, who overwhelmingly voted Apollo 8 the story of the year. *Time* discarded "The Dissenter" in favor of Borman, Lovell, and Anders; and a friend telegraphed Frank Borman, "You have bailed out 1968."

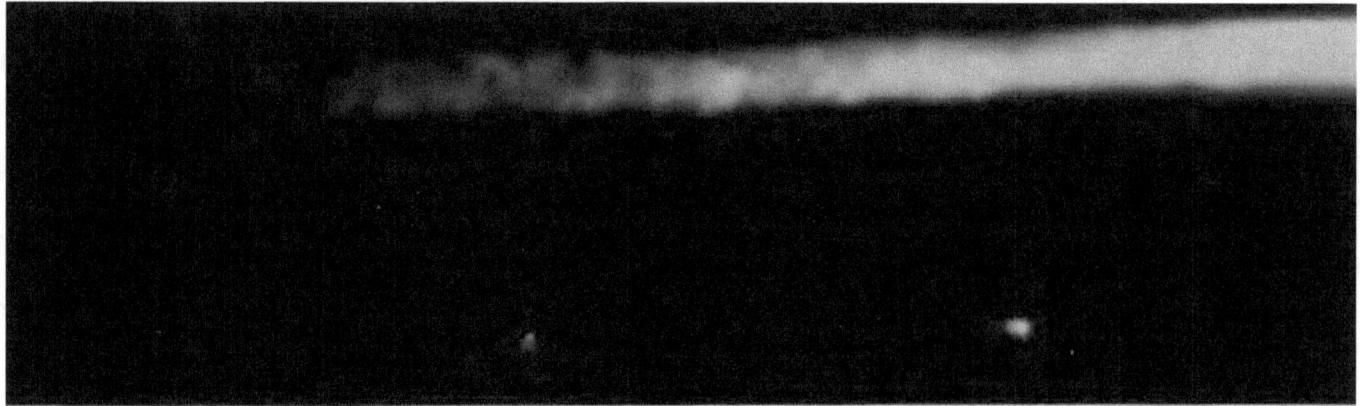

An early-morning reception took place aboard the USS *Yorktown* in the Pacific on December 27, 1968. At left, Apollo 8 Commander Frank Borman thanked the crew for giving up Christmas at home so that the carrier could serve as the prime recovery ship. Astronauts Lovell and Anders stood beside Borman.

A joint session of Congress convened to hear the first men to fly around the Moon. That's Borman at the Speaker's rostrum with Lovell and Anders wearing expressions of somber dignity. Public appearances were a demanding part of the job.

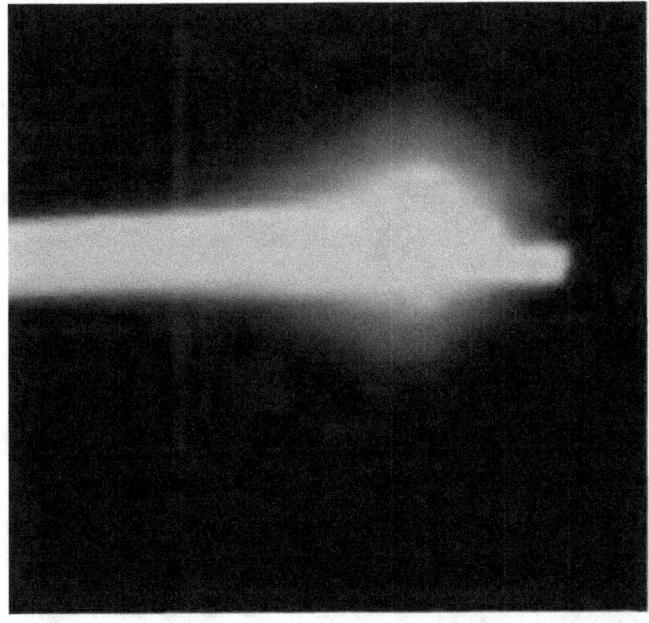

Blazing across the sky, the Apollo 8 command module returns home. A Pan-Am pilot on his way to Honolulu from Fiji glimpsed the sight, and estimated that the fiery streak was as much as 125 miles long. The glowing tail made the spacecraft look like a meteor. This unusual photo was taken by a special USAF camera on a KC-135 aircraft at 40,000 feet.

Charred and dripping, the Apollo 8 command module is hoisted from the Pacific and eased down on its trailer aboard the USS *Yorktown*. Its remaining propellants have to be drained off and its precious film magazines recovered. Then spacecraft engineers will go over it carefully to study the effects of a trip around the Moon.

On a six-city tour, the Apollo 8 astronauts and their wives received a warm welcome and round of tributes. Here New York Governor Nelson Rockefeller is presenting them with a commemorative Steuben glassware plate titled "The Mountains of the Moon."

In the United Nations, Secretary General U Thant introduced the astronauts and their families to members assembled in the Security Council chambers. No simulations in Houston nor the Cape had prepared them for this kind of public attention.

Gleaming in the sunlight, *Spider* and *Gumdrop* are hard-docked while two of the three-man crew venture outside. CM pilot Dave Scott, breathing through an umbilical connection, pokes his head out the command-module hatch. The picture was taken by LM pilot Rusty Schweickart (who wore an independent life-support pack) while perched on the lunar module's "front porch."

CHAPTER TEN

Getting It All Together

By GEORGE E. MUELLER

The Apollo 9 and 10 missions were the learning phase of the program, coming as they did between the first tests of the command module in deep space on Apollo 8 and the lunar landing of Apollo 11.

Apollo 9 and 10 were missions designed to rehearse all the steps and reproduce all the events of the Apollo 11 mission with the single exception of the lunar touchdown, stay, and liftoff. The command and service modules and the lunar module were used in flight procedures identical to those that would later take similar vehicles to the Moon and a landing. The flight mechanics, the mission support systems, the communications, the recording of data—all these techniques and components were tested in a final round of verification.

We learned from those two missions how to develop our flight procedures. They are complex, at best, and for future missions we wanted to know exactly what alternatives we would have under a wide variety of circumstances to assure the safety of the astronauts and the success of the missions, in that order. Fortunately we had very few problems on both missions. Three reasons for that good fortune were the extensive ground tests, the simulation exercises which provided the crews with high-fidelity training for every phase of the flight, and the critical design review procedures.

The crews rehearsed and re-rehearsed their movements in ground simulators and in conditions of inflight weightlessness produced by parabolic flights in a converted Air Force KC–135 tanker and by neutral buoyancy simulations. They had started this kind of training as a crew months before the flight, gradually working toward proficiency and a degree of automatic response to the checklists, the pilots' shorthand notes that were developed simultaneously. By the time they were ready for their actual missions, they had run through all the normal routines many times, and had thoroughly rehearsed emergency procedures for every imaginable trouble or failure.

The critical design reviews were fundamental to the testing and simulation programs. What we actually did was to go over every single part of the spacecraft to make sure that we understood how it worked, how it might go wrong, and what alternative procedure or backup system we had in case it did go wrong. The overriding consideration was, of course, crew safety; there always had to be at least one way back from orbit or from the Moon if something went wrong. These design reviews were done by a number of task groups, because there were too many spacecraft systems for any single group to consider, let alone understand to the required degree.

One of my areas of special interest lay in what we called "software"—the computer programs that provided the intelligence and control functions onboard the spacecraft and at the ground stations. General Phillips asked me to form a special software review group that met for several months and initiated some disciplines which, I believe, finally made it possible for the Apollo 11 lunar landing to take place when it did.

An interesting point about the Apollo 10 flight is that it did fail to do one of the steps that was later done on the Apollo 11 and which caused some brief moments of tension just before the lunar landing. On Apollo 10, the landing radar and the rendezvous radar never were operated at the same time. They are used for two different procedures, and at two different altitudes, so that there didn't seem to be any need for simultaneous operation of the pair. But on Apollo 11, they wanted to check the operation of the high-altitude gate radar as they approached the surface of the Moon, and so it was left on. That radar and the landing radar were feeding information directly into the onboard computer in the Apollo 11 lunar module. Although the data from the altitude radar was not being used, it nevertheless was driving the input registers of the computer and thus forcing the central processor to decide at each cycle that the radar data were not germane.

THOSE COMPUTER ALARMS

Fortunately, in our software review we had insisted on one point: We had to have at least a 10-percent spare capacity in the memory of the onboard computer. We did that to provide for unforeseen contingencies, and that contingency occurred on the Apollo 11 lunar landing. On that mission we used up the 10 percent with those extraneous messages from the high-altitude radar, and in spite of the spare capacity, we nearly overloaded the landing computer. It set off alarms that caused a few seconds of fast thinking, but it was quickly put right and the landing went on to a safe touchdown.

All this critical design review was then followed by a review of all the things that happened during the dry runs before the final countdown, checking to make sure that there were no anomalies that hadn't been identified and traced back to their origins. That was done in the last couple of days before the flight. What we were basically doing was making sure that everybody had done everything that could be done to assure the safety of the mission.

But you can only do so much in ground simulation and analysis, and then you've got to test the articles in actual flight. One example of this was with the docking mechanism, which presented no particular problems in ground tests and in simulation. But the first time it was tried in space, they had considerable difficulty getting the mechanism into place. That led to a very rapid redesign of the mechanism between flights, which was just one of the things accomplished in the very short time between flights. Looking back on it, it seems amazing that we were able to do a number of things on those two-month centers that would have been considered flatly impossible only a few months earlier. The key to that ability, though, was getting ourselves organized and then finding the will to do things. That made all the difference in the world on the Apollo program; there was the highest motivation, and it produced results, time and time again.

Of course, after the flights we had a thorough debriefing and evaluation of the mission, and of the behavior of the spacecraft and other systems, checking the actual results against our predicted performance, analyzing actual anomalies against ones that we had expected and planned for. The rigid discipline of postflight analysis and the preflight reviews were among the most important inputs of Apollo management to the success of the program.

FIRST MANNED FLIGHT OF THE LM

The Apollo 9 mission was to be the first manned flight in the lunar module, and the whole purpose of the mission was to qualify, in flight, that portion of the overall spacecraft system. Further, we wanted to show that the lunar module, in combination with the command and service modules, could perform its assigned tasks in weightless flight. It wasn't necessary to go to the Moon for this work, so the Apollo 9 mission took place in Earth orbit. The conditions were the same, insofar as those qualifying tests were concerned, and we had the further advantage of a more comfortable situation in case any problems developed.

It was planned to fly a ten-day mission, approximating in time a complete trip to the Moon, a lunar landing, stay and ascent, and then return to the Earth. We had developed a mission profile that would put astronauts into the lunar module on three separate occasions during the flight, first of all to check a lot of procedures and other items, and second to do multiple activations and deactivations of the lunar module. This mission was the only one in which the LM was powered up and down more than once; it was done here with the intent of discovering anything that might go wrong, and to refine the procedures worked out in simulations.

The lunar module might well be called the first true spacecraft, since it was designed for flight only in the environment of space. Folded and stowed as it was in the nose of the Saturn V, its frail body caused some concern about its ability to stand up to the stress of a Saturn launch. Some of the earlier Saturn launches had shown what was called a pogo oscillation, named after the pogo-stick phenomenon. We wanted to make certain that the LM would be able to take that longitudinal acceleration and shock. We also wanted to make sure the mechanisms for extending its landing legs and pads would not be jammed by unusual loads in flight, and that the adapter between the LM and the rest of the spacecraft could take those loads.

Other objectives of the mission included checking propulsion system operation in both the docked and undocked conditions, to complete a rendezvous between the LM and the command module, with the LM being the active partner during the maneuver, and to demonstrate extravehicular activity from both spacecraft in order to evaluate the ease or difficulty of that kind of task, and to check out the handholds, handrails, and localized illumination that had been developed for EVA.

The crew for Apollo 9 was commanded by James A. McDivitt. David R. Scott was the command pilot and Russell L. Schweickart was lunar module pilot. There was an initial three-day delay, not because of any equipment failure but because one of the crew had a cold. Liftoff occured on March 3, 1969, at 11 a.m. The normal launch phase followed, and then the S–IVB stage was ignited to place the spacecraft in a nearly circular Earth orbit of about 102 by 104 nautical miles. That orbit was the arena for what was to follow. In a routine manner, the combined command and service modules were separated from the rest of the spacecraft, turned in space, and docked with the lunar module. About one hour later, the docked spacecraft were ejected from the S–IVB, using the ejection mechanism for the first time. From this point on, many of the events were done for the first time.

About two hours after the ejection, the crew fired the service propulsion system for the first time on that mission, in a burst that lasted five seconds. It was the first of several such firings planned to check the service module propulsion system and the resulting maneuvering capability of the docked and undocked spacecraft. That task also completed the list of work for the first day in orbit. It may seem as if there was very little for the astronauts to do on that first day, but it must be realized that much time is spent checking and rechecking onboard systems, the communications and telemetering links to the ground stations, the receipt of data, and all the myriad tasks that confront astronauts in space.

On the second day in Earth orbit, the crew of Apollo 9 also spent much time in systems checks. They fired the service propulsion system three times, each time with about a two-hour interval between firings. Two of the firings were long, lasting almost two and almost five minutes; the third was a shorter burst of less than a half-minute.

DESCENT ENGINE FIRED

McDivitt and Schweickart entered the lunar module on the third day in orbit and powered it up to check out its systems and to fire the descent engine while in the docked condition. The engine burn lasted more than six minutes, during which time the crew controlled the engine manually, and also demonstrated digital-auto-pilot attitude control. After returning to the command module, the full crew made the fifth of the docked maneuvers, using the service propulsion system.

Early on the fourth day, Schweickart and Scott began their extravehicular activity, that spectacular and exhilarating solo performance outside the spacecraft. Unfortunately, Schweickart had beeen affected by a form of motion sickness during the previous days, a difficulty that later led to a regime of zero-*g* flights for all crews during the last days before launch. But this time it affected the planned time line for the EVA, and it was decided to reduce the activities outside the spacecraft.

Schweickart went into the lunar module, which was depressurized, and left it by way of the exit hatch that would later be used by the astronauts who landed on the

First manned flight of an unearthly spacecraft (it couldn't survive atmospheric reentry) took place over the cloud-shrouded Sahara, below. Note the contact probes projecting from the footpads, and the ladder leading from the "front porch" with their "golden slippers" boot retainers. At right, after its landing stage has been jettisoned, the lunar module returns from the successful completion of the first rendezvous. Before docking begins, it rotates in front of CM windows to be inspected for possible damage. The bell of the vital ascent engine protrudes from a nest of foil thermal insulation.

Moon. He stayed around the platform—the "front porch" as the astronauts called it—by that hatch for about three-quarters of an hour, checking the handholds and the lights that had been trained on specific portions of the spacecraft. Scott, meantime, had opened the command module hatch and clambered partially outside, still hooked to the spacecraft through a life-support umbilical system.

Schweickart was using a portable life-support system in his EVA, checking it for the first time outside. Both men detailed their experiences, took some photographs, and retrieved some thermal samples, testing the maneuverability of the space suits and the accessibility of the sample locations on the spacecraft. We had planned to have Schweickart move externally from the LM to the command module, to verify a way of getting from one spacecraft to the other if the tunnel were not available. This step was not completed beyond evaluating the specific handholds that would have been necessary to such a transfer.

THE LM ON ITS OWN

The fifth day of the Apollo 9 mission was the crucial test for the LM. Schweickart and McDivitt entered the LM, powered it up, and prepared to separate from the command module. They released the craft, and the astronauts were flying free in the LM. They backed away and rotated the module so that Scott, in the command module, could see that all the legs were down and extended, and that there were no physical failures in the craft itself. Then they fired the descent engine just enough to move the LM into a parallel orbit about three miles away from the command module. The mechanics of the orbit were such that the LM was on its own, but twice during each swing around the Earth, the two spacecraft were close enough so that Scott could initiate rescue operations should anything have happened to the LM.

For nearly six hours the two astronauts in the LM rehearsed the possible phases of the lunar approach and departure. They exercised the reaction control systems, fired the descent engine again, then jettisoned the descent stage, and then fired the ascent engine for the first time in space. From their adjusted position about 10 miles below and 80 miles behind the command module, they began their approach to a rendezvous and docking, much the same as the actual event to take place later on Apollo 11. The first phase of their rendezvous terminated temporarily with about a 100 ft separation, so that both spacecraft could be photographed. Then they docked, a solid and clean joining that verified the crews' training and the performance of the docking and locking mechanisms. They jettisoned the ascent stage, and remotely fired its engine to insert it into a highly elliptical orbit around the Earth.

From here on, the rest of the mission was almost an anticlimax, because the performance of the spacecraft left no doubt as to its ability to make the lunar trip, complete with landing and departure. But there was still work to be done, and so for the remaining five days in orbit, the crew again and again exercised the service propulsion system, once to lower the perigee of their Earth orbit, once to test the propellant staging system, and once to head home to Earth. On the last four days, they also did a series of landmark tracking experiments, using visual sightings of Earth features which were observed and photographed. They also made a number of photographs with a

The rendezvous radar antenna on the LM, untried in space, was photographed through a CM window by one of the Apollo 9 crew, perhaps in anticipation of the fact that it would soon be unstowed, powered up, and put to the all-important test. A critical and sophisticated part of the rendezvous system, it worked beautifully when tried two days later. The curved metal strap at the extreme left, not part of the antenna, is a handrail to be grasped by an astronaut floating outside the spacecraft.

In the cheerful mood prevailing when the three crew members were back together, Dave Scott mugs for Rusty Schweickart's camera. Here he shows how he, when alone in the command module, had had to peer against the glare to catch his first glimpse of the LM as it flew back in rendezvous.

Five hard days of carefully doing what had never been done before shows in the strained face of Apollo 9 Commander Jim McDivitt, normally a relaxed and equable man. Attempting to describe the cool courage of Mc-Divitt and Schweickart when they went off for the first time over the horizon in the unlandable LM, some observers declared it the bravest act since man first ate a raw oyster.

multiple-camera assembly using different film emulsions to obtain pictures in different portions of the photographic spectrum. These were further augmented by pictures taken almost simultaneously with hand-held cameras, using conventional films.

Once out of orbit, following the eighth successive firing of the service propulsion system, the reentry was normal. They landed right on their target in the Atlantic, 241 hours after takeoff, and were recovered quickly by Navy helicopters.

Apollo 10 was different, because it did go to the Moon—or at least within 47,000 feet of it—in its rehearsal of the Apollo 11 mission. There had been some speculation about whether or not the crew might have landed, having gotten so close. They might have wanted to, but it was impossible for that lunar module to land. It was an early design that was too heavy for a lunar landing, or, to be more precise, too heavy to be able to complete the ascent back to the command module. It was a test module, for the dress rehearsal only, and that was the way it was used. Besides, the discipline on the Apollo program was such that no crew would have made such a decision on its own in any event.

Thomas P. Stafford was the Apollo 10 mission commander, with command module pilot John W. Young and lunar module pilot Eugene A. Cernan. They were launched on schedule at 11:49 a.m., on May 18, 1969, about two and one-half months after Apollo 9 had set out on its successful test flight.

THE DRESS REHEARSAL

Launching was routine, as was the establishment of the Earth parking orbit that followed. After systems checks, and approval from Mission Control, the crew fired the S–IVB stage engine to leave the parking orbit and enter the translunar trajectory. Then they separated the command and service modules from the S–IVB, and turned to dock with the lunar module. Ejection of the docked spacecraft followed, as the crew performed a separation maneuver to increase the distance between the docked spacecraft and the S–IVB stage. Then they eased the S–IVB into a solar orbit by propulsive venting of its excess propellants.

The translunar trajectory had been established so precisely that it was not necessary to make the first midcourse correction, generally a routine step during the early phases of the flight toward the Moon. Finally, after a little more than a day in flight, a single translunar midcourse correction was done to make the flight path of Apollo 10 coincide with the trajectory planned for the Apollo 11 mission. Three days and four hours away from the launch pad, the crew fired the service propulsion system for almost six minutes, inserting the spacecraft into a lunar orbit. Apollo 10 now was orbiting the Moon in a circular flight path about 60 miles above its surface.

Stafford and Cernan entered the lunar module about six hours later to check the systems. They transferred some needed equipment, and moved back into the command module for a normal sleep period. They then reentered the lunar module to go through a complete systems check to prepare for the lunar-orbit rendezvous, the final check of the flight mechanics of the Apollo mission.

Just after two hours into the fourth day, the two spacecraft were undocked and separated. The crew performed its routine communications and radar checks, and then

Stafford and Cernan began their descent toward the lunar surface by firing the descent stage engine. They let down to an altitude of about eight miles above the surface, the closest they were to go in that mission. Over one of the selected landing sites, they checked the landing radar, which worked successfully. After this descent, Stafford and Cernan maneuvered the LM into an elliptical orbit, 11 by 190 miles, to establish the conditions for their rendezvous with Young in the command module. They completed one swing around the Moon, and then staged the lunar module, firing its ascent engine in a simulation of a return from the lunar surface.

These words read rapidly, as if the performance they describe were done swiftly. But it must be appreciated that this rehearsal, which was planned to follow the schedule for the Apollo 11 as closely as possible, actually took more than six and one-half hours from the beginning of the descent toward the Moon until both spacecraft had docked for the second time for crew transfer back to the CSM.

FLAWLESS RENDEZVOUS

The rendezvous was flawless. It began with the usual maneuvering by the LM, using its reaction control system. Young, in the command module, nominally had nothing to do except to wait passively for the docking; the lunar module was expected to be the active seeker and mover in the rendezvous. But in fact he had to be prepared to take over the rendezvous if anything failed on the LM. So he checked the rendezvous every step of the way with sextant observations and ranging with very-high-frequency radio, working out the things he would have to do in the event of a lunar module system failure. The terminal phase of the docking began an hour later, and the lunar module nudged its way into the locking mechanism of the command module. The jubilant crews met back in the command module.

For the final day in lunar orbit, Stafford, Young, and Cernan spent their time in a series of experiments that would add to the general fund of Apollo knowledge. They tracked landmarks, worked on alignment exercises for the inertial platform, and took a series of stereo-pair and sequence photographs of the lunar surface, singling out features that would guide future landing-site selections. And then it was time for the return to Earth.

They fired the service module engine once again to move out of the lunar orbit and back onto a trajectory toward the Earth, using a fast-return flight path that would bring them back again in 54 hours. The flight path again was established with great precision; the only midcourse correction needed was done just about three hours before the reentry, and it changed the velocity by slightly more than two feet per second, or about one part in 20,000.

Fifteen minutes before reentry, they separated the command module from the service module. The command module bearing the crew streaked through the atmosphere; the parachutes deployed and let the spacecraft down on their target coordinates. The astronauts always made much of the accuracy of their landings, with a pool going to the crew landing closest to the target. (In fact, the navigation and guidance accuracy was such that the spacecraft computers pinpointed the target location better than the recovery ships were able to. Later, ships used inertial

Glistening in the sunlight, reflecting the Moon in bright metal as yet undarkened by the savage heat of atmospheric reentry, the Apollo 10 command module ghosts silently within a few yards of its lunar module, occupied by Gene Cernan and Tom Stafford. John Young is alone in the command module at this point. The background is the far side of the Moon, about 60 miles down.

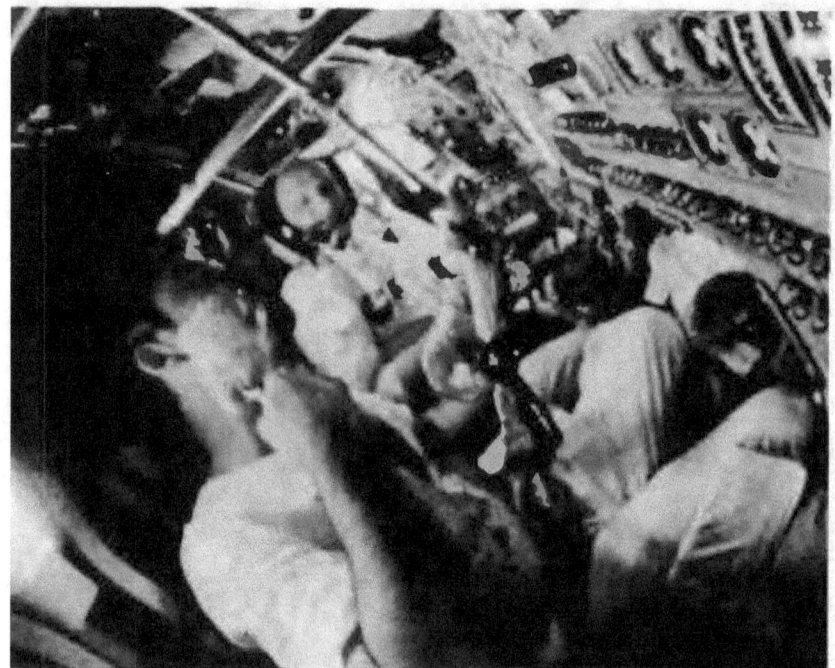

Shaving in space, expected to be a problem (neither astronauts nor delicate mechanism would thrive in a whiskery atmosphere), proved to be no problem at all with an adequately sticky lather. Looking on is Apollo 10 Commander Stafford.

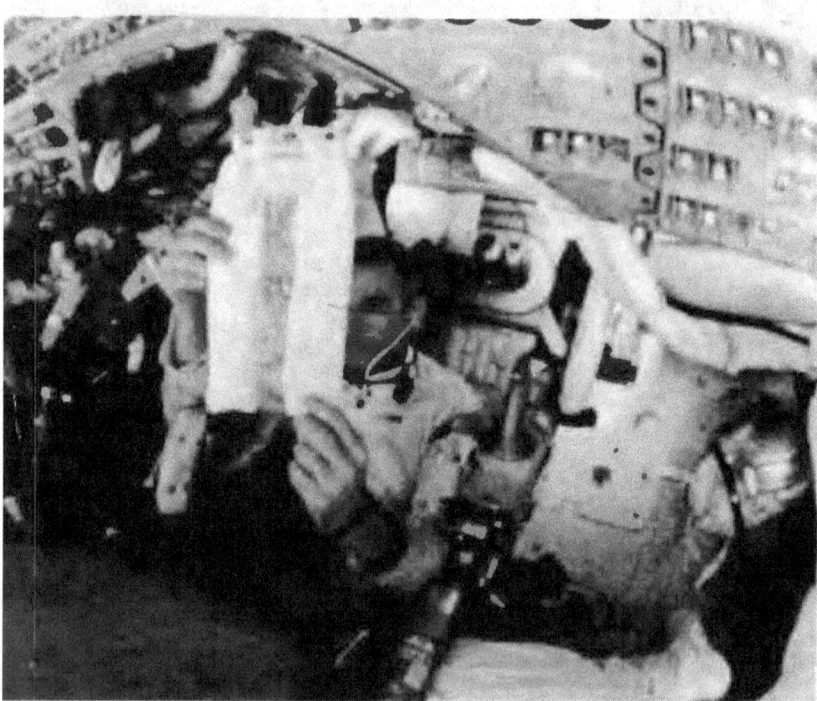

A meal, not a map, is what Gene Cernan is holding up here. It's a plastic envelope containing a chicken and vegetable mix; and with hot water added it made a palatable main course. In the Mercury days space food had been almost as grim as Army survival rations, but during Apollo the eating grew a lot better. By the end of the program, individual astronaut preferences were reflected in the flight menus, and spooned dishes and sandwich spreads were available.

Manning his control console in the Mission Control Center in Houston is George M. Low, then Apollo Spacecraft Program Manager. Behind him is Chris Kraft, Director of Flight Operations at the Manned Spacecraft Center. They and the other men in the photo are viewing a color television transmission from the Apollo 10 during the second day of its lunar orbit mission. The spacecraft at this point was some 112,000 miles from the Earth, about halfway to the Moon. Low's TV monitor is off to the left.

Dawn was just breaking as Apollo 10 gently floated down into the Pacific 395 miles east of Pago Pago. The pinpoint landing was so accurate that the blinking tracking lights on the spacecraft were visible from the USS *Princeton* during the descent.

navigation that was as accurate as that in the spacecraft, but their captains stood off by a mile or two from the landing point to avoid any possibility of a collision.) Navy helicopters swung over their rafts, lifted the crew and took them back to the recovery carrier, the USS *Princeton*, within 39 minutes after the craft had hit the water. About an hour later, the spacecraft itself was hoisted, charred and dripping, onto the deck of the *Princeton*.

These two flights had been successful; more, they had been nearly trouble-free. The confidence they gave to the planned Apollo 11 mission was almost tangible, erasing any doubts about the pace and the direction of the program before Apollo 9 and 10. The thoroughness of the approach, the critical design reviews, and the extensive test and simulation work on the ground had successfully demonstrated the readiness of equipment and crews for the next step.

Everything had been done that would later be done by the crew of Apollo 11, except for the actual touching down on the Moon's surface, the stay there, and the liftoff. Apollo 9 and 10 had done all they could to prepare the entire team—astronauts, flight controllers, ground-support personnel, and management—for the great adventure.

Flotation collar secured, frogmen get ready to assist the Apollo 10 astronauts from the command module. Named *Charlie Brown,* the CM landed three and a half miles from the USS *Princeton.* About one-half hour later the astronauts were aboard the recovery ship, having spent eight days in space.

Returning from the dress rehearsal, Commander Thomas P. Stafford is aided from the command module by frogmen. By demonstrating lunar orbit rendezvous and the LM descent system, this lunar orbit mission set the stage for Apollo 11, which flew two months later and put men on the Moon.

Hoist away! Shortly after they come down from space, the astronauts go back up; this time only briefly as the cage and sling carry them one at a time to the recovery helicopter hovering above (the camera freezes two of its blades).

Glad to be home. Standing in the 'copter doorway the jubilant Apollo 10 crew smile at well-wishers aboard the *Princeton.* From left: LM Pilot Gene Cernan, Commander Tom Stafford, and CM Pilot John Young. The Apollo 10 splashdown was near American Samoa.

The loneliness of space exploration is captured in this picture of Buzz Aldrin standing by *Eagle*'s foil-wrapped footpad. (But a tiny image of Armstrong taking the photograph can be seen on his reflective faceplate.) The *slightly arms-out stance* derives from the pressurized suit. A plaque on the landing stage, which is still on the Moon, is engraved: "Here men from the planet Earth first set foot upon the Moon, July 1969, A.D. We came in peace for all mankind."

CHAPTER ELEVEN

"The Eagle Has Landed"

By MICHAEL COLLINS
and EDWIN E. ALDRIN, JR.

Prelude

All was ready. Everything had been done. Projects Mercury and Gemini. Seven years of Project Apollo. The work of more than 300,000 Americans. Six previous unmanned and manned Apollo flights. Planning, testing, analyzing, training. The time had come.

We had a great deal of confidence. We had confidence in our hardware: the Saturn rocket, the command module, and the lunar module. All flight segments had been flown on the earlier Apollo flights with the exception of the descent to and the ascent from the Moon's surface and, of course, the exploration work on the surface. These portions were far from trivial, however, and we had concentrated our training on them. Months of simulation with our colleagues in the Mission Control Center had convinced us that they were ready.

Although confident, we were certainly not overconfident. In research and in exploration, the unexpected is always expected. We were not overly concerned with our safety, but we would not be surprised if a malfunction or an unforeseen occurrence prevented a successful lunar landing.

As we ascended in the elevator to the top of the Saturn on the morning of July 16, 1969, we knew that hundreds of thousands of Americans had given their best effort to give us this chance. Now it was time for us to give our best.

—Neil A. Armstrong

The splashdown May 26, 1969, of Apollo 10 cleared the way for the first formal attempt at a manned lunar landing. Six days before, the Apollo 11 launch vehicle and spacecraft had crawled from the VAB and trundled at 0.9 mph to Pad 39-A. A successful countdown test ending on July 3 showed the readiness of machines, systems, and people. The next launch window (established by lighting conditions at the landing site on Mare Tranquillitatis) opened at 9:32 a.m. EDT on July 16, 1969. The crew for Apollo 11, all of whom had already flown in space during Gemini, had been intensively training as a team for many months. The following

*mission account makes use of crew members' own words, from books written by two
of them, supplemented by space-to-ground and press-conference transcripts.*

ALDRIN: At breakfast early on the morning of the launch, Dr. Thomas Paine, the Administrator of NASA, told us that concern for our own safety must govern all our actions, and if anything looked wrong we were to abort the mission. He then made a most surprising and unprecedented statement: if we were forced to abort, we would be immediately recycled and assigned to the next landing attempt. What he said and how he said it was very reassuring.

We were up early, ate, and began to suit up—a rather laborious and detailed procedure involving many people, which we would repeat once again, alone, before entering the LM for our lunar landing.

While Mike and Neil were going through the complicated business of being strapped in and connected to the spacecraft's life-support system, I waited near the elevator on the floor below. I waited alone for fifteen minutes in a sort of serene limbo. As far as I could see there were people and cars lining the beaches and highways. The surf was just beginning to rise out of an azure-blue ocean. I could see the massiveness of the Saturn V rocket below and the magnificent precision of Apollo above. I savored the wait and marked the minutes in my mind as something I would always want to remember.

COLLINS: I am everlastingly thankful that I have flown before, and that this period of waiting atop a rocket is nothing new. I am just as tense this time, but the tenseness comes mostly from an appreciation of the enormity of our undertaking rather than from the unfamiliarity of the situation. I am far from certain that we will be able to fly the mission as planned. I think we will escape with our skins, or at least I will escape with mine, but I wouldn't give better than even odds on a successful landing and return. There are just too many things that can go wrong. Fred Haise [the backup astronaut who had checked command-module switch positions] has run through a checklist 417 steps long, and I have merely a half dozen minor chores to take care of—nickel and dime stuff. In between switch throws I have plenty of time to think, if not daydream. Here I am, a white male, age thirty-eight, height 5 feet 11 inches, weight 165 pounds, salary $17,000 per annum, resident of a Texas suburb, with black spot on my roses, state of mind unsettled, about to be shot off to the Moon. Yes, to the Moon.

At the moment, the most important control is over on Neil's side, just outboard of his left knee. It is the abort handle, and now it has power to it, so if Neil rotates it 30° counterclockwise, three solid rockets above us will fire and yank the CM free of the service module and everything below it. It is only to be used in extremis, but I notice a horrifying thing. A large bulky pocket has been added to Neil's left suit leg, and it looks as though if he moves his leg slightly, it's going to snag on the abort handle. I quickly point this out to Neil, and he grabs the pocket and pulls it as far over to the inside of his thigh as he can, but it still doesn't look secure to either one of us. Jesus, I can see the headlines now: "MOONSHOT FALLS INTO OCEAN. Mistake by crew, program officials intimate. Last transmission from Armstrong prior to leaving the pad reportedly was 'Oops.'"

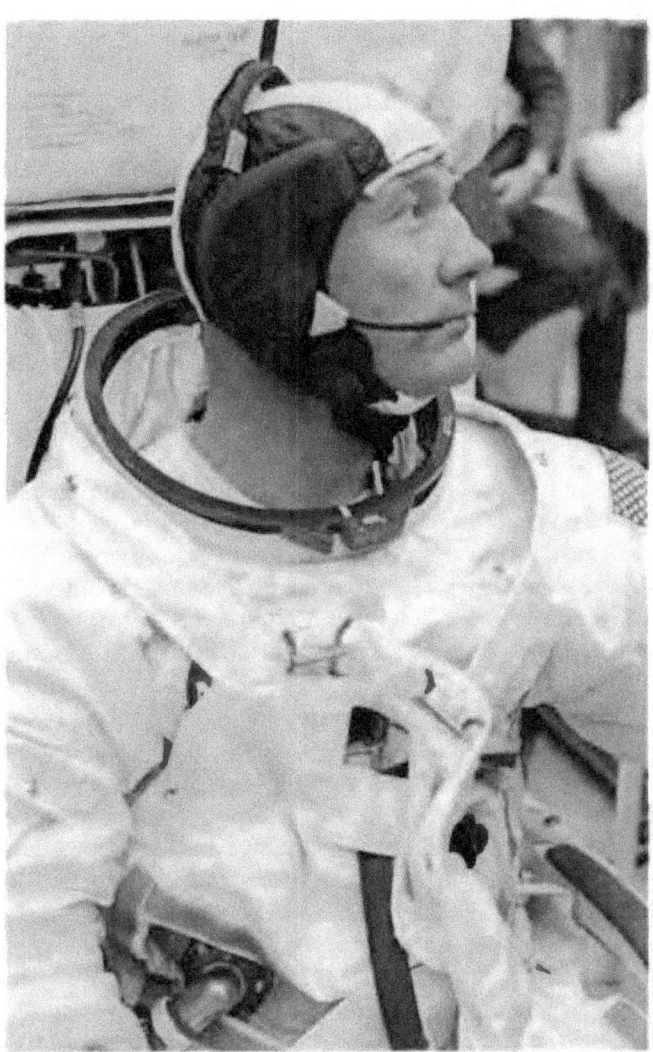

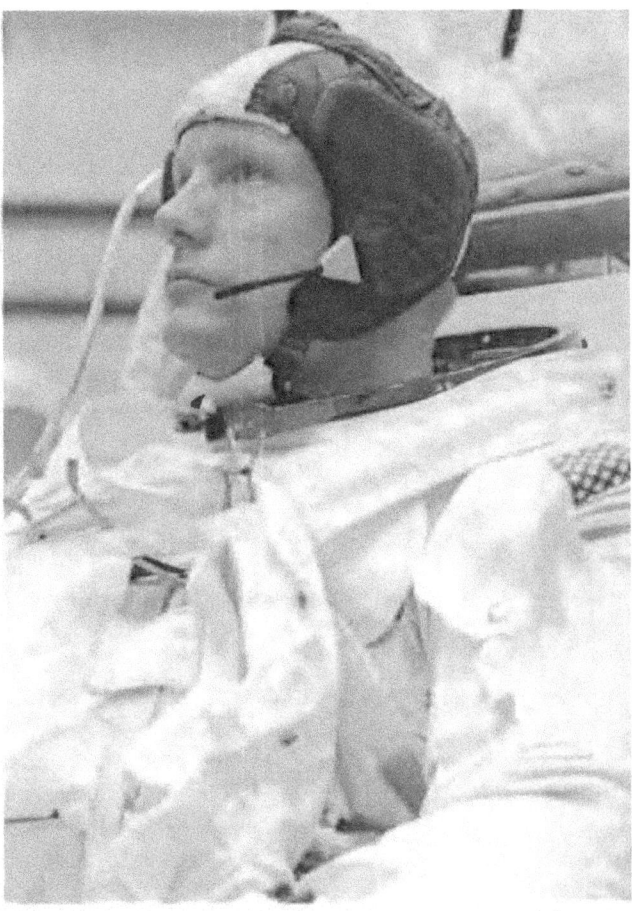

Neil Armstrong, commander of Apollo 11, a civilian, had flown in space in Gemini 8. An engineer and test pilot, he once flew the X–15 above 200,000 feet and at more than 4000 mph.

Edwin E. Aldrin, Jr., then an Air Force colonel, was the lunar module pilot for Apollo 11. He had graduated third in his class at West Point, and earned a Ph.D. in astronautics at MIT. His prior space experience was as pilot for Gemini 12.

Michael Collins was the command module pilot, in orbit above the two others on the surface. A West Pointer and Air Force lieutenant colonel, he had flown in Gemini 10. These unsmiling photos were taken before the mission; but also see page 223.

ARMSTRONG: The flight started promptly, and I think that was characteristic of all events of the flight. The Saturn gave us one magnificent ride, both in Earth orbit and on a trajectory to the Moon. Our memory of that differs little from the reports you have heard from the previous Saturn V flights.

ALDRIN: For the thousands of people watching along the beaches of Florida and the millions who watched on television, our lift-off was ear shattering. For us there was a slight increase in the amount of background noise, not at all unlike the sort one notices taking off in a commercial airliner, and in less than a minute we were traveling ahead of the speed of sound.

COLLINS: This beast is best felt. Shake, rattle, and roll! We are thrown left and right against our straps in spasmodic little jerks. It is steering like crazy, like a nervous lady driving a wide car down a narrow alley, and I just hope it knows where it's going, because for the first ten seconds we are perilously close to that umbilical tower.

ALDRIN: A busy eleven minutes later we were in Earth orbit. The Earth didn't look much different from the way it had during my first flight, and yet I kept looking at it. From space it has an almost benign quality. Intellectually one could realize there were wars underway, but emotionally it was impossible to understand such things. The thought reoccurred that wars are generally fought for territory or are disputes over borders; from space the arbitrary borders established on Earth cannot be seen. After one and a half orbits a preprogrammed sequence fired the Saturn to send us out of Earth orbit and on our way to the Moon.

ARMSTRONG: Hey Houston, Apollo 11. This Saturn gave us a magnificent ride. We have no complaints with any of the three stages on that ride. It was beautiful.

COLLINS: We started the burn at 100 miles altitude, and had reached only 180 at cutoff, but we are climbing like a dingbat. In nine hours, when we are scheduled to make our first midcourse correction, we will be 57,000 miles out. At the instant of shutdown, Buzz recorded our velocity as 35,579 feet per second, more than enough to escape from the Earth's gravitational field. As we proceed outbound, this number will get smaller and smaller until the tug of the Moon's gravity exceeds that of the Earth's and then we will start speeding up again. It's hard to believe that we are on our way to the Moon, at 1200 miles altitude now, less than three hours after liftoff, and I'll bet the launch-day crowd down at the Cape is still bumper to bumper, straggling back to the motels and bars.

ALDRIN: Mike's next major task, with Neil and me assisting, was to separate our command module *Columbia* from the Saturn third stage, turn around and connect with the lunar module *Eagle*, which was stored in the third stage. *Eagle*, by now, was exposed; its four enclosing panels had automatically come off and were drifting away. This of course was a critical maneuver in the flight plan. If the separation and docking did not work, we would return to Earth. There was also the possibility of an in-space collision and the subsequent decompression of our cabin, so we were still in our spacesuits as Mike separated us from the Saturn third stage. Critical as the maneuver is, I felt no apprehension about it, and if there was the slightest inkling of concern it disappeared quickly as the entire separation and docking proceeded perfectly to completion. The nose of *Columbia* was now connected to the top of the *Eagle* and heading for the Moon as we watched the Saturn third stage venting, a propulsive maneuver causing it to move slowly away from us.

Fourteen hours after liftoff, at 10:30 p.m. by Houston time, the three astronauts fasten covers over the windows of the slowly rotating command module and go to sleep. Days 2 and 3 are devoted to housekeeping chores, a small midcourse velocity correction, and TV transmissions back to Earth. In one news digest from Houston, the astronauts are amused to hear that Pravda has referred to Armstrong as "the czar of the ship."

ALDRIN: In our preliminary flight plan I wasn't scheduled to go to the LM until the next day in lunar orbit, but I had lobbied successfully to go earlier. My strongest argument was that I'd have ample time to make sure that the frail LM and its equipment had suffered no damage during the launch and long trip. By that time neither Neil nor I had been in the LM for about two weeks.

THE MOST AWESOME SPHERE

COLLINS: Day 4 has a decidedly different feel to it. Instead of nine hours' sleep, I get seven—and fitful ones at that. Despite our concentrated effort to conserve our energy on the way to the Moon, the pressure is overtaking us (or me at least), and I feel that all of us are aware that the honeymoon is over and we are about to lay our little pink bodies on the line. Our first shock comes as we stop our spinning motion and swing ourselves around so as to bring the Moon into view. We have not been able to see the Moon for nearly a day now, and the change is electrifying. The Moon I have known all my life, that two-dimensional small yellow disk in the sky, has gone away somewhere, to be replaced by the most awesome sphere I have ever seen. To begin with, it is huge, completely filling our window. Second, it is three-dimensional. The belly of it bulges out toward us in such a pronounced fashion that I almost feel I can reach out and touch it. To add to the dramatic effect, we can see the stars again. We are in the shadow of the Moon now, and the elusive stars have reappeared.

As we ease around on the left side of the Moon, I marvel again at the precision of our path. We have missed hitting the Moon by a paltry 300 nautical miles, at a distance of nearly a quarter of a million miles from Earth, and don't forget that the Moon is a moving target and that we are racing through the sky just ahead of its leading edge. When we launched the other day the Moon was nowhere near where it is now: it was some 40 degrees of arc, or nearly 200,000 miles, behind where it is now, and yet those big computers in the basement in Houston didn't even whimper but belched out super-accurate predictions.

As we pass behind the Moon, we have just over eight minutes to go before the burn. We are super-careful now, checking and rechecking each step several times. When the moment finally arrives, the big engine instantly springs into action and reassuringly plasters us back in our seats. The acceleration is only a fraction of one *g* but it feels good nonetheless. For six minutes we sit there peering intent as hawks at our instrument panel, scanning the important dials and gauges, making sure that the proper thing is being done to us. When the engine shuts down, we discuss the matter

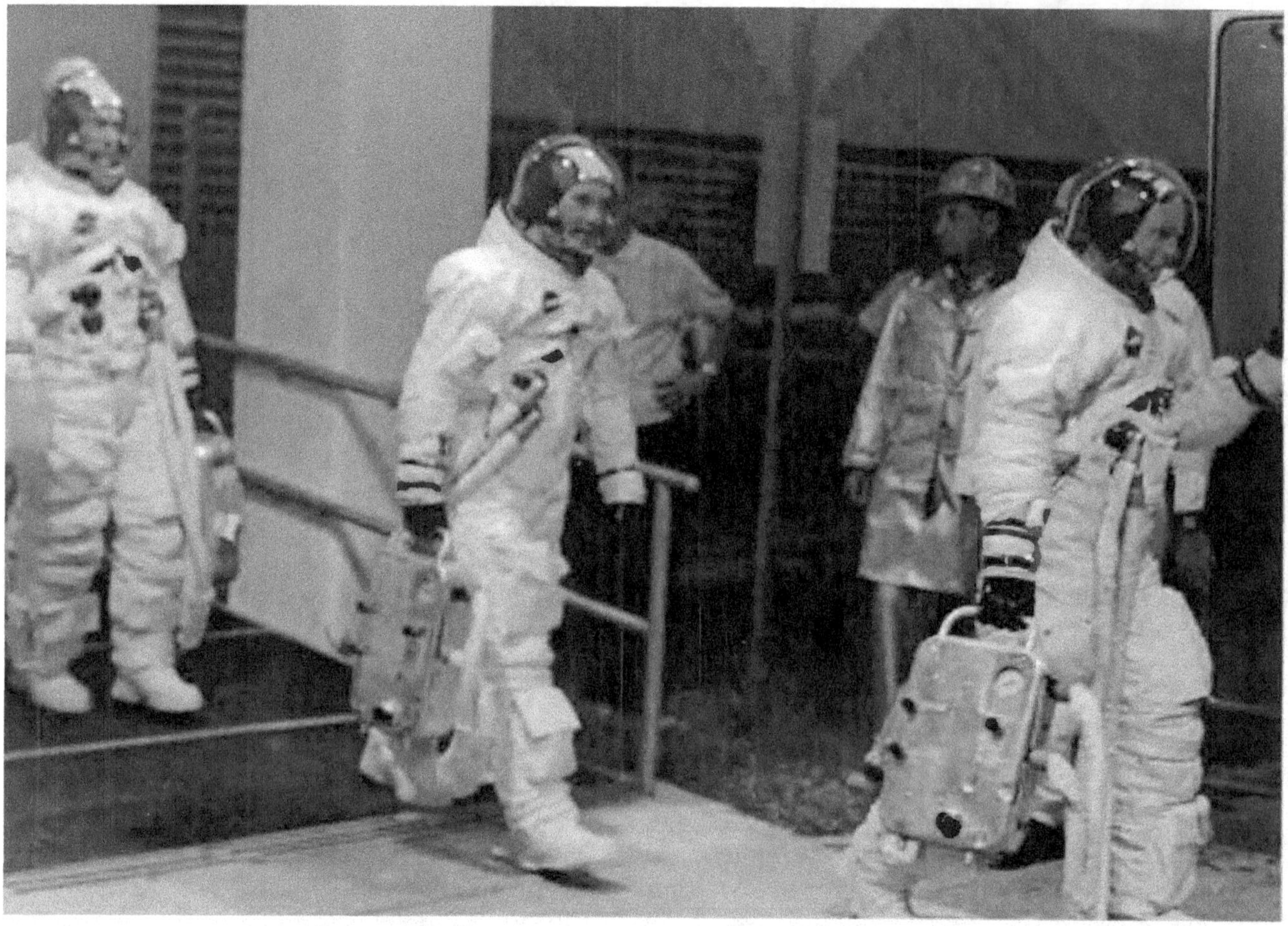

Striding confidently toward the transfer van that will carry them to the launch pad, Apollo 11 Commander Armstrong leads Collins and Aldrin past well-wishers at the start of their historic voyage. Since they are suited up with helmets in place, they carry portable breathing and cooling systems until they can plug into the environmental-control systems aboard their spacecraft.

During the cruise phase there was less work and less tension, although housekeeping and navigational duties still had to be done. Here Aldrin in the lunar module listens to numbers from Houston.

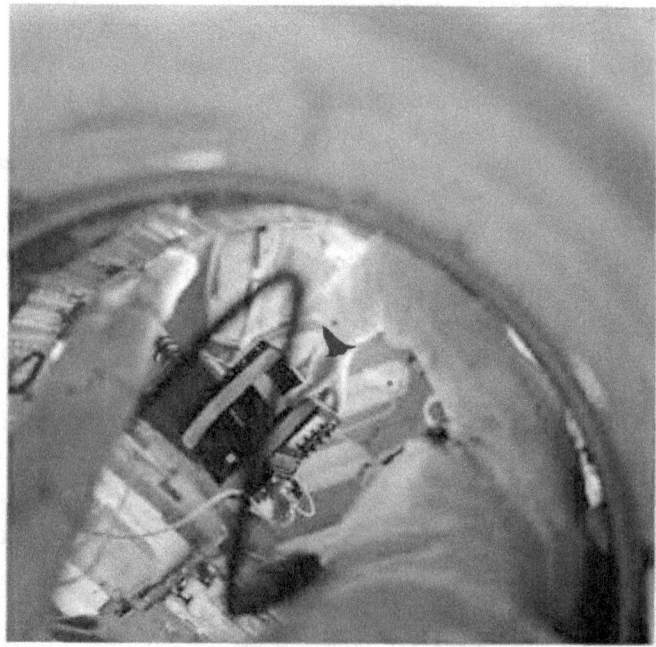

The TV camera with its monitor taped to it was also fired up when work permitted to send back to Earth imagery of itself and of the Moon, as well as homey details in *Columbia*. TV imagery was good, though poorer than on later missions.

with our computer and I read out the results: "Minus one, plus one, plus one." The accuracy of the overall system is phenomenal: out of a total of nearly three thousand feet per second, we have velocity errors in our body axis coordinate system of only a tenth of one foot per second in each of the three directions. That is one accurate burn, and even Neil acknowledges the fact.

ALDRIN: The second burn to place us in closer circular orbit of the Moon, the orbit from which Neil and I would separate from the *Columbia* and continue on to the Moon, was critically important. It had to be made in exactly the right place and for exactly the correct length of time. If we overburned for as little as two seconds we'd be on an impact course for the other side of the Moon. Through a complicated and detailed system of checks and balances, both in Houston and in lunar orbit, plus star checks and detailed platform alignments, two hours after our first lunar orbit we made our second burn, in an atmosphere of nervous and intense concentration. It, too, worked perfectly.

ASLEEP IN LUNAR ORBIT

We began preparing the LM. It was scheduled to take three hours, but because I had already started the checkout, we were completed a half hour ahead of schedule. Reluctantly we returned to the *Columbia* as planned. Our fourth night we were to sleep in lunar orbit. Although it was not in the flight plan, before covering the windows and dousing the lights, Neil and I carefully prepared all the equipment and clothing we would need in the morning, and mentally ran through the many procedures we would follow.

COLLINS: "Apollo 11, Apollo 11, good morning from the Black Team." Could they be talking to me? It takes me twenty seconds to fumble for the microphone button and answer groggily. I guess I have only been asleep five hours or so; I had a tough time getting to sleep, and now I'm having trouble waking up. Neil, Buzz, and I all putter about fixing breakfast and getting various items ready for transfer into the LM. [Later] I stuff Neil and Buzz into the LM along with an armload of equipment. Now I have to do the tunnel bit again, closing hatches, installing drogue and probe, and disconnecting the electrical umbilical. I am on the radio constantly now, running through an elaborate series of joint checks with *Eagle*. I check progress with Buzz: "I have five minutes and fifteen seconds since we started. Attitude is holding very well." "Roger, Mike, just hold it a little bit longer." "No sweat, I can hold it all day. Take your sweet time. How's the czar over there? He's so quiet." Neil chimes in, "Just hanging on—and punching." Punching those computer buttons, I guess he means. "All I can say is, beware the revolution," and then, getting no answer, I formally bid them goodbye. "You cats take it easy on the lunar surface. . . ." "O.K., Mike," Buzz answers cheerily, and I throw the switch which releases them. With my nose against the window and the movie camera churning away, I watch them go. When they are safely clear of me, I inform Neil, and he begins a slow pirouette in place, allowing me a look at his outlandish machine and its four extended legs. "The *Eagle* has wings!" Neil exults.

It doesn't look like any eagle I have ever seen. It is the weirdest-looking contraption ever to invade the sky, floating there with its legs awkwardly jutting out above a body which has neither symmetry nor grace. I make sure all four landing gears are down and locked, report that fact, and then lie a little, "I think you've got a fine-looking flying machine there, *Eagle*, despite the fact you're upside down." "Somebody's upside down," Neil retorts. "O.K., *Eagle*. One minute . . . you guys take care." Neil answers, "See you later." I hope so. When the one minute is up, I fire my thrusters precisely as planned and we begin to separate, checking distances and velocities as we go. This burn is a very small one, just to give *Eagle* some breathing room. From now on it's up to them, and they will make two separate burns in reaching the lunar surface. The first one will serve to drop *Eagle*'s perilune to fifty thousand feet. Then, when they reach this spot over the eastern edge of the Sea of Tranquility, *Eagle*'s descent engine will be fired up for the second and last time, and *Eagle* will lazily arc over into a 12-minute computer-controlled descent to some point at which Neil will take over for a manual landing.

ALDRIN: We were still 60 miles above the surface when we began our first burn. Neil and I were harnessed into the LM in a standing position. [Later] at precisely the right moment the engine ignited to begin the 12-minute powered descent. Strapped in by the system of belts and cables not unlike shock absorbers, neither of us felt the initial motion. We looked quickly at the computer to make sure we were actually functioning as planned. After 26 seconds the engine went to full throttle and the motion became noticeable. Neil watched his instruments while I looked at our primary computer and compared it with our second computer, which was part of our abort guidance system.

I then began a computer read-out sequence to Neil which was also being transmitted to Houston. I had helped develop it. It sounded as though I was chattering like a magpie. It also sounded as though I was doing all the work. During training we had discussed the possibility of making the communication only between Neil and myself, but Mission Control liked the idea of hearing our communications with each other. Neil had referred to it once as "that damned open mike of yours," and I tried to make as little an issue of it as possible.

A YELLOW CAUTION LIGHT

At six thousand feet above the lunar surface a yellow caution light came on and we encountered one of the few potentially serious problems in the entire flight, a problem which might have caused us to abort, had it not been for a man on the ground who really knew his job.

COLLINS: At five minutes into the burn, when I am nearly directly overhead, *Eagle* voices its first concern. "Program Alarm," barks Neil, "It's a 1202." What the hell is that? I don't have the alarm numbers memorized for my own computer, much less for the LM's. I jerk out my own checklist and start thumbing through it, but before I can find 1202, Houston says, "Roger, we're GO on that alarm." No problem, in other words. My checklist says 1202 is an "executive overflow," meaning simply that the computer has been called upon to do too many things at once and is forced to postpone some of them. A little farther along, at just three thousand feet above the

Its legs folded up for launch, the lunar module looked like this as the command module eased in to dock and draw it free from the third stage of the Saturn launch vehicle.

An offset docking target on the LM lined up with the pilot's window on the CSM if alignment was right. Then an array of powerful latches locked the two spacecraft together.

The unknown that is soon to be known: this picture shows the Apollo 11 landing site one orbit before descent was begun. Tranquility Base is near the shadow line, a little to the right of center. The big jagged shape to the left is not a shadow but an out-of-focus LM thruster.

surface, the computer flashes 1201, another overflow condition, and again the ground is superquick to respond with reassurances.

ALDRIN: Back in Houston, not to mention on board the *Eagle*, hearts shot up into throats while we waited to learn what would happen. We had received two of the caution lights when Steve Bales, the flight controller responsible for LM computer activity, told us to proceed, through Charlie Duke, the capsule communicator. We received three or four more warnings but kept on going. When Mike, Neil, and I were presented with Medals of Freedom by President Nixon, Steve also received one. He certainly deserved it, because without him we might not have landed.

ARMSTRONG: In the final phases of the descent after a number of program alarms, we looked at the landing area and found a very large crater. This is the area we decided we would not go into; we extended the range downrange. The exhaust dust was kicked up by the engine and this caused some concern in that it degraded our ability to determine not only our altitude in the final phases but also our translational velocities over the ground. It's quite important not to stub your toe during the final phases of touchdown.

From the space-to-ground tapes:

EAGLE: 540 feet, down at 30 [feet per second] . . . down at 15 . . . 400 feet down at 9 . . . forward . . . 350 feet, down at 4 . . . 300 feet, down 3½ . . . 47 forward . . . 1½ down . . . 13 forward . . . 11 forward, coming down nicely . . . 200 feet, 4½ down . . . 5½ down . . . 5 percent . . . 75 feet . . . 6 forward . . . lights on . . . down 2½ . . . 40 feet, down 2½, kicking up some dust . . . 30 feet, 2½ down . . . faint shadow . . . 4 forward . . . 4 forward . . . drifting to right a little . . . O.K. . . .

HOUSTON: 30 seconds [fuel remaining].

EAGLE: Contact light! O.K., engine stop . . . descent engine command override off . . .

HOUSTON: We copy you down, *Eagle.*

EAGLE: Houston, Tranquility Base here. The *Eagle* has landed!

HOUSTON: Roger, Tranquility. We copy you on the ground. You've got a bunch of guys about to turn blue. We're breathing again. Thanks a lot.

TRANQUILITY: Thank you . . . That may have seemed like a very long final phase. The auto targeting was taking us right into a football-field-sized crater, with a large number of big boulders and rocks for about one or two crater-diameters around it, and it required flying manually over the rock field to find a reasonably good area.

HOUSTON: Roger, we copy. It was beautiful from here, Tranquility. Over.

TRANQUILITY: We'll get to the details of what's around here, but it looks like a collection of just about every variety of shape, angularity, granularity, about every variety of rock you could find.

HOUSTON: Roger, Tranquility. Be advised there's lots of smiling faces in this room, and all over the world.

TRANQUILITY: There are two of them up here.

COLUMBIA: And don't forget one in the command module.

ARMSTRONG: Once [we] settled on the surface, the dust settled immediately

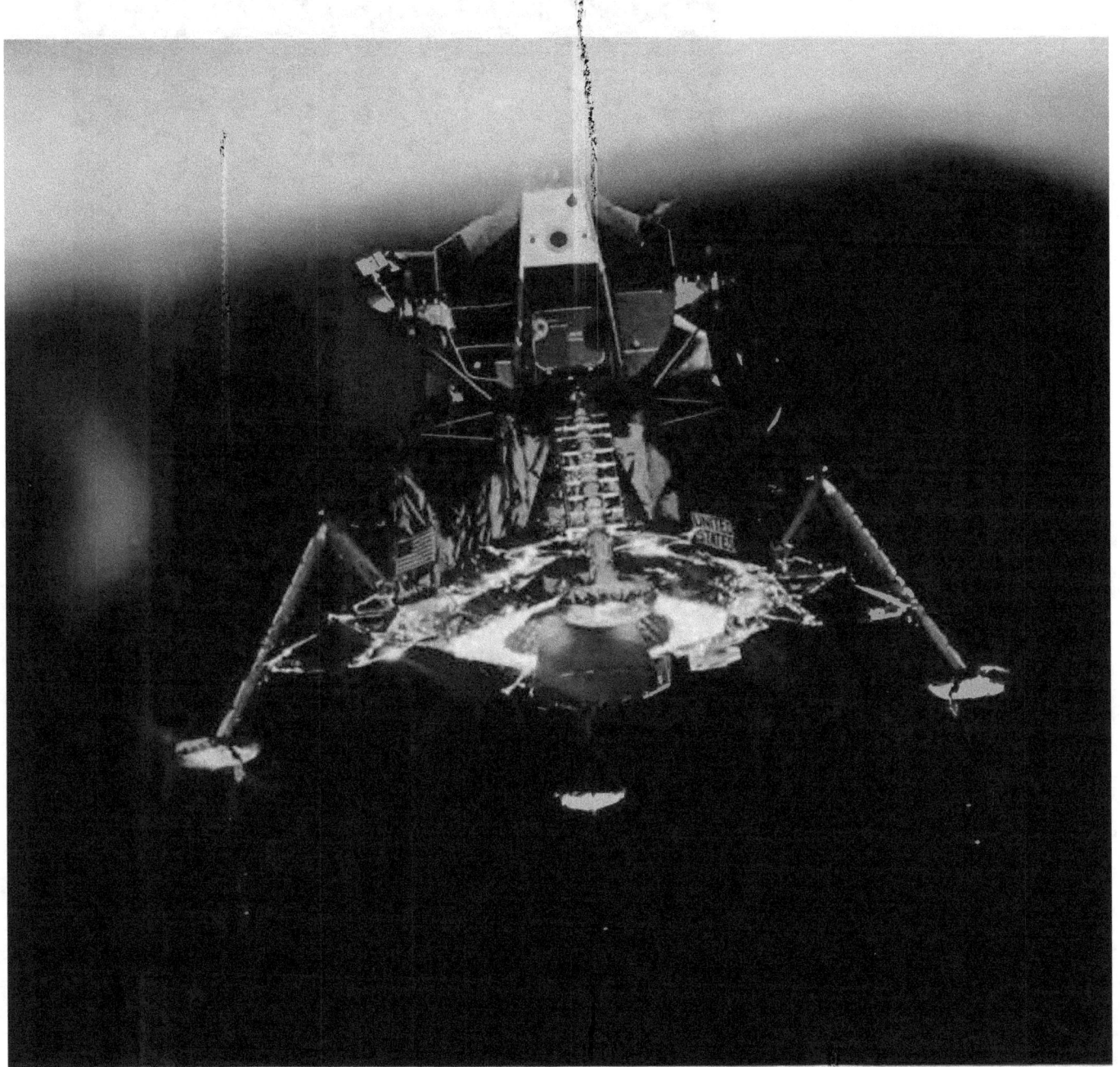

"You cats take it easy," Collins radioed in farewell as the lunar module separated for its historic descent to the surface of the Moon. The three probes extending below the footpads were to turn on a contact light reporting when the LM was within a few feet of the surface. On the nearest landing leg note the ladder giving access to the surface from the cabin at the top. The exhaust bell of the big descent engine can be seen in the center. It made two burns, finally settling the LM gently on the surface with only seconds of fuel remaining.

Leaving the ninth step of the ladder, Aldrin jumps down to the Moon. Earlier on the "porch" he had radioed, "Now I want to partially close the hatch, making sure not to lock it on my way out." Armstrong's dry response was: "A good thought." On Earth his weight, including the spacesuit and mechanism-filled portable life-support system, would have totaled 360 lb; but here the gross came only to a bouncy 60 lb. The descent-engine exhaust bell (extreme right) came to rest about a foot above the surface.

The dusty surface took footprints like damp sand. Although superficially soft, it proved remarkably resistant to penetration by coring tubes, which generally hung up after being driven a few inches.

The flag of Tranquility Base was not a symbol of territorial claim so much as identification of the nation that had carried out the first manned landing. Aldrin's forward-leaning stance here was the normal resting position of an astronaut wearing the big life-support pack. Note eroded, half-buried rock in right foreground.

and we had an excellent view of the area surrounding the LM. We saw a crater surface, pockmarked with craters up to 15, 20, 30 feet, and many smaller craters down to a diameter of 1 foot and, of course, the surface was very fine-grained. There were a surprising number of rocks of all sizes.

A number of experts had, prior to the flight, predicted that a good bit of difficulty might be encountered by people due to the variety of strange atmospheric and gravitational characteristics. This didn't prove to be the case and after landing we felt very comfortable in the lunar gravity. It was, in fact, in our view preferable both to weightlessness and to the Earth's gravity.

When we actually descended the ladder it was found to be very much like the lunar-gravity simulations we had performed here on Earth. No difficulty was encountered in descending the ladder. The last step was about 3½ feet from the surface, and we were somewhat concerned that we might have difficulty in reentering the LM at the end of our activity period. So we practiced that before bringing the camera down.

ALDRIN: We opened the hatch and Neil, with me as his navigator, began backing out of the tiny opening. It seemed like a small eternity before I heard Neil say, "That's one small step for man . . . one giant leap for mankind." In less than fifteen minutes I was backing awkwardly out of the hatch and onto the surface to join Neil, who, in the tradition of all tourists, had his camera ready to photograph my arrival.

I felt buoyant and full of goose pimples when I stepped down on the surface. I immediately looked down at my feet and became intrigued with the peculiar properties of the lunar dust. If one kicks sand on a beach, it scatters in numerous directions with some grains traveling farther than others. On the Moon the dust travels exactly and precisely as it goes in various directions, and every grain of it lands nearly the same distance away.

THE BOY IN THE CANDY STORE

ARMSTRONG: There were a lot of things to do, and we had a hard time getting them finished. We had very little trouble, much less trouble than expected, on the surface. It was a pleasant operation. Temperatures weren't high. They were very comfortable. The little EMU, the combination of spacesuit and backpack that sustained our life on the surface, operated magnificently. The primary difficulty was just far too little time to do the variety of things we would have liked. We had the problem of the five-year-old boy in a candy store.

ALDRIN: I took off jogging to test my maneuverability. The exercise gave me an odd sensation and looked even more odd when I later saw the films of it. With bulky suits on, we seemed to be moving in slow motion. I noticed immediately that my inertia seemed much greater. Earth-bound, I would have stopped my run in just one step, but I had to use three of four steps to sort of wind down. My Earth weight, with the big backpack and heavy suit, was 360 pounds. On the Moon I weighed only 60 pounds.

At one point I remarked that the surface was "Beautiful, beautiful. Magnificent desolation." I was struck by the contrast between the starkness of the shadows and the desertlike barrenness of the rest of the surface. It ranged from dusty gray to light tan and was unchanging except for one startling sight: our LM sitting there with its black, silver, and bright yellow-orange thermal coating shining brightly in the otherwise colorless landscape. I had seen Neil in his suit thousands of times before, but on the Moon the unnatural whiteness of it seemed unusually brilliant. We could also look around and see the Earth, which, though much larger than the Moon the Earth was seeing, seemed small—a beckoning oasis shining far away in the sky.

As the sequence of lunar operations evolved, Neil had the camera most of the time, and the majority of pictures taken on the Moon that include an astronaut are of me. It wasn't until we were back on Earth and in the Lunar Receiving Laboratory looking over the pictures that we realized there were few pictures of Neil. My fault perhaps, but we had never simulated this in our training.

COAXING THE FLAG TO STAND

During a pause in experiments, Neil suggested we proceed with the flag. It took both of us to set it up and it was nearly a disaster. Public Relations obviously needs practice just as everything else does. A small telescoping arm was attached to the flagpole to keep the flag extended and perpendicular. As hard as we tried, the telescope wouldn't fully extend. Thus the flag, which should have been flat, had its own unique permanent wave. Then to our dismay the staff of the pole wouldn't go far enough into the lunar surface to support itself in an upright position. After much struggling we finally coaxed it to remain upright, but in a most precarious position. I dreaded the possibility of the American flag collapsing into the lunar dust in front of the television camera.

COLLINS: [On his fourth orbital pass above] "How's it going?" "The EVA is progressing beautifully. I believe they're setting up the flag now." Just let things keep going that way, and no surprises, please. Neil and Buzz sound good, with no huffing and puffing to indicate they are overexerting themselves. But one surprise at least is in store. Houston comes on the air, not the slightest bit ruffled, and announces that the President of the United States would like to talk to Neil and Buzz. "That would be an honor," says Neil, with characteristic dignity.

The President's voice smoothly fills the air waves with the unaccustomed cadence of the speechmaker, trained to convey inspiration, or at least emotion, instead of our usual diet of numbers and reminders. "Neil and Buzz, I am talking to you by telephone from the Oval Office at the White House, and this certainly has to be the most historic telephone call ever made . . . Because of what you have done, the heavens have become a part of man's world. As you talk to us from the Sea of Tranquility, it inspires us to redouble our efforts to bring peace and tranquility to Earth . . ." My God, I never thought of all this bringing peace and tranquility to anyone. As far as I am concerned, this voyage is fraught with hazards for the three of us—and especially two of us—and that is about as far as I have gotten in my thinking.

Neil, however, pauses long enough to give as well as he receives. "It's a great honor and privilege for us to be here, representing not only the United States but men of peace of all nations, and with interest and a curiosity and a vision for the future."

Scientific experiments, stowed compactly for their trip to the Moon, are unpacked for deployment by Aldrin. Note the spread-leg stability of the landed LM, its sturdy legs foil-wrapped for thermal insulation. Beyond the right leg can be seen the solar-wind experiment, an exposed foil sheet that will be brought back to Earth for careful analysis; and beyond it the television camera. At right, Aldrin drops off the retroreflector for laser ranging of the Earth-Moon distance, and takes the seismometer experiment 15 feet farther out. The former gave new accuracy to measurement of the Moon's orbit. The seismometer was the first of an array of seismic stations now emplaced on the Moon.

Ghosting up to its crucial rendezvous with *Columbia*, its legs and landing stage left behind on the surface as a launching platform, *Eagle*'s historic voyage is almost done. Once its film, rock boxes, and two exhilarated astronauts have come aboard, *Eagle* will be left in lunar orbit while the three men set out for the distant half-planet shown here that is man's home.

[Later] Houston cuts off the White House and returns to business as usual, with a long string of numbers for me to copy for future use. My God, the juxtaposition of the incongruous: roll, pitch, and yaw; prayers, peace, and tranquility. What will it be like if we really carry this off and return to Earth in one piece, with our boxes full of rocks and our heads full of new perspectives for the planet? I have a little time to ponder this as I zing off out of sight of the White House and the Earth.

ALDRIN: We had a pulley system to load on the boxes of rocks. We found the process more time-consuming and dust-scattering than anticipated. After the gear and both of us were inside, our first chore was to pressure the LM cabin and begin stowing the rock boxes, film magazines, and anything else we wouldn't need until we were connected once again with the *Columbia*. We removed our boots and the big backpacks, opened the LM hatch, and threw these items onto the lunar surface, along with a bagful of empty food packages and the LM urine bags. The exact moment we tossed every thing out was measured back on Earth—the seismometer we had put out was even more sensitive than we had expected.

Before beginning liftoff procedures [we] settled down for our fitful rest. We didn't sleep much at all. Among other things we were elated—and also cold. Liftoff from the Moon, after a stay totaling twenty-one hours, was exactly on schedule and fairly uneventful. The ascent stage of the LM separated, sending out a shower of brilliant insulation particles which had been ripped off from the thrust of the ascent engine. There was no time to sightsee. I was concentrating on the computers, and Neil was studying the attitude indicator, but I looked up long enough to see the flag fall over . . . Three hours and ten minutes later we were connected once again with the *Columbia*.

COLLINS: I can look out through my docking reticle and see that they are steady as a rock as they drive down the center line of that final approach path. I give them some numbers. "I have 0.7 mile and I got you at 31 feet per second." We really *are* going to carry this off! For the first time since I was assigned to this incredible flight, I feel that it *is* going to happen. Granted, we are a long way from home, but from here on it should be all downhill. Within a few seconds Houston joins the conversation, with a tentative little call. "*Eagle* and *Columbia,* Houston standing by." They want to know what the hell is going on, but they don't want to interrupt us if we are in a crucial spot in our final maneuvering. Good heads! However, they needn't worry, and Neil lets them know it. "Roger, we're stationkeeping."

ALL SMILES AND GIGGLES

[After docking] it's time to hustle down into the tunnel and remove hatch, probe, and drogue, so Neil and Buzz can get through. Thank God, all the claptrap works beautifully in this its final workout. The probe and drogue will stay with the LM and be abandoned with it, for we will have no further need of them and don't want them cluttering up the command module. The first one through is Buzz, with a big smile on his face. I grab his head, a hand on each temple, and am about to give him a smooch on the forehead, as a parent might greet an errant child; but then, embarrassed, I think better of it and grab his hand, and then Neil's. We cavort about a little bit, all smiles and giggles over our success, and then it's back to work as usual.

We Walk On Moon

Armstrong, Aldrin Step Onto Surface

Daily Press

Luna 15 Darts To Lower Moon Orbit

World Cheers Landing

'Tranquillity Base Here; The Eagle Has Landed'

The Evening Bulletin

NIGHT EXTRA

Man's 'Giant Leap'

Kennedy Faces Auto Charge In Fatal Crash

Apollo 11 Lands Crew on Moon, Kennedy Goal

Huntsville News

Morning Edition

Souvenir Issue

One small step for man, one giant leap for mankind'

Millions on planet earth watch lunar surface feat

THE HOUSTON POST

MOON SPECIAL

NEIL WALKS THE MOON!

'A Great Moment of Our Time'

U.S. Flag Is Planted Near Ship

Eagle Crew All Calm as History Made

Old Man Moon Not The Same

Best Words Of Descent: Eagle Lands

...ounts ...on TV

Chicago Tribune

'O. K., EAGLE HAS LANDED'

Russ Craft in New Orbit

Apollo 11 Down Safely at 3:17 on Moon Flight

LUNA CIRCLES ONLY 10 MILES FROM MOON

Aldrin Plans Communion on the Moon

Probe Is Closer in Apollo Site

Nixon Tells Confidence in Apollo 11's Success

After Church

The Philadelphia Inquirer

PUBLIC LEDGER

CITY EDITION

AN INDEPENDENT NEWSPAPER FOR ALL THE PEOPLE

MAN LANDS ON MOON

'Eagle Has Landed,' Astronauts Say

Crater Avoided In Touchdown; 2 Stay in Ship

THE PLAIN DEALER

FINAL

Astronauts Walk Moon

One Giant Leap for Mankind

Touchdown Words From Moon

Birmingham Post-Herald
SUNRISE EDITION

Americans Walk On Moon!

It's A Banner Day For U.S.!

'1 Small Step For Man, One Giant Leap'

Whew!

Chance Of Collision 'Infinitesimal'

Luna 15 Drops To Lower Orbit

Wheeler Sees No Full Viet Pullout

Hot Days, Warm Nights Continue

Moon Landing React

St. Louis Globe-Democrat

Our Men Are on the Moon

Crater

The New York Times

LATE CITY EDITION

MEN WALK ON MOON

ASTRONAUTS LAND ON A PLAIN AFTER STEERING PAST CRATER

Voice From Moon: 'Eagle Has Landed'

A Powdery Surface Found by Armstrong

Apollo 11 Heads Home

Daily Press

Splashdown Is Thursday

Gas, Oil Tax Hike Is Voted

Soviet Union Moon Probe Goes Down

Eagle Begins Trip Home

LUNA DOWN ON THE M

The Evening

The Washington Post

Apollo Sets Course For Home After Perfect Lunar Takeoff

Ascent, Docking Flawless

Luna 15 Crashes On Moon

TO THE MOON...AND BACK
7th DAY

MOONSHIP DOIN'S TODAY

HERALD EXAMINER
LATEST NEWS SPORTS

COMING HOME

ck to Orbit

Rocks May Prove Moon Is Alive, Hot

No one knew it when Columbus first stepped on Watlings Island, but every headline writer was challenged by the first manned Moon landing. The full newspaper coverage suggested by these front pages was just a part of it. Considering the nearly worldwide radio and television coverage, it has been estimated that more than half the population of the planet was aware of the events of Apollo 11.

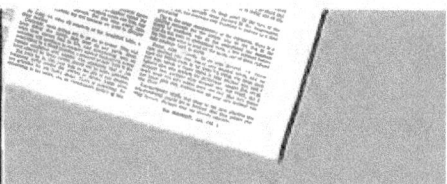

American Airlines

Excerpts from a TV program broadcast by the Apollo 11 astronauts on the last evening of the flight, the day before splashdown in the Pacific:

COLLINS: ". . . The Saturn V rocket which put us in orbit is an incredibly complicated piece of machinery, every piece of which worked flawlessly. This computer above my head has a 38,000-word vocabulary, each word of which has been carefully chosen to be of the utmost value to us. The SPS engine, our large rocket engine on the aft end of our service module, must have performed flawlessly or we would have been stranded in lunar orbit. The parachutes up above my head must work perfectly tomorrow or we will plummet into the ocean. We have always had confidence that this equipment will work properly. All this is possible only through the blood, sweat, and tears of a number of people. First, the American workmen who put these pieces of machinery together in the factory. Second, the painstaking work done by various test teams during the assembly and retest after assembly. And finally, the people at the Manned Spacecraft Center, both in management, in mission planning, in flight control, and last but not least, in crew training. This operation is somewhat like the periscope of a submarine. All you see is the three of us, but beneath the surface are thousands and thousands of others, and to all of those, I would like to say, 'Thank you very much.' "

ALDRIN: ". . . This has been far more than three men on a mission to the Moon; more, still, than the efforts of a government and industry team; more, even, than the efforts of one nation. We feel that this stands as a symbol of the insatiable curiosity of all mankind to explore the unknown. Today I feel we're really fully capable of accepting expanded roles in the exploration of space. In retrospect, we have all been particularly pleased with the call signs that we very laboriously chose for our spacecraft, *Columbia* and *Eagle*. We've been pleased with the emblem of our flight, the eagle carrying an olive branch, bringing the universal symbol of peace from the planet Earth to the Moon. Personally, in reflecting on the events of the past several days, a verse from Psalms comes to mind. 'When I consider the heavens, the work of Thy fingers, the Moon and the stars, which Thou hast ordained; What is man that Thou art mindful of him?' "

ARMSTRONG: "The responsibility for this flight lies first with history and with the giants of science who have preceded this effort; next with the American people, who have, through their will, indicated their desire; next with four administrations and their Congresses, for implementing that will; and then, with the agency and industry teams that built our spacecraft, the Saturn, the *Columbia,* the *Eagle,* and the little EMU, the spacesuit and backpack that was our small spacecraft out on the lunar surface. We would like to give special thanks to all those Americans who built the spacecraft; who did the construction, design, the tests, and put their hearts and all their abilities into those craft. To those people tonight, we give a special thank you, and to all the other people that are listening and watching tonight, God bless you. Good night from Apollo 11."

[*Portions of the text of this chapter have been excerpted with permission from* Carrying the Fire, © *1974 by Michael Collins, and* Return to Earth, © *1973 by Aldrin-Warga Associates.*]

Relief and jubilation greeted the safe splashdown of Apollo 11. "Many of us still can't believe that the goal we set out to achieve in 1961 has been achieved," said George Low; and some of the parties that night have entered folk-lore. From right above: Gilruth, Low, Kraft, Phillips (looking down), and Mueller (partly behind flag).

Aboard the Hornet in their quarantine trailer on the hangar deck, three buoyant astronauts chaff with the President of the United States. Compare their expressions here with those in the photographs before launch on page 205.

Sombreros replaced space helmets when crowds stalled a motorcade in Mexico City. In a 45-day world tour aboard Air Force One to 27 cities in 24 countries, the Apollo 11 astronauts received exhausting acclaim.

Apollo 12 Astronaut Alan Bean examines Surveyor III's television camera. The two astronauts walked down to the spacecraft from their own lunar module, which they had landed about 600 feet away. They removed the TV camera and the scoop so that scientists could study the effects on well-known materials of a 31-month lunar sojourn. A third spacecraft, Lunar Orbiter III, made the pinpoint landing possible by its earlier feat of photographing the site in exquisite detail.

CHAPTER TWELVE

Ocean of Storms and Fra Mauro

By CHARLES CONRAD, JR.,
and ALAN B. SHEPARD, JR.

*S*cientific exploration of the Moon began in earnest with the Apollo 12 and 14 *missions. Four astronauts worked on the Moon in four-hour shifts, walking from site to site. The Apollo 12 astronauts carried everything: experiments, equipment, tools, sample bags, cameras. The Apollo 14 team had a small equipment cart, and some of the time it was a help. But the missions showed that a man in a pressurized suit had definite limitations on the rugged and perplexing lunar surface. It was more work than it seemed, and in the case of Apollo 14, medical advice from Earth ended one phase of activity. But the two missions produced a wealth of new scientific data and lunar samples, and both laid a firm foundation for the great voyages of lunar exploration to follow.*

The sky was cloudy and rain was falling on November 14, 1969, as the Apollo 12 crew prepared for launch. Half a minute after liftoff a lightning strike opened the main circuit breakers in the spacecraft. Quick action by the crew and Launch Control restored power, and Astronauts Charles "Pete" Conrad, Jr., Richard F. Gordon, and Alan L. Bean sped into sunlight above the clouds. "We had everything in the world drop out," Conrad reported. "We've had a couple of cardiac arrests down here too," Launch Control radioed back.

Their destination was the Ocean of Storms. Four and a half days later, Conrad and Bean entered the lunar module Intrepid *and separated from Gordon in the command module* Yankee Clipper. *Their landing site was about 1300 miles west of where Apollo 11 had landed, on a surface believed covered by debris splashed out from the crater Copernicus some 250 miles away. The exact site was a point where, 31 months before, the unmanned lunar scout Surveyor III had made a precarious automatic landing. The Surveyor site was a natural choice: it was a geologically different surface, it would demonstrate pinpoint landing precision, and it would offer a chance to bring back metal, electronic, and optical materials that had soaked for many months in the lunar environment.*

Here is Pete Conrad's account of the mission:

It was really pioneering in lunar exploration. We had planned our traverses carefully, we covered them, and we stayed on the time line. We had a real-time link with the ground, to help guide our work on the surface. Of course we had practiced a lot beforehand, working with geologists in the field to learn techniques from them while they learned what we could and couldn't do in the lunar environment.

Our first important task was the precision landing near Surveyor III. When we pitched over just before the landing phase, there it was, looking as if we would land practically on target. The targeting data were just about perfect, but I maneuvered around the crater, landing at a slightly different spot than the one we had planned. In my judgment, the place we had prepicked was a little too rough. We touched down about 600 feet from the Surveyor. They didn't want us to be nearer than 500 feet because of the risk that the descent engine might blow dust over the spacecraft.

Our second important task—and of course the real reason for going to the Moon in the first place—was to accomplish useful scientific work on the surface. We had to set up the ALSEP and its experiments; we had to do a lot of geologizing; and finally we had to bring back some pieces of the Surveyor, so they could be analyzed for the effects of their exposure.

Al Bean and I made two EVAs, each lasting just under four hours; and we covered the planned traverses as scheduled. We learned things that we could never have found out in a simulation. A simple thing like shoveling soil into a sample bag, for instance, was an entirely new experience. First, you had to handle the shovel differently, stopping it before you would have on Earth, and tilting it to dump the load much more steeply, after which the whole sample would slide off suddenly.

LITTLE CLOUDS AROUND YOUR FEET

And the dust! Dust got into everything. You walked in a pair of little dust clouds kicked up around your feet. We were concerned about getting dust into the working parts of the spacesuits and into the lunar module, so we elected to remain in the suits between our two EVAs. We thought that it would be less risky that way than taking them off and putting them back on again.

On the first EVA, the first thing I did was to take the contingency sample. When Al joined me on the surface, we started with the experimental setups. We set out the solar wind experiment and the ALSEP items. We planted the passive seismic experiment, deployed and aligned antennas, laid out the lunar surface magnetometer, and took core samples. Some of the experiments started working right away as planned, sending data back. Others weren't set to start operating until after we had left.

We were continually describing what we were doing; we kept up a stream of chatter so that people on the ground could follow what was going on if we were to lose the video signal. And we did lose it, too, soon after we landed. That was hard to take.

One strange surface phenomenon was a group of conical mounds, looking for all the world like small volcanoes. They were maybe five feet tall and about fifteen feet in diameter at the base. Both of us really enjoyed working on the surface; we took a

Going its separate way for a landing, the Apollo 12 Lunar Module *Intrepid* gleams in the sunlight as it pulls ahead of *Yankee Clipper*, the command module. The view is westward, from a circular orbit 69 miles above the surface, with *Intrepid* very nearly as high. With the Sun above and behind the camera, the very rough lunar terrain below appears greatly subdued. The circular crater in the middle distance on the right is Herschel. The smooth-floored giant crater Ptolemaeus occupies much of the area to its left.

Tire tracks trace the path of the Apollo 14 astronauts from their lunar module *Antares* to the site, some 200 yards to the west, where they set up the Apollo Lunar Surface Experiments Package (ALSEP). On this mission, they had a two-wheeled, light, hand-pulled cart (shown on page 237) to carry their equipment and samples. The Modular Equipment Transporter, or MET, had pneumatic tires, which compacted the soil as they rolled. In this photo- taken in the direction of the Sun, the tracks are brightly backlighted. In general, however, where astronauts worked, the soil scuffed up by their boots was distinctly darker than the undisturbed surface material.

Astronaut Alan Bean unloads equipment from the Apollo 12 lunar module *Intrepid* in preparation for the walk to the ALSEP site. The lunar module—surely the clumsiest-looking flying machine ever built—consisted of a descent stage, destined to remain on the Moon, and an ascent stage that later carried the crew and samples into lunar orbit. Scientific equipment and gear for use on the lunar surface was stowed in four bays of the descent stage. The panel that covered the bay facing Bean folded down to provide a work table.

lot of kidding later about the way we reacted. But it *was* exciting; there we were, the third and fourth people on the Moon, doing what we were supposed to do, what we had planned to do, and keeping within schedule. Add to that the excitement of just being there, and I think we could be forgiven for reacting with enthusiasm.

Our second EVA was heavily scheduled. We were to make visual observations, collect a lot more samples, document photographically the area around the Ocean of Storms, and—if we could—bring back pieces of the Surveyor III spacecraft. We had rehearsed that part with a very detailed mockup before the flight, and were well prepared.

We moved on a traverse, picking up samples and describing them and the terrain around them, as well as documenting the specific sites with photography. We rolled a rock into a crater so that scientists back on Earth could see if the seismic experiment was working. (It was sensitive enough to pick up my steps as I walked nearby.) Anyway, we rolled the rock and they got a jiggle or two, indicating that experiment was off and running.

TAN DUST ON SURVEYOR

We found some green rocks, and some gray soil that maintained its light color even below the surface, which is not common, and we finally reached the Surveyor crater. I was surprised by its size and its hard surface. We could have landed right there, I believe now, but it would have been a scary thing at the time. The Surveyor was covered with a coating of fine dust, and it looked tan or even brown in the lunar light, instead of the glistening white that it was when it left Earth more than two years earlier. It was decided later that the dust was kicked up by our descent onto the surface, even though we were 600 feet away.

We cut samples of the aluminum tubing, which seemed more brittle than the same material on Earth, and some electrical cables. Their insulation seemed to have gotten dry, hard, and brittle. We managed to break off a piece of glass, and we unbolted the Surveyor TV camera. Then Al suggested that we cut off and take back the sampling scoop, and so we added that to the collection.

Then we headed back to the *Intrepid*. We retrieved the solar-wind experiment, stowed it and the sample bags in the *Intrepid*, got in, buttoned it up, and started repressurization. Altogether we brought back about 75 pounds of rocks, and 15 pounds of Surveyor hardware. We also brought back the 25-pound color TV camera from *Intrepid* so that its failure could be investigated.

While we were busy on the surface, Dick Gordon was busy in lunar orbit. The *Yankee Clipper* was a very sophisticated observation and surveying spacecraft. One of the experiments that Dick performed was multispectral photography of the lunar surface, which gave scientists new data with which to interpret the composition of the Moon.

After Al and I got back to *Yankee Clipper* following lunar liftoff and rendezvous, all three of us worked on the photography schedule. We were looking specifically for good coverage of proposed future landing sites, especially Fra Mauro, which was then scheduled for Apollo 13. That's a rough surface, and we wanted to get the highest-resolution photos we could so that the crew of the Apollo 13 mission would have the best training information they could get.

We changed the plane of our lunar orbit to cover the sites better, and we also elected to stay an extra day in lunar orbit so that we could complete the work without feeling pressured. We took hundreds of stills, and thousands of feet of motion-picture film of the Fra Mauro site, and of the Descartes and Lalande craters, two other proposed landing sites.

Meantime the experiments we had left on the lunar surface were busy recording and transmitting data. They all worked well, with one exception, and were really producing useful data. One unexpected result came from the seismic experiment recording the impact of *Intrepid* on the surface after we had jettisoned it. The entire Moon rang like a gong, vibrating and resonating for almost on hour after the impact. The best guess was that the Moon was composed of rubble a lot deeper below its surface than anybody had assumed. The internal structure, being fractured instead of a solid mass, could bounce the seismic energy from piece to piece for quite a while.

The same phenomenon was observed at two ALSEP stations when the Apollo 14 crew jettisoned their lunar module *Antares* and programmed it to crash between the Apollo 12 and 14 sites.

With every mission after Apollo 12, additional seismic calibrations were obtained by aiming the Saturn S-IVB stage to impact a selected point on the Moon after separation from the spacecraft. The seismic vibrations from these impacts lasted about three hours.

Apollo 13 was supposed to land in the Fra Mauro area. The explosion on board wiped out that mission, and it became instead a superb example of a crew's ability to turn a very risky situation into a safe return to Earth.

So the Fra Mauro site was reassigned to Apollo 14, because scientists gave that area a high priority. The following account is by Alan B. Shepard, Jr., the first American into space and one of the original seven astronauts.

The Fra Mauro hills stand a couple of hundred of miles to the east of the Apollo 12 landing site. I was selected to command this mission, my first since the original Mercury flight in 1961. With me to the lunar surface went Edgar D. Mitchell in *Antares,* while Stuart A. Roosa was the command module pilot of *Kitty Hawk.*

CHOOSING A SMOOTHER SPOT

The targeting data for the Apollo 14 landing site were every bit as good as the data for Apollo 12; but we had to fly around for a little while for the same reason they had to. The landing site was rougher, on direct observation, than the photos had been able to show. So I looked for a smoother area, found one, and landed there.

Our first EVA was similar to those before; we got out, set up the solar-wind experiment and the flag, and deployed the ALSEP. The latter had two new experiments. One was called the "thumper." Ed Mitchell set up an array of geophones, and then walked out along a planned survey line with a device that could be placed against the surface and fired, to create a local impact of known size. Thirteen of the 21 charges went off, registering good results. The other different experiment we had was a grenade

Color telecasts, live from the Apollo 14 site, came by way of the erectable S-band antenna shown here. The S-band of radio frequencies (between 1550 and 5200 megahertz) was used for high-data-rate space transmissions. The gold-colored parabolic reflector, which opened just like an umbrella, provided a higher gain than the lunar module's own steerable antenna. Note how featureless the lunar surface appears in the area just above the astronaut's shadow. This illustrates the visibility problem that the astronauts faced in walking down-Sun.

launcher, with four grenades to be fired off by radio command some time after we had left the Moon. They were designed to impact at different distances from the launcher, to get a pattern of seismic response to the impact explosions.

While Ed and I were working on our first EVA, Stu was doing the photographic part of the orbital science experiments. One job was to get detailed photographic coverage of the proposed site for the Apollo 15 mission, near the Descartes crater.

He was asked also to get a number of other photos of the lunar surface, in areas that had not been well-covered in earlier missions. Stu produced some great photos of the surface, rotating the command module *Kitty Hawk* to compensate for the motion of the image. He photographed the area around Lansberg B, which had been the predicted impact site of the Apollo 13 S–IVB stage. It was calculated that the impact could have produced a crater about 200 feet in diameter, and scientists wanted good pictures of the area so they could search for the brand-new crater on the Moon.

Stu also found them another new crater on the back side of the Moon. It was serendipity; he was shooting other pictures and suddenly this very bright, young crater came into view directly under *Kitty Hawk*. So he swung the camera around, pushed the button, and then went back to his original assignment.

A LUNAR RICKSHAW

Ed and I worked on the surface for 4 hours and 50 minutes during our first EVA; after the return to *Antares*, a long rest period, and then resuiting, we began the second EVA. This time we had the MET—modularized equipment transporter, although we called it the lunar rickshaw—to carry tools, cameras, and samples so we could work more effectively and bring back a larger quantity of samples.

Our planned traverse was to take us from *Antares* more or less due east to the rim of Cone crater. That traverse had been chosen because scientists wanted samples and rocks from the crater's rim. The theory is that the oldest rocks from deep under the Moon's surface were thrown up and out of the crater by the impact, and that the ones from the extreme depth of the crater were to be found on the rim.

On our way to the crater, one of the first things Ed did was to take a magnetometer reading at the first designated site. When he read the numbers over the air, there was some excitement back at Houston because the readings were about triple the values gotten on Apollo 12. They were also higher than the values Stu was reading in the *Kitty Hawk*, and so it seemed that the Moon's magnetic field varied spatially.

Our first sampling began a little further on, in a rock field with boulders about two or three feet along the major dimension. These were located in the centers of a group of three craters, each about sixty feet across. Like the bulk of the samples brought back, these were documented samples. That means photographing the soil or rocks, describing them and their position over the voice link to Mission Control, and then putting the sample in a numbered bag, identifying the bag at the same time on the voice hookup.

Apollo 14 tried an experiment to do something constructive about the dust that plagued all of the missions. NASA engineers wanted to check out some of the finishes proposed for the Rover and other pieces of operating equipment. I had a group of

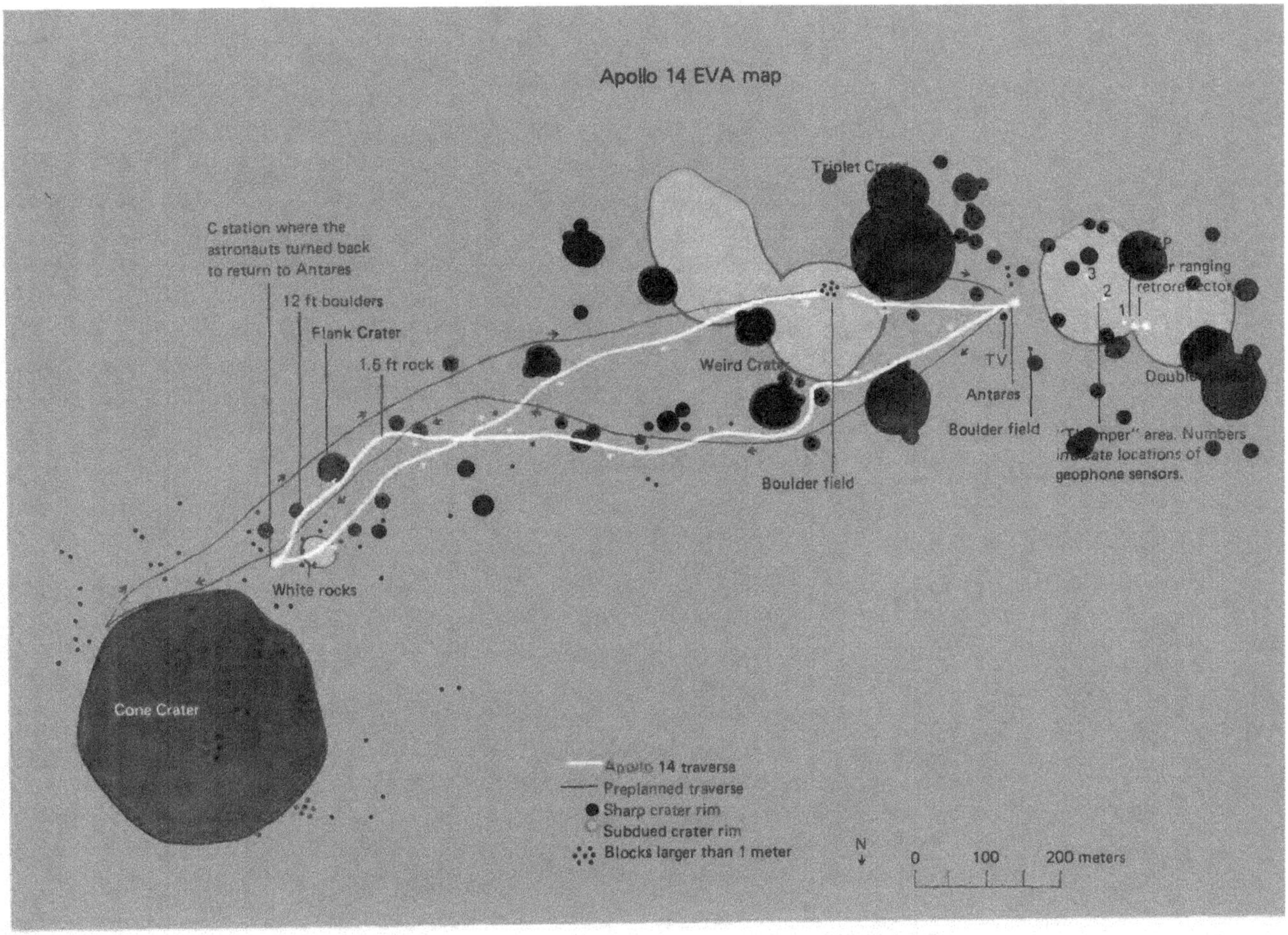

Apollo 14 EVA map

C station where the
astronauts turned back
to return to Antares

12 ft boulders

Flank Crater

1.5 ft rock

White rocks

Cone Crater

Triplet Crater

Weird Crater

Boulder field

Boulder field

TV

Antares

Boulder field

3

2

1

ALSEP ranging
retroreflector

Double Crater

"Thumper" area. Numbers
indicate locations of
geophone sensors.

Apollo 14 traverse
Preplanned traverse
● Sharp crater rim
○ Subdued crater rim
Blocks larger than 1 meter

N
↓

0   100   200 meters

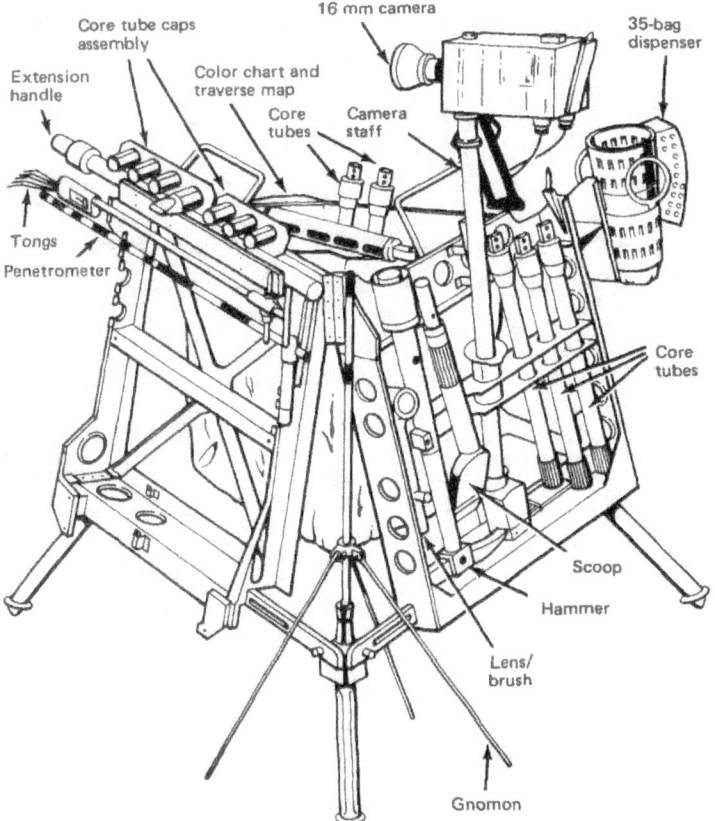

Extension
handle

Core tube caps
assembly

Color chart and
traverse map

Core
tubes

Tongs

Penetrometer

16 mm camera

Camera
staff

35-bag
dispenser

Core
tubes

Scoop

Hammer

Lens/
brush

Gnomon

The planned traverse route for the second EVA is shown by a fine black line on this map of the Apollo 14 site. The heavier white line is the traverse actually covered. The craters and boulders encountered are plotted, as are the locations of the emplaced experiments. Such maps are essential for an understanding of the sample sources and the experiment data.

The well-stocked tool rack at left, which fitted neatly on the rickshaw, was at least better than traipsing about carrying everything, including samples already collected. But it proved to be a drag in deep dust, easier to carry than to tow. The problem of doing on-the-spot lunar geologising in an efficient way awaited the electric Rover.

samples—material chips with different finishes—and I dusted them with the surface dust, shook them off, and then brushed one set to try to determine the abrasive effects, if any, of such dust removal. The other set was left unbrushed as a control sample. All this was of course recorded with the closeup camera.

The mapped traverse was to take us nearly directly to the rim of Cone crater, a feature about 1000 feet in diameter. As we approached, the boulders got larger, up to four and five feet in size. And at this time, the going started to get rough for us. The terrain became more steep as we approached the rim, and the increased grade accentuated the difficulty of walking in soft dust.

THE HUNT FOR THE RIM OF CONE

Another problem was that the ruggedness and unevenness of the terrain made it very hard to navigate by landmarks, which is the way a man on foot gets around. Ed and I had difficulty in agreeing on the way to Cone, just how far we had traveled, and where we were. We did some more sampling, and then moved on toward Cone, into terrain that had almost continuous undulations, and very small flat areas. Soon after that, the surface began to slope upward even more steeply, and it gave us the feeling that we were starting the last climb to the rim of Cone. We passed a rock which had a lot of glass in it, and reported to Houston that it was too big to pick up.

We continued, changing our suit cooling rates to match our increased work output as we climbed, and stopping a couple of times briefly to rest. For a while, we picked up the cart and carried it, preferring to move this way because it was a little faster.

And then came what had to be one of the most frustrating experiences on the traverse. We thought we were nearing the rim of Cone, only to find we were at another and much smaller crater still some distance from Cone. At that point, I radioed Houston that our positions were doubtful, and that there was probably quite a way to go yet to reach Cone.

About then, there was a general concurrence that maybe that was about as far as we should go, even though Ed protested that we really ought to press on and look into Cone crater. But in the end, we stopped our traverse short of the lip and turned for the walk back to *Antares*.

Later estimates indicated we were perhaps only 30 feet or so below the rim of the crater, and yet we were just not able to define it in that undulating and rough country.

One of the rocks we sampled in that area was a white breccia (a rock made up of pieces of stone embedded in a matrix). The white coloring came from the very high percentage of feldspar that was in the breccia. That rock, and others in the area, were believed to approach 4.6 billion years in age.

We stopped at Weird crater, for more sampling and some panoramic photography, and then continued the return traverse. At the Triplet craters, more than three-quarters of the way back to *Antares*, we stopped again. Ed's job there was to drive some core tubes; I was to dig a trench to check the stratification of the surface. But the core material was granular and slipped out of the tube every time Ed lifted it clear of the surface. I wasn't having any better luck with my trenching, because the side

Apollo 14 astronaut Shepard fits a core tube section to the extension handle in preparation for taking a vertical sample of the subsurface material. Core tubes were among the handtools carried on the MET.

Bean cautiously removes hot fuel capsule from its graphite cask in order to insert it into the Radioisotope Thermoelectric Generator (RTG) at his right. The temperature of the capsule, which was filled with plutonium-238, is about 1350° F.

The RTG was the powerhouse for the entire experiment package. The temperature difference between the fuel capsule and the finned outer housing was converted into electrical power by 442 lead telluride thermocouples. Starting at about 74 watts, the output to the central station will continue for years at a slowly diminishing rate.

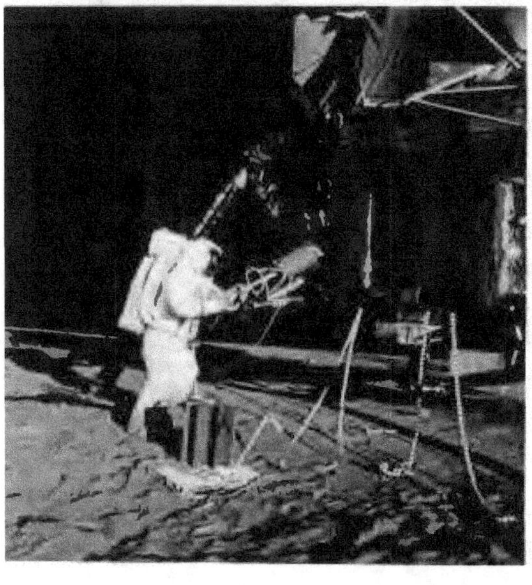

walls kept collapsing. I did get enough of a trench dug so that I could observe some stratification of the surface materials, seeing their color shift into the darker browns and near blacks, and then into a surprisingly light-colored layer underneath the darkest one.

That was it. *Antares* was in sight, as it had been throughout much of the traverse, and our long Moon walk was almost over. I went on past *Antares* to the ALSEP site to check antenna alignment because of reports from Houston that a weak signal was being received. Ed took some more samples from a nearby field of boulders.

At that, our surface tasks were done, with the exception of recovering the solar-wind experiment and getting back into *Antares* for the return flight. We had covered a

ALSEP: Scientific Station on the Moon

Although the Apollo astronauts could stay on the lunar surface for only a few days, scientists wished to make some kinds of observations over a period of weeks or even years, if possible. The solution was to have the astronauts set up an unmanned, automatic scientific station called ALSEP (Apollo Lunar Surface Experiments Package). An ALSEP was emplaced at each landing site, beginning with Apollo 12.

An ALSEP is a group of geophysical instruments arrayed about a central station, as in the accompanying sketch. Each ALSEP has a different set of experiments. Power is supplied by a Radio-isotope Thermoelectric Generator. Radio communication for the transmission of experiment data and the receipt of instrument adjustment commands is maintained through a rod-shaped antenna pointed in the Earth's general direction.

Each ALSEP can send about 9 million instrument readings a day. With five ALSEPs operating simultaneously, a staggering amount of information has already accumulated. An unexpected bonus has been the unusually long lifetimes of the ALSEP units. Originally designed for one year of reliable operation, all were still sending useful data five years after Apollo 12.

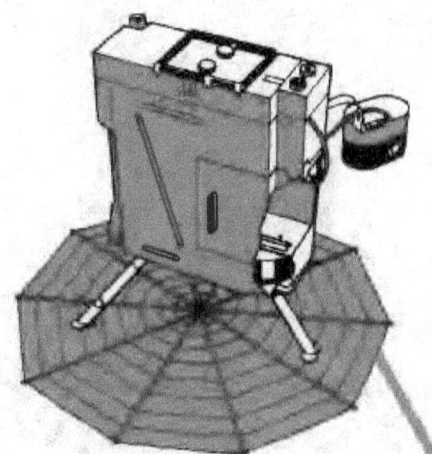

The **Suprathermal Ion Detector** Experiment (SIDE) measures the energy and mass of the positive ions that result from the ionization of gases near the lunar surface by the solar wind or ultraviolet radiation. The **Cold Cathode Gauge Experiment** (CCGE) measures changes in the extremely low concentrations of gas in the lunar atmosphere. The electronics for the CCGE are housed in the SIDE.

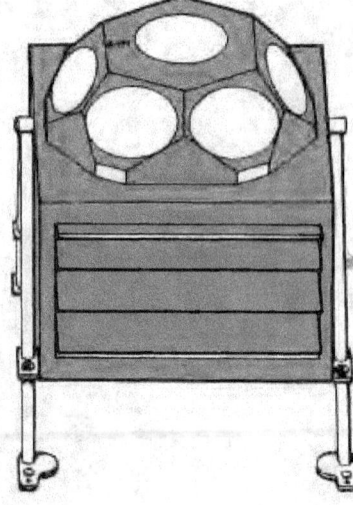

The **Solar-Wind Spectrometer** experiment uses seven Faraday-cup sensors to measure the energy spectra of charged particles that strike it from various directions. Because the Moon, unlike the Earth, is not protected from the solar-wind plasma by a magnetic shield, the instrument can detect subtle variations in the wind's intensity and direction.

distance of about two miles and collected many samples during four and one-half hours on the surface in the second EVA. I also threw a makeshift javelin, and hit a couple of golf shots.

After liftoff there were still experiments left to do. The first of these was another seismic event, generated by the impact of the jettisoned *Antares* on the Moon. Again the Moon responded with that resonant ringing for some time after the event. Once we were on our way back to Earth, we did a series of four experiments in weightlessness. One was a simple metal casting experiment, to see what the effects of zero gravity would be on the purity or the homogeneity of the mass. The materials included some pure samples, and others with crystals or fibers for strengthening. As you might expect,

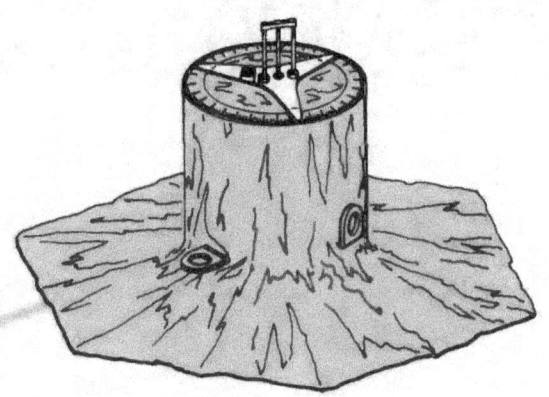

The Passive Seismic Experiment uses four extremely sensitive seismometers to measure lunar surface vibrations, free oscillations, and tidal variations in surface tilt. Three long-period seismometers are mounted orthogonally to measure wave motions with periods between 1/2 and 250 seconds, while the short-period seismometer measures vertical motions with periods between 1/20 and 20 seconds. The electronics are housed in the ALSEP central station. The thermal shroud isolates the sensor and a patch of ground 5 feet in diameter from the temperature extremes of the lunar day and night.

Lunar Surface Magnetometers, operating at three ALSEP stations, have simultaneously measured the global response of the Moon to fluctuations in large-scale solar and terrestrial magnetic fields. By considering these responses in conjunction with the free-space magnetic data from the lunar satellite, Explorer 35, scientists have estimated rock temperatures (which affect electrical conductivity) deep in the lunar interior.

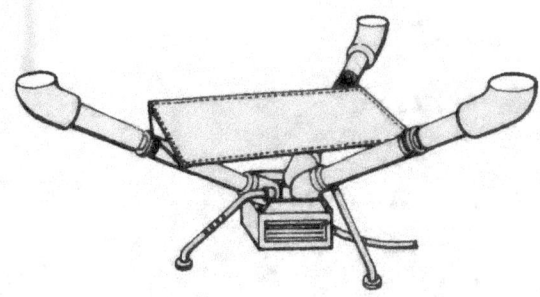

With its three gold-covered booms outspread, the Lunar Surface Magnetometer can measure the three orthogonal components of the magnetic field. Periodically, the fluxgate sensors at the ends of the booms are flipped over mechanically to check the calibration. An astronaut initially oriented the instrument by means of the shadowgraph shown at the base of the X-axis (right) boom and the bubble level on the sunshade.

The Laser Ranging Retroreflector (LRRR) is a completely passive array of small fused-silica corner cubes that reflect incident light precisely back toward its sources. When the source is a pulsed ruby laser at a large telescope, the distance from the LRRR to the ground station can be routinely measured within 6 inches. The three LRRR arrays on the Moon permit long-term studies of subtle Earth and Moon motions.

The top surface of the central station in an aluminum honeycomb sunshield. Before deployment, the antenna and several ALSEP experiments were attached to the brackets atop the sunshield with quick-release bolts. When raised, the sunshield and insulating side curtains provide thermal protection for the electronics. A leveling head on the antenna mast permitted the astronaut to aim the helical S-band antenna earthward.

A wavy golden ribbon connects the Apollo 14 Suprathermal Ion Detector and its accompanying Cold Cathode Gauge with the ALSEP central station some 50 feet away. This pair of instruments was also emplaced at the Apollo 12 and 16 sites. The wide range of the three Ion Detector look angles permits study of the directional characteristics of the flow of ions on both sides of the Earth's magnetospheric tail.

The Passive Seismic Experiment is completely hidden by its many-layered shroud of aluminized Mylar. The top of the thermal shroud is the platform for the bubble level and Sun compass that the astronaut used to orient the experiment initially. An internal set of leveling motors keeps the seismometers constantly level within a few seconds of arc. Seismic motions are recorded on Earth with a magnification factor of 10 million. The network created by the four ALSEPs that have this experiment enables seismologists to locate moonquakes in three dimensions, and to study the seismic velocities and propagation characteristics of subsurface materials.

Like any tourist in a strange place, Ed Mitchell consults a map on his way to Cone crater. He was photographed by his companion, Alan Shepard, during the second Apollo 14 EVA. During their 9 hours on the lunar surface, these tourists collected 95 pounds of lunar samples to bring home. Their main complaint during their stay was the way the lunar dust stuck to their suits almost up to their knees.

the materials turned out to be more homogeneous under zero-gravity conditions. We measured heat flow and convection in some samples and, sure enough, zero gravity changed those characteristics also. We did some electrophoretic separations, which are techniques used by the pharmaceutical industry to make vaccines, in the belief that maybe zero-gravity conditions could simplify a complex and expensive process. Finally, we did some fluid transfer experiments, simply trying to pour a fluid from one container to another in zero gravity. The surface tension works against you there, and so it was much easier when the containers being used were equipped with baffles that the fluid could cling to, as it were.

That was our mission. Our return was routine, our landing on target, and our homecoming as joyous as those before.

I look back now on the flights carrying Pete's crew and my crew as the real pioneering explorations of the Moon. Neil, Buzz, and Mike in Apollo 11 proved that man could get to the Moon and do useful scientific work, once he was there. Our two flights—Apollo 12 and 14—proved that scientists could select a target area and define a series of objectives, and that man could get there with precision and carry out the objectives with relative ease and a very high degree of success. And both of our flights, as did earlier and later missions, pointed up the advantage of manned space exploration. We all were able to make minor corrections or major changes at times when they were needed, sometimes for better efficiency, and sometimes to save the mission.

Apollo 12 and 14 were the transition missions. After us came the lunar rover, wheels to extend greatly the distance of the traverse and the quantity of samples that could be carried back to the lunar module. And on the last flight, a trained scientist who was also an astronaut went along on the mission.

I'd like to look on that last flight as just a temporary hold in the exploration of space.

The flag flutters on the Moon in the genuine wind of a rocket exhaust as the ascent stage of the Apollo 14 lunar module *Antares* lifts off from the Moon. Pieces of the gold-coated insulating foil torn off the descent stage by the blast were also sent flying. Who knows how many thousands of years will pass before a wind of vaporized rock from some nearby meteorite impact once more sets this flag flapping?

In the blackness of space, the Apollo 14 command-service module *Kitty Hawk* gleams brilliantly as it draws near the camera in the lunar module *Antares*. The single-orbit rendezvous procedure, used for the first time in lunar orbit on this mission, brought the two craft together in two hours. After crew transfer, *Antares* was guided to lunar impact at a point between the Apollo 12 and 14 sites. The resulting seismic signal, recorded by both instruments, lasted 1½ hours.

At its journey's end, the Apollo 14 command module splashes down into the sparkling South Pacific, some 900 miles south of Samoa. The parachutes collapse as they are freed of their load. On this occasion, the command module remained right side up in the water after landing. Like a kayak, a command module was just as stable in the water when it was upside-down (stable two). If it toppled over to an inverted position, as happened on other splashdowns, the crew could right it by means of inflatable airbags.

Astronauts Mitchell, Shepard, and Roosa, and a recovery team frogman wait aboard the raft *Lily Pad* for a helicopter pickup. With the hatch open, the command module was vulnerable to swamping, along with its priceless load of lunar samples and film, which is why frogmen routinely lashed an inflated flotation collar around a spacecraft.

The veritable pay dirt of the Apollo expeditions is the collection of lunar samples that is now available for the most detailed examination and analysis. Scientists have long been aware that our understanding of the nature and history of the solar system has been biased in unknown ways by the fact that all of the study material comes from one planet. Although meteorites are fascinating samples of the material of the solar system at large, there is never any direct evidence of the source of an individual meteorite. Now, within a few years, mankind has assembled the material of another world, recording where each piece came from and what was nearby. Here, scientists at the Lunar Receiving Laboratory work with an Apollo 14 sample in a sterile nitrogen atmosphere.

In Mission Control the Gold Team, directed by Gerald Griffin (seated, back of head to camera), prepares to take over from Black Team (Glynn Lunney, seated, in profile) during a critical period. Seven men with elbows on console are Deke Slayton, Joe Kerwin (Black CapCom), Vance Brand (Gold CapCom), Phil Shaffer (Gold FIDO), John Llewellyn (Black RETRO), Charles Deiterich (Gold RETRO), and Lawrence Canin (Black GNC). Standing at right is Chester Lee, Mission Director from NASA's Washington headquarters, and broad back at right belongs to Rocco Petrone, Apollo Program Director. Apollo 13 had two other "ground" teams, the White and the Maroon. All devised heroic measures to save the mission from disaster.

CHAPTER THIRTEEN

"Houston, We've Had a Problem"

By JAMES A. LOVELL

Since Apollo 13 many people have asked me, "Did you have suicide pills on board?" We didn't, and I never heard of such a thing in the eleven years I spent as an astronaut and NASA executive.

I did, of course, occasionally think of the possibility that the spacecraft explosion might maroon us in an enormous orbit about the Earth—a sort of perpetual monument to the space program. But Jack Swigert, Fred Haise, and I never talked about that fate during our perilous flight. I guess we were too busy struggling for survival.

Survive we did, but it was close. Our mission was a failure but I like to think it was a successful failure.

Apollo 13, scheduled to be the third lunar landing, was launched at 1313 Houston time on Saturday, April 11, 1970; I had never felt more confident. On my three previous missions, I had already logged 572 hours in space, beginning with Gemini 7, when Frank Borman and I stayed up 14 days—a record not equaled until Skylab.

Looking back, I realize I should have been alerted by several omens that occurred in the final stages of the Apollo 13 preparation. First, our command module pilot, Ken Mattingly, with whom Haise and I had trained for nearly two years, turned out to have no immunity to German measles (a minor disease the backup LM pilot, Charlie Duke, had inadvertently exposed us to). I argued to keep Ken, who was one of the most conscientious, hardest working of all the astronauts. In my argument to Dr. Paine, the NASA Administrator, I said, "Measles aren't that bad, and if Ken came down with them, it would be on the way home, which is a quiet part of the mission. From my experience as command module pilot on Apollo 8, I know Fred and I could bring the spacecraft home alone if we had to." Besides, I said, Ken doesn't have the measles now, and he may never get them. (Five years later, he still hadn't.)

Dr. Paine said no, the risk was too great. So I said in that case we'll be happy to accept Jack Swigert, the backup CMP, a good man (as indeed he proved to be, though he had only two days of prime-crew training).

The second omen came in ground tests before launch, which indicated the possibility of a poorly insulated supercritical helium tank in the LM's descent stage. So we modified the flight plan to enter the LM three hours early, in order to obtain an on-board readout of helium tank pressure. This proved to be lucky for us because it gave us a chance to shake down this odd-shaped spacecraft that was to hold our destiny in its spidery hands. It also meant the LM controllers were in Mission Control when they would be needed most.

Then there was the No. 2 oxygen tank, serial number 10024X-TA0009. This tank had been installed in the service module of Apollo 10, but was removed for modification (and was damaged in the process of removal). I have to congratulate Tom Stafford, John Young, and Gene Cernan, the lucky dogs, for getting rid of it.

This tank was fixed, tested at the factory, installed in our service module, and tested again during the Countdown Demonstration Test at the Kennedy Space Center beginning March 16, 1970. The tanks normally are emptied to about half full, and No. 1 behaved all right. But No. 2 dropped to only 92 percent of capacity. Gaseous oxygen at 80 psi was applied through the vent line to expel the liquid oxygen, but to no avail. An interim discrepancy report was written, and on March 27, two weeks before launch, detanking operations were resumed. No. 1 again emptied normally, but its idiot twin did not. After a conference with contractor and NASA personnel, the test director decided to "boil off" the remaining oxygen in No. 2 by using the electrical heater within the tank. The technique worked, but it took eight hours of 65-volt dc power from the ground-support equipment to dissipate the oxygen.

With the wisdom of hindsight, I should have said, "Hold it. Wait a second. I'm riding on this spacecraft. Just go out and replace that tank." But the truth is, I went along, and I must share the responsibility with many, many others for the $375 million failure of Apollo 13. On just about every spaceflight we have had some sort of failure, but in this case, it was an accumulation of human errors and technical anomalies that doomed Apollo 13.

At five and a half minutes after liftoff, Swigert, Haise, and I felt a little vibration. Then the center engine of the S-II stage shut down two minutes early. This caused the remaining four engines to burn 34 seconds longer than planned, and the S-IVB third stage had to burn nine seconds longer to put us in orbit. No problem: the S-IVB had plenty of fuel.

The first two days we ran into a couple of minor surprises, but generally Apollo 13 was looking like the smoothest flight of the program. At 46 hours 43 minutes Joe Kerwin, the CapCom on duty, said, "The spacecraft is in real good shape as far as we are concerned. We're bored to tears down here." It was the last time anyone would mention boredom for a long time.

At 55 hours 46 minutes, as we finished a 49-minute TV broadcast showing how comfortably we lived and worked in weightlessness, I pronounced the benediction: "This is the crew of Apollo 13 wishing everybody there a nice evening, and we're just about ready to close out our inspection of *Aquarius* (the LM) and get back for a pleasant evening in *Odyssey* (the CM). Good night."

On the tapes I sound mellow and benign, or some might say fat, dumb, and happy. A pleasant evening, indeed! Nine minutes later the roof fell in; rather, oxygen tank No. 2 blew up, causing No. 1 tank also to fail. We came to the slow conclusion that our normal supply of electricity, light, and water was lost, and we were about 200,000 miles from Earth. We did not even have power to gimbal the engine so we could begin an immediate return to Earth.

The message came in the form of a sharp bang and vibration. Jack Swigert saw a warning light that accompanied the bang, and said, "Houston, we've had a problem here." I came on and told the ground that it was a main B bus undervolt. The time was 2108 hours on April 13.

Next, the warning lights told us we had lost two of our three fuel cells, which were our prime source of electricity. Our first thoughts were ones of disappointment, since mission rules forbade a lunar landing with only one fuel cell.

With warning lights blinking on, I checked our situation; the quantity and pressure gages for the two oxygen tanks gave me cause for concern. One tank appeared to be completely empty, and there were indications that the oxygen in the second tank was rapidly being depleted. Were these just instrument malfunctions? I was soon to find out.

Thirteen minutes after the explosion, I happened to look out of the left-hand window, and saw the final evidence pointing toward potential catastrophe. "We are venting something out into the—into space," I reported to Houston. Jack Lousma, the CapCom replied, "Roger, we copy you venting." I said, "It's a gas of some sort."

It was a gas—oxygen—escaping at a high rate from our second, and last, oxygen tank. I am told that some amateur astronomers on top of a building in Houston could actually see the expanding sphere of gas around the spacecraft.

ARRANGING FOR SURVIVAL

The knot tightened in my stomach, and all regrets about not landing on the Moon vanished. Now it was strictly a case of survival.

The first thing we did, even before we discovered the oxygen leak, was to try to close the hatch between the CM and the LM. We reacted spontaneously, like submarine crews, closing the hatches to limit the amount of flooding. First Jack and then I tried to lock the reluctant hatch, but the stubborn lid wouldn't stay shut! Exasperated, and realizing that we didn't have a cabin leak, we strapped the hatch to the CM couch.

In retrospect, it was a good thing that we kept the tunnel open, because Fred and I would soon have to make a quick trip to the LM in our fight for survival. It is interesting to note that days later, just before we jettisoned the LM, when the hatch had to be closed and locked, Jack did it—easy as pie. That's the kind of flight it was.

The pressure in the No. 1 oxygen tank continued to drift downward; passing 300 psi, now heading toward 200 psi. Months later, after the accident investigation was complete, it was determined that, when No. 2 tank blew up, it either ruptured a line on the No. 1 tank, or caused one of the valves to leak. When the pressure reached 200 psi, it was obvious that we were going to lose all oxygen, which meant that the last fuel cell would also die.

At 1 hour and 29 seconds after the bang, Jack Lousma, then CapCom, said after instructions from Flight Director Glynn Lunney: "It is slowly going to zero, and we are starting to think about the LM lifeboat." Swigert replied, "That's what we have been thinking about too."

A lot has been written about using the LM as a lifeboat after the CM has become disabled. There are documents to prove that the lifeboat theory was discussed just before the Lunar Orbit Rendezvous mode was chosen in 1962. Other references go back to 1963, but by 1964 a study at the Manned Spacecraft Center concluded: "The LM [as lifeboat] . . . was finally dropped, because no single reasonable CSM failure

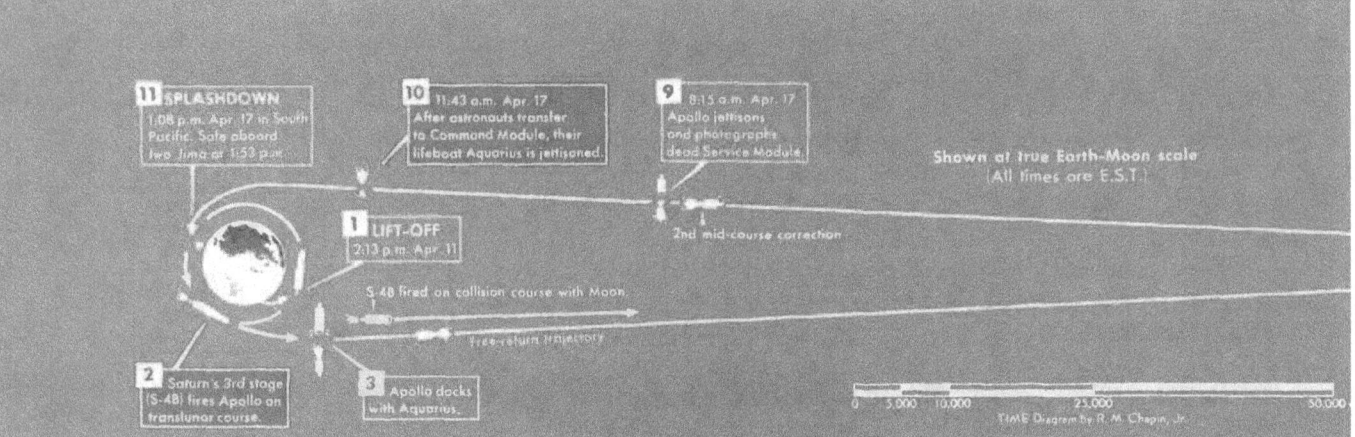

Detailed chronology from 2.5 minutes before

G.E.T.*

55:52:31 Master caution and warning triggered by low hydrogen pressure in tank No. 1.

55:52:58 CapCom (Charlie Duke): "13, we've got one more item for you, when you get a chance. We'd like you to stir up your cryo tanks. In addition, I have shaft and trunnion—"

55:53:06 Swigert: "Okay."

55:53:07 "—for looking at the Comet Bennett, if you need it."

55:53:12 Swigert: "Okay. Stand by."

55:53:18 Oxygen tank No. 1 fans on.

55:53:19 Oxygen tank No. 2 pressure decreases 8 psi.

55:53:20 Oxygen tank No. 2 fans turned on.

55:53:20 Stabilization control system electrical disturbance indicates a power transient.

55:53:21 Oxygen tank No. 2 pressure decreases 4 psi.

55:53:22.718 Stabilization control system electrical disturbance indicates a power transient.

55:53:22.757 1.2-volt decrease in ac bus 2 voltage.

55:53:22.772 11.1-amp rise in fuel cell 3 current for one sample.

* Ground Elapsed Time (hours, minutes, seconds)

55:53:36 Oxygen tank No. 2 pressure begins rise lasting for 24 seconds.

55:53:38.057 11-volt decrease in ac bus 2 voltage for one sample.

55:53:38.085 Stabilization control system electrical disturbance indicates a power transient.

55:53:41.172 22.9-amp rise in fuel cell 3 current for one sample.

55:53:41.192 Stabilization control system electrical disturbance indicates a power transient.

55:54:00 Oxygen tank No. 2 pressure rise ends at a pressure of 953.8 psia.

55:54:15 Oxygen tank No. 2 pressure begins to rise.

55:54:30 Oxygen tank No. 2 quantity drops from full scale for 2 seconds and then reads 75.3 percent.

55:54:31 Oxygen tank No. 2 temperature begins to rise rapidly.

55:54:43 Flow rate of oxygen to all three fuel cells begins to decrease.

55:54:45 Oxygen tank No. 2 pressure reaches maximum value of 1008.3 psia.

55:54:51 Oxygen tank No. 2 quantity jumps to off-scale high and then begins to drop until the time of telemetry loss, indicating

failed sensor.

55:54:52 Oxygen tank No. 2 temperature reads − 151.3° F.

55:54:52.703 Oxygen tank No. 2 temperature suddenly goes off-scale low, indicating failed sensor.

55:54:52.763 Last telemetered pressure from oxygen tank No. 2 before telemetry loss is 995.7 psia.

55:54:53.182 Sudden accelerometer activity on X, Y, Z axes.

55:54:53.220 Stabilization control system rate changes begin.

55:54:53.323 Oxygen tank No. 1 pressure drops 4.2 psi.

55:54:53.500 2.8-amp rise in total fuel cell current.

55:54:53.542 X, Y, and Z accelerations in CM indicate 1.17g, 0.65g, and 0.65g.

55:54:53.555 Loss of telemetry.

55:54:53.555+ Master caution and warning triggered by dc main bus B undervoltage. Alarm is turned off in 6 seconds. All indications are that the cryogenic oxygen tank No. 2 lost pressure in this time period and the panel separated.

55:54:54.741 Nitrogen pressure in fuel cell 1 is off-scale low indicating failed sensor.

55:54:55.350 Telemetry recovered.

could be identified that would prohibit use of the SPS." Naturally, I'm glad that view didn't prevail, and I'm thankful that by the time of Apollo 10, the first lunar mission carrying the LM, the LM as a lifeboat was again being discussed. Fred Haise, fortunately, held the reputation as the top astronaut expert on the LM—after spending fourteen months at the Grumman plant on Long Island, where the LM was built.

Fred says: "I never heard of the LM being used in the sense that we used it. We had procedures, and we had trained to use it as a backup propulsion device, the rationale being that the thing we were really covering was the failure of the command

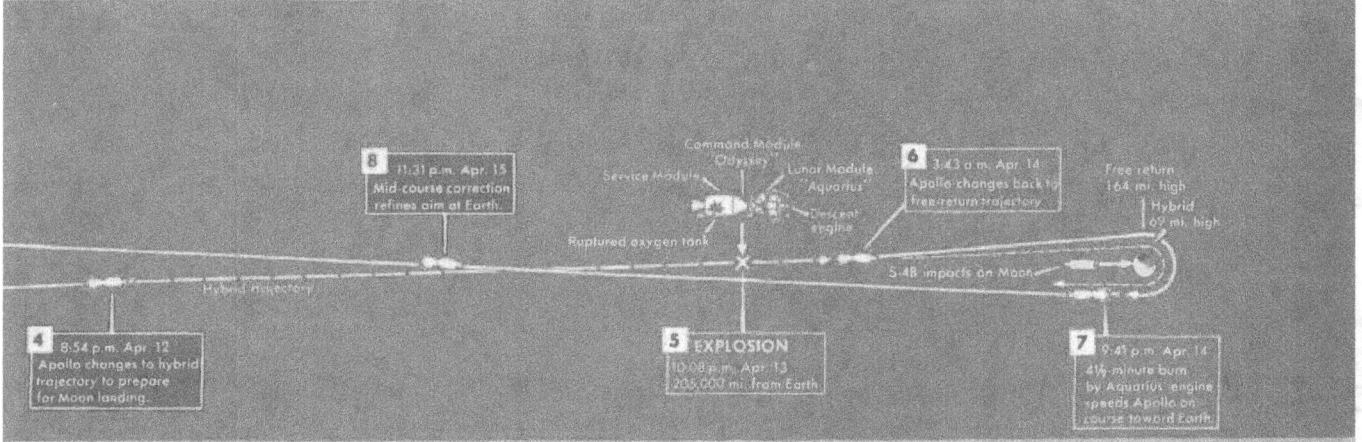

the accident to about 5 minutes after

55:54:56 Service propulsion system engine valve body temperature begins a rise of 1.65° F in 7 seconds. Dc main bus A decreases 0.9 volt to 28.5 volts and dc main bus B decreases 0.9 volt to 29.0 volts. Total fuel cell current is 15 amps higher than the final value before telemetry loss. High current continues for 19 seconds. Oxygen tank No. 2 temperature reads off-scale high after telemetry recovery, probably indicating failed sensors. Oxygen tank No. 2 pressure reads off-scale low following telemetry recovery, indicating a broken supply line, a tank pressure below 19 psi, or a failed sensor. Oxygen tank No. 1 pressure reads 781.9 psia and begins to drop.

55:54:57 Oxygen tank No. 2 quantity reads off-scale high following telemetry recovery indicating failed sensor.

55:55:01 Oxygen flow rates to fuel cells 1 and 3 approached zero after decreasing for 7 seconds.

55:55:02 The surface temperature of the service module oxidizer tank in bay 3 begins a 3.8° F increase in a 15-second period. The service propulsion system helium tank temperature begins a 3.8° F increase in a 32-second period.

55:55:09 Dc main bus A voltage recovers to

29.0 volts; dc main bus B recovers to 28.8.

55:55:20 Swigert: "Okay, Houston, we've had a problem here."

55:55:28 Duke: "This is Houston. Say again, please."

55:55:35 Lovell: "Houston, we've had a problem. We've had a main B bus undervolt."

55:55:42 Duke: "Roger. Main B undervolt."

55:55:49 Oxygen tank No. 2 temperature begins steady drop lasting 59 seconds, indicating failed senor.

55:56:10 Haise: "Okay. Right now, Houston, the voltage is—is looking good. And we had a pretty large bang associated with the caution and warning there. And as I recall, main B was the one that had an amp spike on it once before."

55:56:30 Duke: "Roger, Fred."

55:56:38 Oxygen tank No. 2 quantity becomes erratic for 69 seconds before assuming an off-scale-low state, indicating failed sensor.

55:56:54 Haise: "In the interim here, we're starting to go ahead and button up the tunnel again."

55:57:04 Haise: "That jolt must have rocked the sensor on—see now—oxygen quantity 2. It was oscillating down around 20 to 60 percent. Now it's full-scale high."

55:57:39 Master caution and warning triggered

by dc main bus B undervoltage. Alarm is turned off in 6 seconds.

55:57:40 Dc main bus B drops below 26.25 volts and continues to fall rapidly.

55:57:44 Lovell: "Okay. And we're looking at our service module RCS helium 1. We have —B is barber poled and D is barber poled, helium 2, D is barber pole, and secondary propellants, I have A and C barber pole." Ac bus fails within 2 seconds.

55:57:45 Fuel cell 3 fails.

55:57:59 Fuel cell current begins to decrease.

55:58:02 Master caution and warning caused by ac bus 2 being reset.

55:58:06 Master caution and warning triggered by dc main bus A undervoltage.

55:58:07 Dc main bus A drops below 26.25 volts and in the next few seconds levels off at 25.5 volts.

55:58:07 Haise: "Ac 2 is showing zip."

55:58:25 Haise: "Yes, we got a main bus A undervolt now, too, showing. It's reading about 25½. Main B is reading zip right now."

56:00:06 Master caution and warning triggered by high hydrogen flow rate to fuel cell 2.

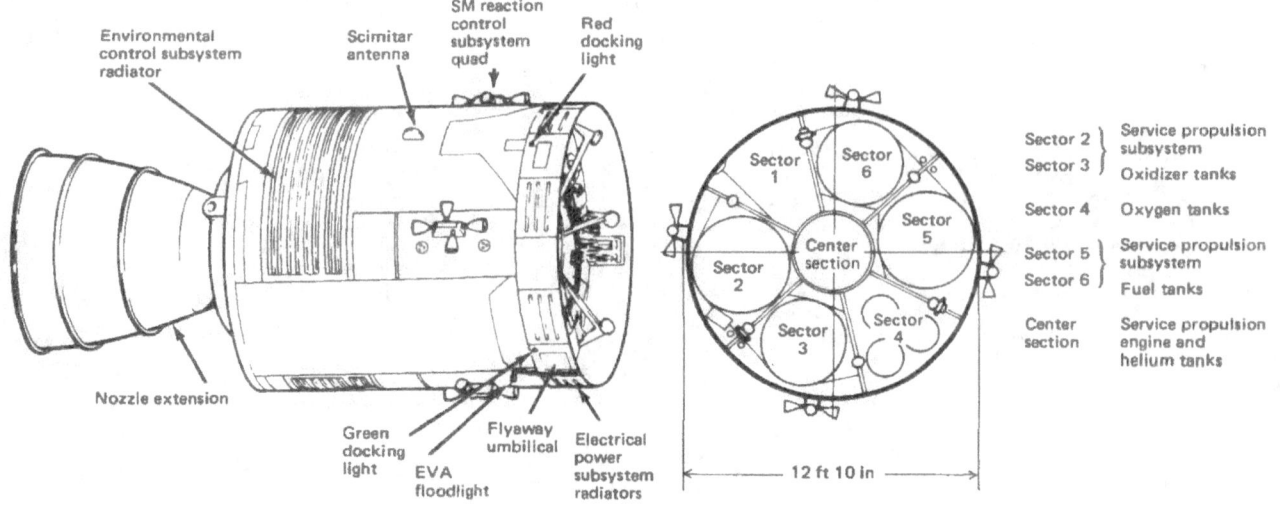

"There's one whole side of that spacecraft missing," said Lovell in astonishment. About five hours before splashdown the service module was jettisoned in a manner that would permit the astronauts to assess its condition. Until then, nobody realized the extent of the damage.

Vital stores of oxygen, water, propellant, and power were lost when the side of the service module blew off. The astronauts quickly moved into the lunar module which had been provided with independent supplies of these space necessities for the landing on the Moon. Years before, Apollo engineers had talked of using the lunar module as a lifeboat.

Environmental control subsystem radiator   Scimitar antenna   SM reaction control subsystem quad   Red docking light

Nozzle extension

Green docking light   EVA floodlight   Flyaway umbilical   Electrical power subsystem radiators

Sector 1   Sector 6   Center section   Sector 5   Sector 2   Sector 3   Sector 4

12 ft 10 in

Sector 2 } Sector 3 } Service propulsion subsystem Oxidizer tanks

Sector 4   Oxygen tanks

Sector 5 } Sector 6 } Service propulsion subsystem Fuel tanks

Center section   Service propulsion engine and helium tanks

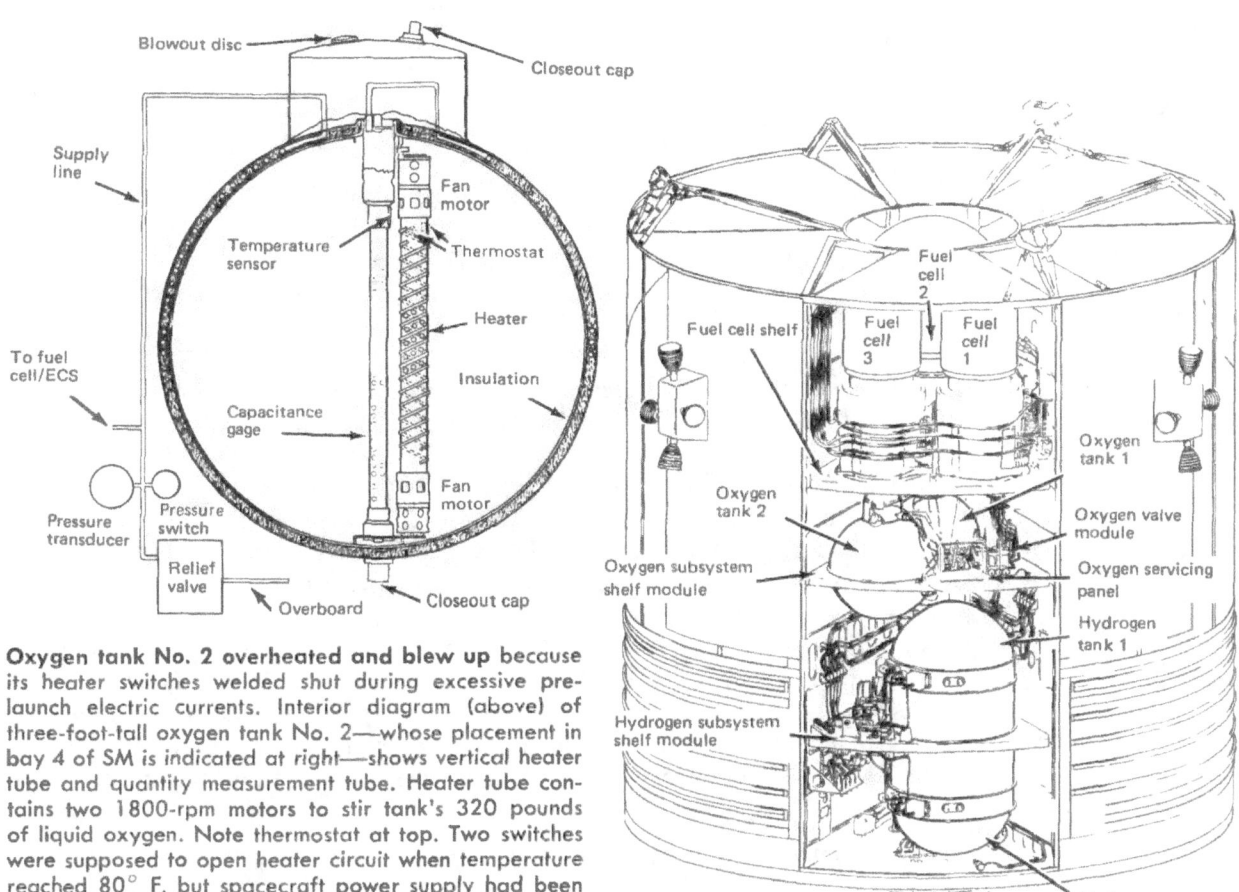

Oxygen tank No. 2 overheated and blew up because its heater switches welded shut during excessive pre-launch electric currents. Interior diagram (above) of three-foot-tall oxygen tank No. 2—whose placement in bay 4 of SM is indicated at right—shows vertical heater tube and quantity measurement tube. Heater tube contains two 1800-rpm motors to stir tank's 320 pounds of liquid oxygen. Note thermostat at top. Two switches were supposed to open heater circuit when temperature reached 80° F, but spacecraft power supply had been changed from 28 to 65 Vdc—while contractors and NASA test teams nodded—so switches welded shut and heater tube temperature probably reached 1000° F.

Top of Apollo 13's fuel tank No. 2 (bottom part is below shelf), photographed before it left North American Rockwell plant. Tank was originally installed in Apollo 10's SM, but was removed for modification and in process was dropped two inches (skin of tank is only 0.02 inch thick). Then it was installed on Apollo 13 and certified, despite test anomalies. In raging heat, it burst and the explosion was ruinous to the SM.

Nestled amid crinkled metal foil used for thermal insulation, oxygen tank No. 2 was mounted above and close to a pair of hydrogen tanks in spacecraft bay. (Diagram above and photo below.)

module's main engine, the SPS engine. In that case, we would have used combinations of the LM descent engine, and in some cases, for some lunar aborts, the ascent engine as well. But we never really thought and planned, and obviously, we didn't have the procedures to cover a case where the command module would end up fully powered down."

To get Apollo 13 home would require a lot of innovation. Most of the material written about our mission describes the ground-based activities, and I certainly agree that without the splendid people in Mission Control, and their backups, we'd still be up there.

They faced a formidable task. Completely new procedures had to be written and tested in the simulator before being passed up to us. The navigation problem was also theirs; essentially how, when, and in what attitude to burn the LM descent engine to provide a quick return home. They were always aware of our safety, as exemplified by the jury-rig fix of our environmental system to reduce the carbon dioxide level.

However, I would be remiss not to state that it really was the teamwork between the ground and flight crew that resulted in a successful return. I was blessed with two shipmates who were very knowledgeable about their spacecraft systems, and the disabled service module forced me to relearn quickly how to control spacecraft attitude from the LM, a task that became more difficult when we turned off the attitude indicator.

FIFTEEN MINUTES OF POWER LEFT

With only 15 minutes of power left in the CM, CapCom told us to make our way into the LM. Fred and I quickly floated through the tunnel, leaving Jack to perform the last chores in our forlorn and pitiful CM that had seemed such a happy home less than two hours earlier. Fred said something that strikes me as funny as I read it now: "Didn't think I'd be back so soon." But nothing seemed funny in real time on that 13th of April, 1970.

There were many, many things to do. In the first place, did we have enough consumables to get home? Fred started calculating, keeping in mind that the LM was built for only a 45-hour lifetime, and we had to stretch that to 90. He had some data from previous LMs in his book—average rates of water usage related to amperage level, rate of water needed for cooling. It turned out that we had enough oxygen. The full LM descent tank alone would suffice, and in addition, there were two ascent-engine oxygen tanks, and two backpacks whose oxygen supply would never be used on the lunar surface. Two emergency bottles on top of those packs had six or seven pounds each in them. (At LM jettison, just before reentry, 28.5 pounds of oxygen remained, more than half of what we started with.)

We had 2181 ampere hours in the LM batteries. We thought that was enough if we turned off every electrical power device not absolutely necessary. We could not count on the precious CM batteries, because they would be needed for reentry after the LM was cast off. In fact, the ground carefully worked out a procedure where we charged the CM batteries with LM power. As it turned out, we reduced our energy consumption to a fifth of normal, which resulted in our having 20 percent of our LM electrical

Blast-gutted service module was set adrift from the combined command module and lunar module just four hours before Earth re-entry. Mission Control had insisted on towing the wrecked service module for 300,000 miles because its bulk protected the command module's heat shield from the intense cold of space. The astronauts next revived the long-dormant command module and prepared to leave their lunar module lifeboat.

The jettisoning of elements during the critical last hours of the Apollo 13 mission is shown in this sequence drawing. When the lifesaving LM was shoved off by tunnel pressure about an hour before splashdown, everyone felt a surge of sentiment as the magnificent craft peeled away. Its maker, Grumman, later jokingly sent a bill for more than $400,000 to North American Rockwell for "towing" the CSM 300,000 miles.

Carbon dioxide would poison the astronauts unless scrubbed from the lunar module atmosphere by lithium hydride canisters. But the lunar module had only enough lithium hydride for 4 man-days—plenty for the lunar landing but not the 12 man-day's worth needed now. Here Deke Slayton (center) explains a possible fix to (left to right) Sjoberg, Kraft, and Gilruth. At left is Flight Director Glynn Lunney.

"Backroom" experts at Mission Control worked many hours to devise the fix that possibly kept the astronauts from dying of carbon dioxide. CapCom Joe Kerwin led astronaut Swigert, step by step, for an hour to build a contraption like the one the experts had constructed on Earth. It involved stripping the hose from a lunar suit and rigging the hose to the taped-over CM double canister, using the suit's fan to draw carbon dioxide from the cabin through the canister and expel it back into the LM as pure oxygen.

Emergency scrubbers were built by Swigert and Haise. Photo at right shows Swigert sitting next to a taped-over double canister and holding one end of a suit nozzle while Fred Haise used both his hands to manipulate the long hose. Underneath the canister is a "mailbox" built of arched cardboard, which was covered by a plastic bag. When Lovell later saw the rig put together on Earth, he said, "it looks just like the one we made." It saved the astronauts from possible death from carbon dioxide.

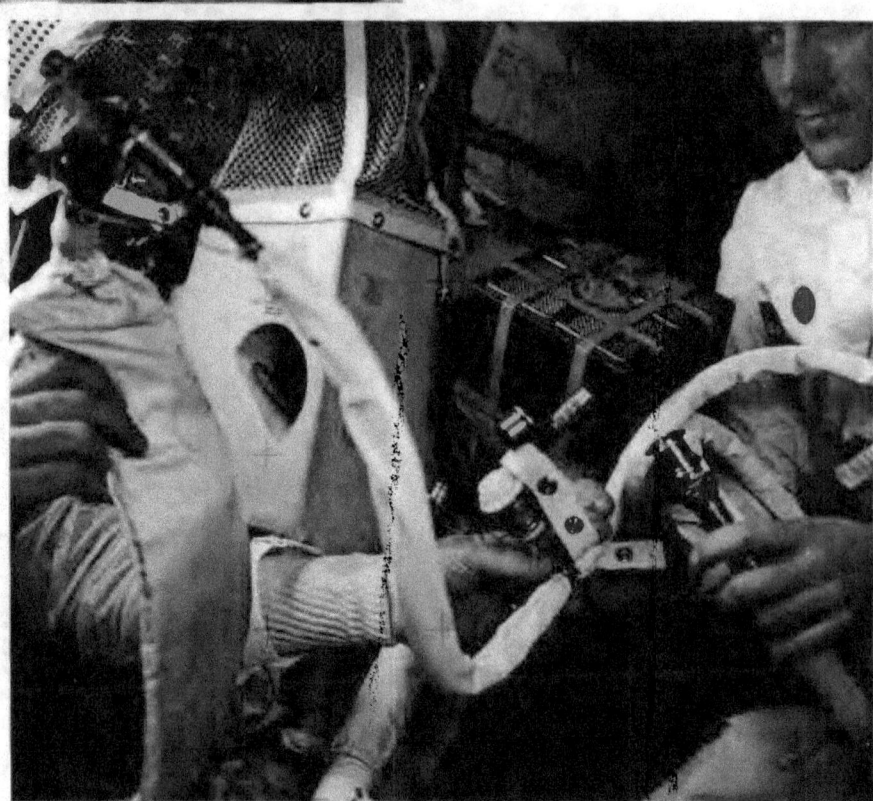

power left when we jettisoned *Aquarius*. We did have one electrical heart-stopper during the mission. One of the CM batteries vented with such force that it momentarily dropped off the line. We knew we were finished if we permanently lost that battery.

Water was the real problem. Fred figured that we would run out of water about five hours before we got back to Earth, which was calculated at around 151 hours. But even there, Fred had an ace in the hole. He knew we had a data point from Apollo 11, which had not sent its LM ascent stage crashing into the Moon, as subsequent missions did. An engineering test on this vehicle showed that its mechanisms could survive seven or eight hours in space without water cooling, until the guidance system rebelled at this enforced toasting. But we did conserve water. We cut down to six ounces each per day, a fifth of normal intake, and used fruit juices; we ate hot dogs and other wet-pack foods when we ate at all. (We lost hot water with the accident and dehydratable food is not palatable with cold water.) Somehow, one doesn't get very thirsty in space, and we became quite dehydrated. I set one record that stood up throughout Apollo: I lost fourteen pounds, and our crew set another by losing a total of 31.5 pounds, nearly 50 percent more than any other crew. Those stringent measures resulted in our finishing with 28.2 pounds of water, about 9 percent of the total.

Fred had figured that we had enough lithium hydroxide canisters, which remove carbon dioxide from the spacecraft. There were four cartridges from the LM, and four from the backpacks, counting backups. But he forgot that there would be three of us in the LM instead of the normal two. The LM was designed to support two men for two days. Now it was being asked to care for three men nearly four days.

A SQUARE PEG IN A ROUND HOLE

We would have died of the exhaust from our own lungs if Mission Control hadn't come up with a marvelous fix. The trouble was the square lithium hydroxide canisters from the CM would not fit the round openings of those in the LM environmental system. After a day and a half in the LM a warning light showed us that the carbon dioxide had built up to a dangerous level, but the ground was ready. They had thought up a way to attach a CM canister to the LM system by using plastic bags, cardboard, and tape—all materials we had on board. Jack and I put it together: just like building a model airplane. The contraption wasn't very handsome, but it worked. It was a great improvisation—and a fine example of cooperation between ground and space.

The big question was, "How do we get back safely to Earth?" The LM navigation system wasn't designed to help us in this situation. Before the explosion, at 30 hours and 40 minutes, we had made the normal midcourse correction, which would take us out of a free-return-to-Earth trajectory and put us on our lunar landing course. Now we had to get back on that free-return course. The ground-computed 35-second burn, by an engine designed to land us on the Moon, accomplished that objective 5 hours after the explosion.

As we approached the Moon, the ground informed us that we would have to use the LM descent engine a second time; this time a long 5-minute burn to speed up our return home. The maneuver was to take place 2 hours after rounding the far side of

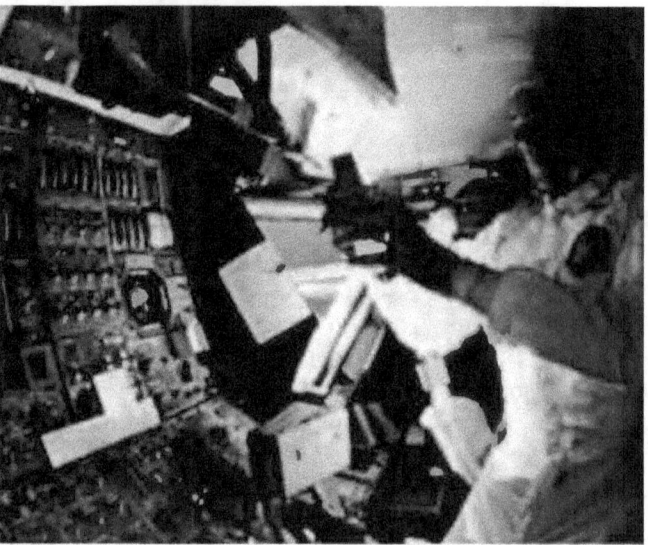

Without a course correction, the returning spacecraft would have missed the Earth completely. Inertial references were stored in the guidance platform, but sightings on the Sun gave the astronauts confidence that the crucial burn would be properly oriented.

In the darkened, power-short spacecraft, temperature dropped to 38° F. Lovell and Haise donned their lunar boots, Swigert an extra suit of underwear. Water condensed on the cabin walls and windows frosted. Food was refrigerator-cold.

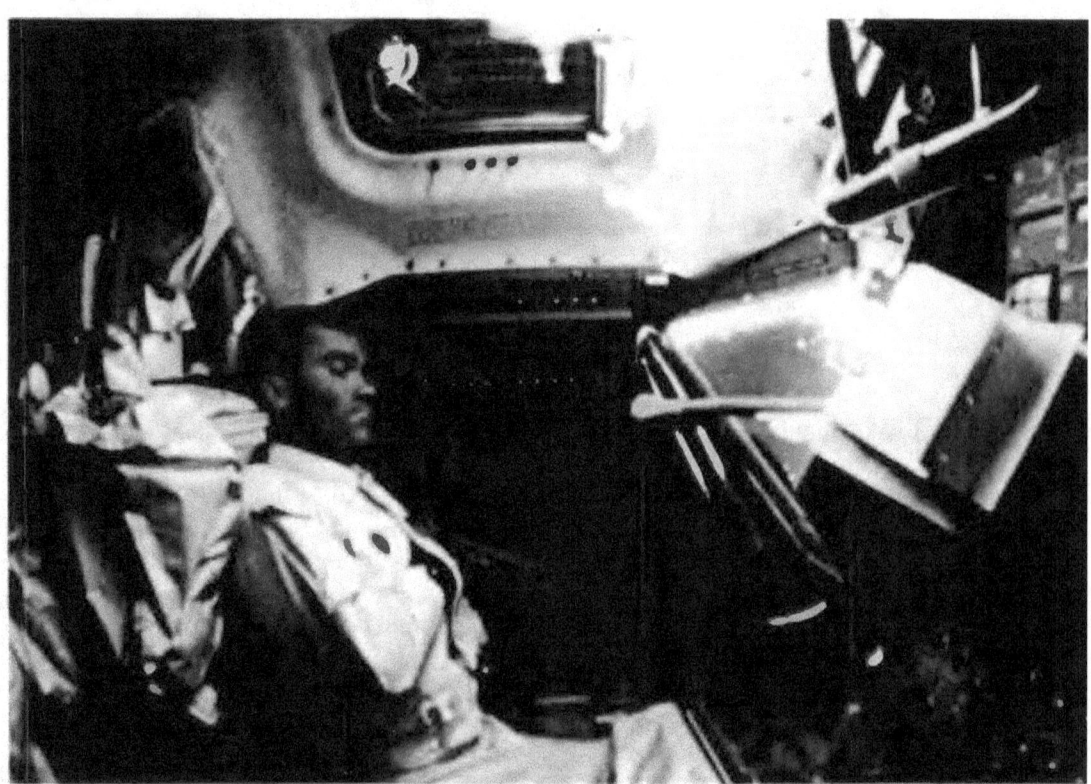

Haise dozes with his hands restrained to keep his weightless arms from flopping about. In the cold gloom, one crew member stayed on watch at all times. Stay-awake pills helped during the final hours as the Earth came whistling at them like a freight train.

the Moon, and I was busy running down the procedures we were to use. Suddenly, I noticed that Swigert and Haise had their cameras out and were busy photographing the lunar surface. I looked at them incredulously and said, "If we don't make this next maneuver correctly, you won't get your pictures developed!" They said, "Well, you've been here before and we haven't." Actually, some of the pictures these tourists took turned out to be very useful.

It was about this time that I said, "Boys, take a good look at the Moon. It's going to be a long time before anybody gets up here again." Later on I was accused of sabotaging Apollo; poor Dr. Paine had to explain that I didn't really mean it, and the space program would go on. The Senate Space Committee asked me about it a week after we got back. Actually, I didn't mean that remark to be public. (I later learned that, unknown to us, we had had a hot mike for about 45 minutes.) Nonetheless, it was 9 months before Apollo 14 was launched.

We had many crises on Apollo 13, but the biggest heart-stopper has hardly been noticed, partly because the transcription released to the press was garbled, and partly because there wasn't much point in talking about a crisis that had been averted earlier. It occurred prior to the second maneuver I mentioned earlier; we called it P.C. + 2 (pericynthian · 2 hours).

We had transferred the CM platform alignment to the LM, but we had to make sure that this alignment was accurate before we made the long P.C. + 2 burn. Ordinarily it is simple to look through the sextant device, called the Alignment Optical Telescope, find a suitable navigation star, and with the help of our computer verify the guidance platform's alignment. But traveling with us was a swarm of debris from the ruptured service module. The sunlight glinting on these bits of junk—I called them false stars—made it impossible to sight a real star.

So what to do? If we couldn't verify the accuracy of the alignment, we didn't have a way to make an accurate burn, or to align the CM platform for reentry. In other words, the ground would have no accurate way to tell us the correct attitude to make the proper maneuvers to return home.

A genius in Mission Control came up with the idea of using the Sun to check the accuracy of our alignment. No amount of debris could blot out that star! Its large diameter could result in considerable error, but nobody had a better plan.

I rotated the spacecraft to the attitude Houston had requested. If our alignment was accurate, the Sun would be centered in the sextant.

When I looked through the AOT, the Sun just had to be there. It really had to be. And it was. At 73:46 hours the air-to-ground transcript sounds like a song from "My Fair Lady":

Lovell: O.K. We got it. I think we got it. What diameter was it?

Haise: Yes. It's coming back in. Just a second.

Lovell: Yes, yaw's coming back in. Just about it.

Haise: Yaw is in. . . .

Lovell: What have you got?

Haise: Upper right corner of the Sun. . . .

Lovell: We've got it!

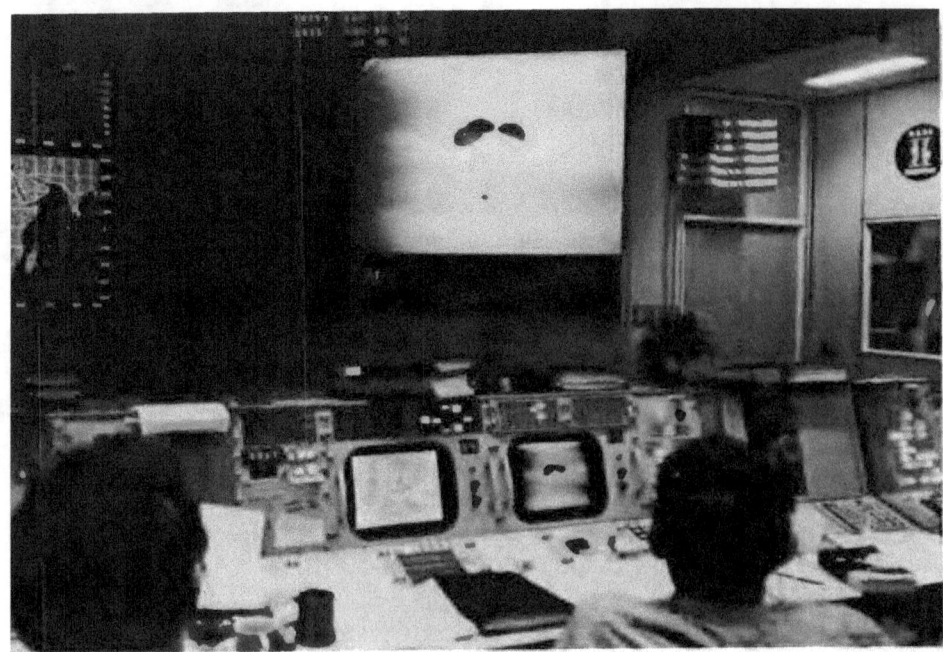

A beautiful sight! Two flight controllers at the Mission Control Center watch the parachute deployment as *Odyssey* floats down toward a gentle landing in the Pacific near American Samoa. Splashdown, at 1:07 p.m. EST, brought down the curtain on the most harried and critical emergency of the entire manned space program.

The charred command module splashed down less than four miles from the recovery ship USS *Iwo Jima*. Three very tired, hungry, cold, dehydrated astronauts await a ride up into the recovery helicopter. They were aboard the recovery ship less than one hour after touching down in the Pacific.

Haise, Lovell, and Swigert step off the recovery helicopter to the *Iwo Jima* in the South Seas. The crew lost a total of 31.5 pounds; Lovell alone 14 pounds—records in both cases. Dehydrated and exhausted, Haise was invalided three weeks by infection.

In Honolulu Lovell is joyously united with wife Marilyn after she and Mary Haise and bachelor Swigert's parents had flown from Houston with President Nixon. During the Apollo 8 mission sixteen months earlier, Lovell had nicknamed a crater on Moon "Mount Marilyn."

Fred and Mary Haise draped with Hawaiian leis. A physician had accompanied pregnant Mary Haise to Honolulu in the event Air Force One should have its first airborne birth. However, the Haise's fourth child Thomas did not arrive until ten weeks later.

"We didn't realize the complete magnitude of this flight," said Lovell, "until we got back home and started reading about it." *Christian Science Monitor* said: "Never in recorded history has a journey of such peril been watched and waited-out by almost the entire human race."

In Houston the 37th President pays tribute to the men who performed miracles in Mission Control to save Apollo 13. To President's right are NASA Administrator Thomas Paine and Mrs. Nixon. To his left: Flight directors Eugene Kranz, Gerald Griffin, Milton Windler (the fourth, Glynn Lunney, is behind lectern), then Chief Astronaut Donald K. Slayton, James A. Lovell III (in uniform), and Sigurd Sjoberg, Director of Flight Operations, who in behalf of "the ground" received the Nation's highest award, Medal of Freedom.

If we raised our voices, I submit it was justified.

I'm told the cheer of the year went up in Mission Control. Flight Director Gerald Griffin, a man not easily shaken, recalls: "Some years later I went back to the log and looked up that mission. My writing was almost illegible I was so damned nervous. And I remember the exhilaration running through me: My God, that's kinda the last hurdle—if we can do that, I know we can make it. It was funny, because only the people involved knew how important it was to have that platform properly aligned." Yet Gerry Griffin barely mentioned the alignment in his change-of-shift briefing— "That check turned out real well" is all he said an hour after his penmanship failed him. Neither did we, as crew members, refer to it as a crisis in our press conference nor in later articles.

The alignment with the Sun proved to be less than a half a degree off. Hallelujah! Now we knew we could do the 5-minute P.C. $+ 2$ burn with assurance, and that would cut the total time of our voyage to about 142 hours. We weren't exactly home free: we had a dead service module, a command module with no power, and a lunar module that was a wonderful vehicle to travel home in, but unfortunately didn't have a heat shield required to enter the Earth's atmosphere. But all we needed now was a continuation of the expertise we seemed blessed with, plus a little luck.

TIRED, HUNGRY, WET, COLD, DEHYDRATED

The trip was marked by discomfort beyond the lack of food and water. Sleep was almost impossible because of the cold. When we turned off the electrical systems, we lost our source of heat, and the Sun streaming in the windows didn't much help. We were as cold as frogs in a frozen pool, especially Jack Swigert, who got his feet wet and didn't have lunar overshoes. It wasn't simply that the temperature dropped to 38° F: the sight of perspiring walls and wet windows made it seem even colder. We considered putting on our spacesuits, but they would have been bulky and too sweaty. Our teflon-coated inflight coveralls were cold to the touch, and how we longed for some good old thermal underwear.

The ground, anxious not to disturb our homeward trajectory, told us not to dump any waste material overboard. What to do with urine taxed our ingenuity. There were three bags in the command module; we found six little ones in the LM, then we connected a PLSS condensate tank to a long hose, and finally we used two large bags designed to drain remaining water out of the PLSS's after the first lunar EVA. I'm glad we got home when we did, because we were just about out of ideas for stowage.

A most remarkable achievement of Mission Control was quickly developing procedures for powering up the CM after its long cold sleep. They wrote the documents for this innovation in three days, instead of the usual three months. We found the CM a cold, clammy tin can when we started to power up. The walls, ceiling, floor, wire harnesses, and panels were all covered with droplets of water. We suspected conditions were the same behind the panels. The chances of short circuits caused us apprehension, to say the least. But thanks to the safeguards built into the command module after the disastrous fire in January 1967, no arcing took place. The droplets furnished one sensation as we decelerated in the atmosphere: it rained inside the CM.

Four hours before landing, we shed the service module; Mission Control had insisted on retaining it until then because everyone feared what the cold of space might do to the unsheltered CM heat shield. I'm glad we weren't able to see the SM earlier. With one whole panel missing, and wreckage hanging out, it was a sorry mess as it drifted away.

Three hours later we parted with faithful *Aquarius*, rather rudely, because we blasted it loose with pressure in the tunnel in order to make sure it completely cleared. Then we splashed down gently in the Pacific Ocean near Samoa, a beautiful landing in a blue-ink ocean on a lovely, lovely planet.

Nobody believes me, but during this six-day odyssey we had no idea what an impression Apollo 13 made on the people of Earth. We never dreamed a billion people were following us on television and radio, and reading about us in banner headlines of every newspaper published. We still missed the point on board the carrier *Iwo Jima*, which picked us up, because the sailors had been as remote from the media as we were. Only when we reached Honolulu did we comprehend our impact: there we found President Nixon and Dr. Paine to meet us, along with my wife Marilyn, Fred's wife Mary (who being pregnant, also had a doctor along just in case), and bachelor Jack's parents, in lieu of his usual airline stewardesses.

In case you are wondering about the cause of it all, I refer you to the report of the Apollo 13 Review Board, issued after an intensive investigation. In 1965 the CM had undergone many improvements, which included raising the permissible voltage to the heaters in the oxygen tanks from 28 to 65 volts dc. Unfortunately, the thermostatic switches on these heaters weren't modified to suit the change. During one final test on the launch pad, the heaters were on for a long period of time. "This subjected the wiring in the vicinity of the heaters to very high temperatures (1000° F), which have been subsequently shown to severely degrade teflon insulation . . . the thermostatic switches started to open while powered by 65 volts dc and were probably welded shut." Furthermore, other warning signs during testing went unheeded and the tank, damaged from 8 hours overheating, was a potential bomb the next time it was filled with oxygen. That bomb exploded on April 13, 1970, 200,000 miles from Earth.

Made it!

A collective sigh of relief rose from the millions following the drama of Apollo 13 when the *Odyssey* splashed down. The CM-shaped bandage on this eagle cleverly depicts the flight home. "Only in a formal sense will Apollo 13 go into history as a failure," editorialized *The New York Times*.

Chicago Sun-Times, Sat., Apr. 18, 1970

CHAPTER FOURTEEN

The Great Voyages of Exploration

By HARRISON H. SCHMITT

First I want to share a new view of Earth, using the corrected vision of space. Like our childhood home, we really see the Earth only as we prepare to leave it. There are the basically familar views from the now well-traveled orbits: banded sunrises and sunsets changing in seconds from black to purple to red to yellow to searing daylight and then back; tinted oceans and continents with structural patterns wrought by aging during four and a half billion years; shadowed clouds and snows ever-varying in their mysteries and beauty; and the warm fields of lights and homes, now seen without the boundaries in our minds.

Again like the childhood home that we now only visit—changing in time but unchanged in the mind—we see the full Earth revolve beneath us. All the tracks of man's earlier greatness and folly are displayed in the window: the Roman world, the explorers' paths around the continents, the trails across older frontiers, the great migrations of peoples. The strange perspective is that of the entire Earth filling only one window, and gradually not even doing that. No longer is it the Earth of our past, but only a delicate blue globe in space. With something of the sadness felt as loved ones age, we see the full Earth change to half and then to a crescent and then to a faint moonlit hole in space. The line of night crosses water, land, and cloud, sending its armies of shadows ahead. We see that night, like time itself, masks but does not destroy beauty.

In sunlight, the sparkling sea shows its ever-changing character in the Sun's reflection, in varying hues of blue and green around the turquoise island beads, and in its icy competition with polar lands. The arcing, changing sails of clouds, following whirling, streaking pathways of wind, mark the passage of the airy lifeblood of the planet.

The revolving equatorial view concentrates our attention. There is the vast unbroken expanse of the Indian Ocean, south of the even more vast green and tan continent of Asia. In another complete view there are all of the blending masses of

Vistas without parallel in human experience surrounded the crews on the great voyages of exploration. Mount Hadley, rising 2¾ miles above the plain, is Apollo 15's backdrop as Jim Irwin sets up the first Lunar Roving Vehicle on the Moon.

(Photo captions for this chapter by the author.)

greens, reds, and yellows of Africa from the Mediterranean to the Cape of Good Hope, from Cap Vert to the Red Sea. Then we see across the great Atlantic from matching coast to matching coast. Scanning all of South America with one glance, we seemingly cease to move as the planet turns beneath us. And then there is the South Pacific. At one point only the brilliant ranges and plains of Antarctica remind a viewer that land still exists. The red continent of Australia finally conquers the illusion that the Earth is ocean alone, becoming the Earth's natural desert beacon.

When at last we are held to our own cyclic wandering about the Moon, we see Earthrise, that first and lasting symbol of a generation's spirit, imagination, and daring. That lonesome, marbled bit of blue with ancient seas and continental rafts is our planet, our home as men travel the solar system. The challenge for all of us is to guard and protect that home, together, as people of Earth.

A NEW VIEW OF THE MOON

What will historians write many years from now about the Apollo expeditions to the Moon? Perhaps they will note that it was a technological leap not undertaken under the threat of war; competition, yes, but not war. Surely they will say that Apollo marked man's evolution into the solar system, an evolution no longer marked by the slow rates of biological change but from then paced only by his intellect and collective will. Finally, I believe that they will record that it was then that men first acquired an understanding of a second planet.

What then is the nature of this understanding, and how did the visits of Apollo 15, 16, and 17 to Hadley-Apennines, Descartes, and Taurus-Littrow relate to it?

The origins of the Moon and the Earth remain obscure, although the boundaries of possibility are now much more limited. The details of the silicate chemistry of the rocks of the Moon and Earth now make us reasonably confident that these familiar bodies were formed about 4.6 billion years ago in about the same part of the youthful solar system. However, the two bodies evolved separately.

As many scientists now view the results of our Apollo studies, the Moon, once formed, evolved through six major phases. Of great future importance is the strong possibility that the first five of these phases also occurred on Earth, although other processes have obscured their effects. Thus, the Moon appears to be an ever more open window into our past.

The known phases of lunar evolution are as follows:

1. The existence of a *melted shell* from about 4.6 to 4.4 billion years ago.

2. Bombardment to form the *cratered highlands* from about 4.4 to 4.1 billion years ago.

3. The creation of the *large basins* from about 4.1 to 3.9 billion years ago.

4. A brief period of formation of *light-colored plains* about 3.9 billion years ago.

5. The eruption of the *basaltic maria* from about 3.8 to about 3.1 billion years ago.

6. The gradual transition to a *quiet crust* from about 3.0 billion years ago until the present.

Comparable to the Grand Canyon in scale and grandeur, the Valley of Taurus-Littrow extends some 20 miles through the ring of massifs surrounding the plains of the Serenitatis basin. In this westward-looking view from Apollo 17 LM *Challenger*, CSM *America* is the small central speck below and ahead, approaching the neck of the valley between the 1½ mile high massifs.

The detail by which we understand these six phases of lunar evolution is quite great. It derives from analysis of returned samples and observations of their geologic setting on the Moon, from the interpretation of geophysical and geochemical data from stations that still operate on the Moon or that previously operated in lunar orbit, and from our experience on Earth.

During the *melted shell phase* from about 4.6 to 4.4 billion years ago, at least the outer 200 miles of the Moon was molten or partially molten. As this shell cooled, the formation and settling of crystals of differing composition resulted in the creation of major chemical differences between various layers tens to hundreds of miles thick. A crust, mantle, and core apparently were formed at this time. The crust consisted of light-colored minerals rich in calcium and aluminum (largely the mineral plagioclase); the mantle contained dark minerals rich in magnesium and iron (largely the minerals pyroxene and olivine); and the core probably was composed of dense, molten material rich in iron and sulfur.

INCONCEIVABLE VIOLENCE

The *cratered highland phase* that followed was extremely, almost inconceivably violent. The debris left over from the creation of the planets bombarded the light-colored crust. These highland surfaces have survived as the bright portions of the full Moon we see today. They were pulverized, remelted, reaggregated, and, finally, saturated with craters at least 30 to 60 miles in diameter. The sheer violence of those times is difficult to comprehend.

The *large basin phase* was the time when very large basins were formed. This appears to have been the result of a distinctly more massive scale of bombardment than that which preceded their formation. These large basins dominate the surface character of the front side of the Moon and are responsible for the major chemical differences we have measured between various large surface regions.

The *light-colored plains phase* that followed was a brief, still controversial period in which most old basins appear to have been partially filled with debris largely derived from the surrounding light-colored crust. The events that created these plains are poorly understood partly because several different processes related to both meteor impact and internal vulcanism may have produced similar plains.

The *basaltic maria phase* was the main period during which the accumulation of heat from radioactive elements within the Moon produced melting and volcanic eruptions. Those eruptions filled all of the large basins with thick masses of dark-colored basalt called the maria. (These sea-like regions are the dark portions of the full Moon.) The lunar basalts are very different from basalts on Earth; they contain much less sodium, carbon, and water and commonly have much more titanium, iron, and heavy elements. At least the upper parts of the maria are ancient lava flows up to 300 feet thick. Many flows differ significantly from each other in chemical and mineral characteristics, differences that vary with both the age and the region.

The *quiet crust phase* from about 3.0 billion years ago to the present was largely just that—quiet. Compared to the past, very little happened except for the formation of scattered, very bright craters like Tycho and Copernicus, the creation of regional

Landing sites for the great voyages of exploration, Apollos 15, 16, and 17, were chosen primarily to expand knowledge of light-colored highland areas, and the ancient crust of the Moon that is 4.5 billion years old. Choices of Hadley-Apennines for 15, Descartes for 16, and Taurus-Littrow for 17 were compromises between this goal and constraints imposed by availability of high-quality imagery for planning, ideal distribution of geophysical instruments, landing-safety considerations, and propulsive energy of the Apollo/Saturn system. Note that later missions were not targeted so closely to the equator. Full-Moon photo here, taken from space, is not precisely congruent with the projection used in the chart.

Field camps on the Moon created by the last three LMs were provisioned with oxygen, water, food, and power for about 70 hours plus some reserves. The Rover had battery power sufficient for about 55 miles, although 22 miles was the most that one was driven. Here Gene Cernan prepares the Rover for our second day of work on Apollo 17. The weird vehicle at left is test vehicle for a Mobile Laboratory, a super-rover that never was built.

Mounting the Rover when spacesuited takes a bit of doing. You stand facing forward by the side of the vehicle, jump upward about two feet with a simultaneous sideways push, kick your feet out ahead, and wait as you slowly settle into the seat, ideally in the correct one. Here I'm completing the job.

The television camera on the Rover, which could be remotely controlled from Earth during traverse stops, was the eye of the science team. It gave them much of the information needed to radio advice up to us.

Broken fenders slung lunar dust about wickedly, but could be repaired by a field fix consisting of spare maps held by clamps. By chance the commanders of the last three missions each somehow managed to break a fragile fender.

fault systems like the Hyginus Rille, and the appearance of mysterious light-colored swirls like Reiner Gamma. Eruptions of basaltic maria also seem to have continued along a ridge and volcanic system that stretches for 1200 miles along the north-south axis of Mare Procellarum. Some of the events may be indications of continuing internal activity and stress beneath a now strong crust, such as the slow, solid convection of the lunar mantle.

For the most part, the surface of the Moon appears to have completed recording its history about three billion years ago. It has been largely unchanged except for the continued eroding rain of small meteors and now by the first primitive probings of men.

The Moon is as chemically and structurally differentiated as the Earth, lacking only the continued refinements of internal melting, solid convection, surficial weather-

Folded up to fit within its storage bay in the LM descent stage, the little car was designed so that it almost assembled itself.

Touring the Moon

Encumbered by a spacesuit, an astronaut on foot could not venture very far from the LM; carrying tools and samples made his forays more difficult. On the last three lunar missions a lightweight electric car greatly increased the productivity of the scientific traverses. Mission rules restricted us from going more than 6 miles from the LM—the distance we could walk back in a pinch—but even so the area that could be investigated was ten times greater than before. The Rover's mobility was quite high; it could climb and descend slopes above 25°. Crossing a steep slope was uneasy for the man on the downhill side, but there were no rollovers. On the level we averaged close to our top speed of 7 mph. Once, going down the Lee-Lincoln scarp, we set an informal lunar speed record for four-wheeled vehicles of 11 mph.

Deploying the Lunar Rover

Carried to the Moon in a nose-down, floorpan-out position, the Rover could be deployed by an astronaut paying out two nylon tapes. In the first stage the car swings out from its storage bay. Then the rear part of the chassis unfolds and locks, and the rear wheels unfold. In the third stage the front chassis and wheels snap out.

Finally, the astronaut lowers it to the surface, and unfolds the seats and footrests. Torsion-bar springs and latches made assembly semiautomatic. Power for the Rover came from two 36-volt silver-zinc batteries driving an independent ¼-hp motor in each wheel. A navigation system kept track of the bearing and range to the LM.

ing, and recycling of the crust. It moves through space as an ancient text, related to the history of the Earth only through the interpretations of our minds. It also exists as an archive of our Sun, possibly preserving in its soils much information of importance to man's future.

If we are to continue to read the text, we must continue to go there and beyond.

THE MISSIONS OF UNDERSTANDING

The last three Apollo journeys were great missions of understanding during which our interpretation of the evolution of the Moon evolved. In July 1971 the first of these missions, Apollo 15, visited Hadley Rille at the foot of the Apennine Mountains. Apollo 15 gave lunar exploration a new scale in duration and complexity. Col. David R. Scott, Col. James B. Irwin, and Lt. Col. Alfred M. Worden looked

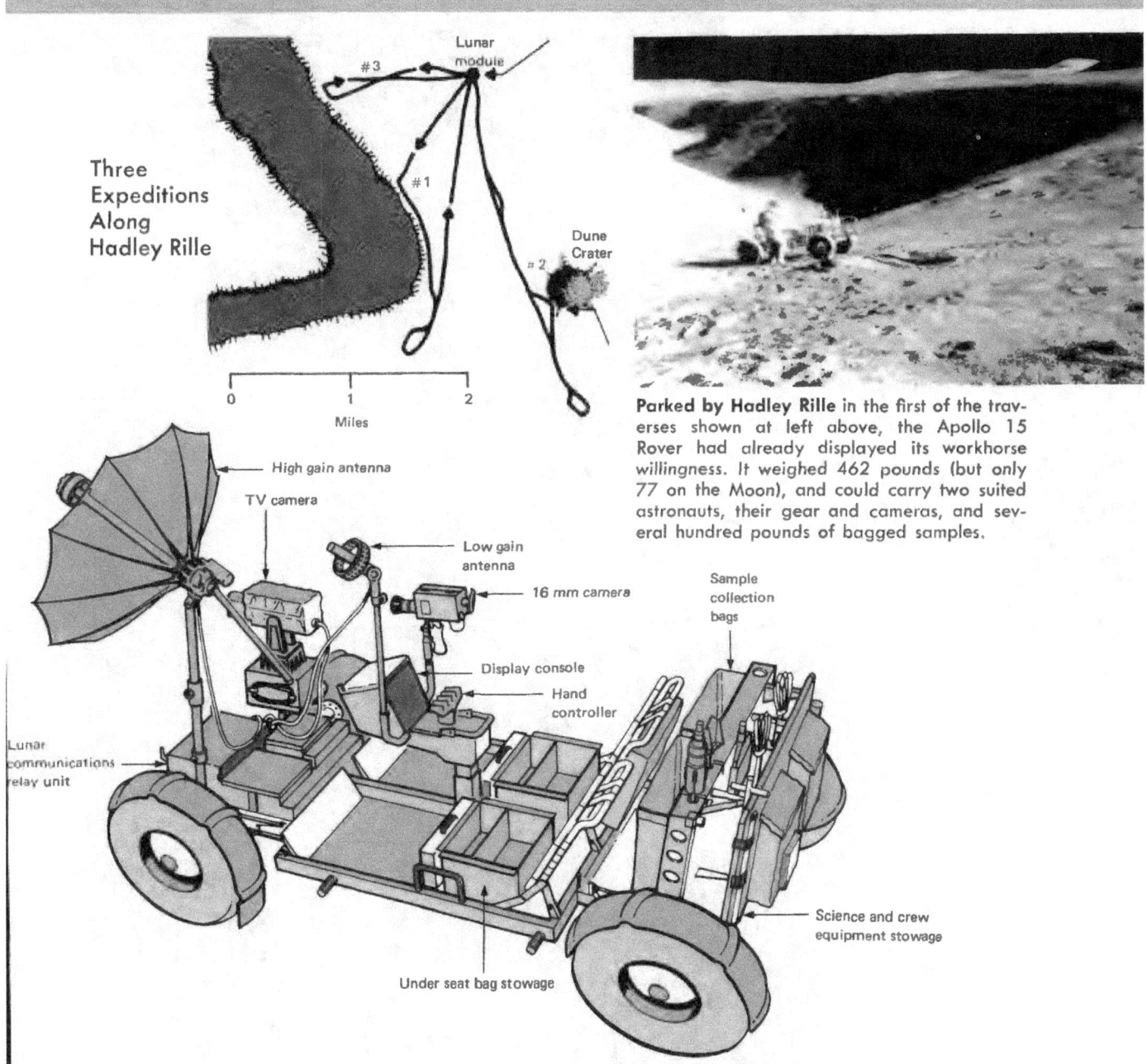

Parked by Hadley Rille in the first of the traverses shown at left above, the Apollo 15 Rover had already displayed its workhorse willingness. It weighed 462 pounds (but only 77 on the Moon), and could carry two suited astronauts, their gear and cameras, and several hundred pounds of bagged samples.

Sampling scoop in hand, I go questing at Station 5 by Camelot Crater (see traverse diagram below). At this point I had already collected a load of samples, and will shortly curve back to the Rover, off-camera at left, to unload. This was our second EVA, covering some 11 miles.

Our three traverses on Apollo 17 came very close to those we had preplanned, differing only because of unexpected findings. The first run was a mile and a quarter to Steno Crater and back, in the 4 o'clock position below. The second was the longest, at 5 o'clock. At Station 4 near Shorty Crater we found the orange soil (see page 276). The third run, at 12 to 1 o'clock, was more than six miles. The great fractured boulder shown on page 285 is on the slope near Station 6.

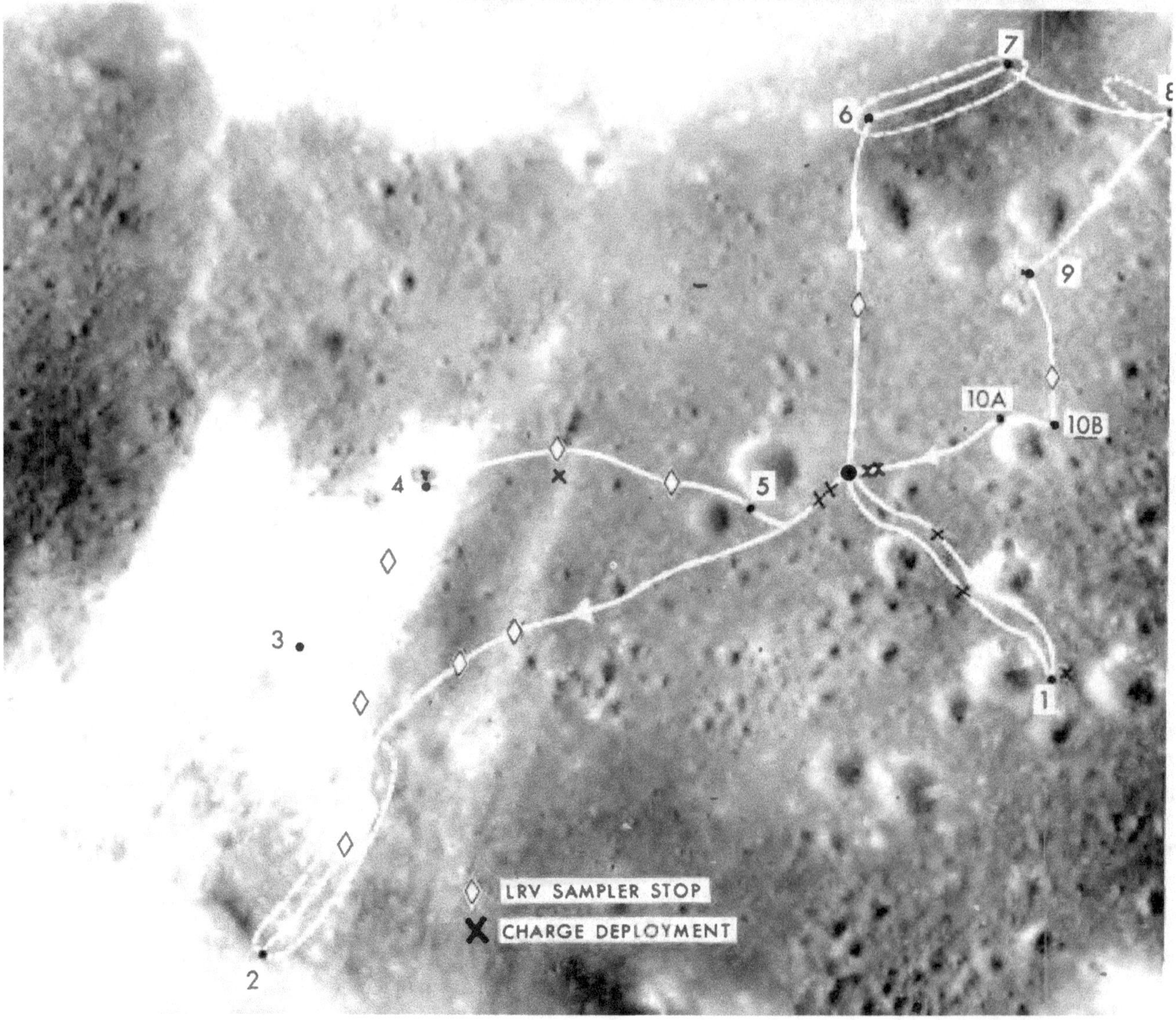

◇ LRV SAMPLER STOP

✕ CHARGE DEPLOYMENT

Sampling by scoop was the main way we obtained the large numbers of small samples that provide good statistical information about the composition of the surface. That instrument to the left of Apollo 15's Jim Irwin is a gnomon. It provides a vertical-seeking rod of known length, a color chart, and a shadow—all useful for calibrating pictures.

Sampling tongs and coring tubes gave us other means of collecting special samples. The spring-loaded tongs below let us pick up small rocks and fragments without getting down on our knees or otherwise reaching way down with clumsy gloves. The core tubes were hammered down into the soil and then drawn back out and capped. They gave us a way to collect sections of soil that preserved the relative relationships undisturbed.

Finding orange soil near Station 4 on Apollo 17 at a time when oxygen was running low kept us on the jump. We dug a trench 8 inches deep and 35 inches long, took samples of the orange soil and nearby gray soil, drove a core tube into the deposit, sampled surrounding rocks, described and photographed the crater site in detail, and packed the samples—all in 35 minutes. The effort gave scientists a most unusual lunar sample: very small beads of orange volcanic glass, formed in a great eruption of fire fountains over 3.5 billion years ago.

Rocks too big to bring back were studied where they were, described and photographed, and sampled by chipping pieces from their corners. If we could roll the rock over, as at right, we could take soil samples underneath that had been shielded from the effects of solar and cosmic radiation.

The basaltic lavas of the lunar maria, like this sample photographed after return to a laboratory on Earth, tell much about the partial melting of the Moon between three and four billion years ago. This sample, 70017, is identified in its documentation photograph by the number on the counter and by the B orientation code.

Ancient beads of orange volcanic glass in the photomicrograph below have revealed secrets of the Moon's deep interior. Produced most likely by the partial melting of the lunar mantle, and discovered by Apollo 17, these beads are unusually rich in such volatile elements as lead, zinc, tellurium, and sulfur. This indicates not only volcanic origin but also derivation from rocks possibly as deep as 200 miles in the Moon. Similar glass beads, green in color, were discovered by Apollo 15.

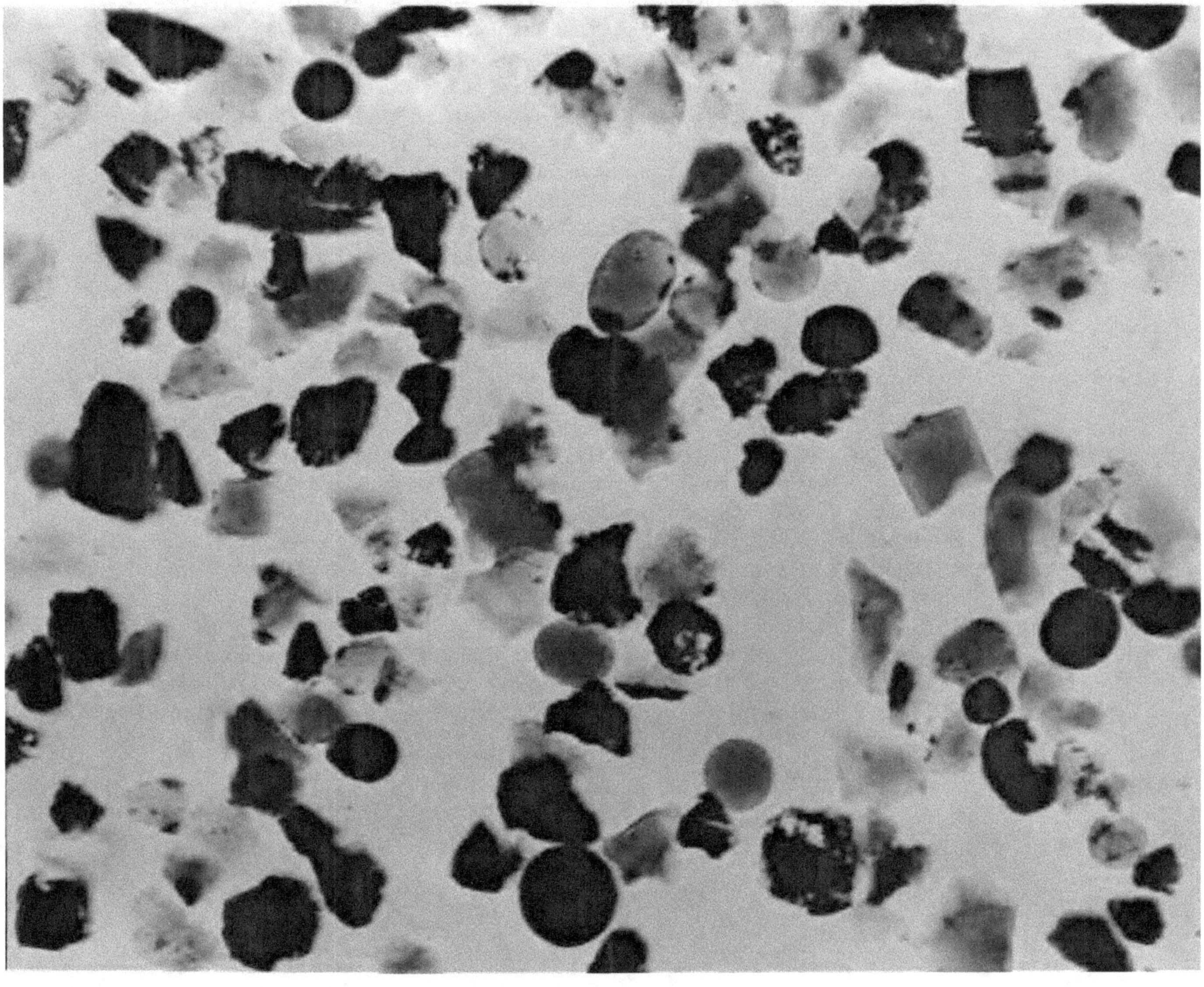

at the whole planet for 13 days through the eyes of precision cameras and electronics as well as the eyes of men. Scott and Irwin spent nearly 67 hours on the Moon's surface, and were the first to use a wheeled surface vehicle, the Rover, to inspect a wide variety of geological features. Finally, before returning to Earth, they placed a small satellite in lunar orbit that greatly expanded our knowledge of the distribution and geological correlation of gravitational and magnetic variations within the Moon's crust.

The varied samples and observations from the vicinity of Hadley Rille and the mountain ring of Imbrium called the Apennines pushed knowledge of lunar processes back past the four-billion-year barrier we had seemed to see on previous missions. We also discovered that lunar history behind this barrier was partially masked by multiple cycles of impact melting and fragmentation. Nevertheless, the rock fragments we sampled gave vague glimpses into the first half-billion years of lunar evolution and into some details of the nature of the melted shell. Part of this view into the past was provided by the well-known "Genesis Rock" of anorthosite (a plagioclase-rich rock). In addition, we expanded our understanding of the complex volcanic processes that created the present surfaces of the maria. These processes were now seen to have included not only the internal separation of minerals within lava flows but possible processes of volcanic erosion and fracturing that could have created the rilles.

The Apollo 15 astronauts placed instruments on the Moon which, in conjunction with earlier missions, finally established a geophysical net of stations. Of particular importance was a net of seismometers by which we began to decipher the inner structure of the Moon. Correlations of information from these stations with other facts enabled us to interpret several major portions of the interior. The Moon's crustal rocks, rich in the calcium and aluminum silicate plagioclase, are broken extensively near the surface but more coherent at depths from 15 to 40 miles. The crust rests on an upper mantle 125 to 200 miles thick that contains the magnesium and iron silicates, pyroxene and olivine. From about 200 or 250 to about 400 miles deep, the lower mantle is possibly similar to some types of stony meteorites called chondrites. From about 400 miles to about 700 miles deep, the chondrite material appears to be locally melted and seismically active. There are also many reasons now to believe that the Moon has an iron-rich core from about 700 miles deep to its center at 1080 miles that produced a global magnetic field until only recent times.

The geophysical station at Hadley-Apennines also told us that the flow of heat from the Moon was possibly two times that expected for a body having approximately the same radioisotopic composition as the Earth's mantle. If true, this tended to confirm earlier suggestions that much of the radioisotopic material in the Moon was concentrated in its crust. Otherwise, the interior of the Moon would be more fluid and show greater activity than we sense with the seismometers.

We began with Apollo 15 to be able to correlate our landing areas around the whole Moon by virtue of very-high-quality photographs and geochemical x-ray and gamma-ray mapping from orbit. The x-ray remote sensing investigations disclosed the provincial nature of lunar chemistry, particularly by highlighting differences in aluminum-to-silicon and magnesium-to-silicon ratios within the maria and the highlands. By outlining variations in the distribution of uranium, thorium, and potassium, the gamma-ray information suggested that large basin-forming events were capable of creating geochemical provinces by the ejection of material from depths of six or more miles.

Possibly of equal importance with all these findings by Apollo 15 was the discovery—shared through television by millions of people—that there existed beauty and majesty in views of nature that had previously been outside human experience.

The mission of Apollo 16 to Descartes in April 1972 revealed that we were not yet ready to understand the earliest chapters of lunar history exposed in the southern highlands. In the samples that Capt. John W. Young, Comdr. Thomas K. Mattingly, and Col. Charles M. Duke, Jr., obtained in the Descartes area, the major central events of that history seemed to be compressed in time far more than we had guessed. There are indications that the formation of the youngest major lunar basins, the eruption of light-colored plains materials, and the earliest extrusions of mare basalts required only about 100 million years of time around 3.9 billion years ago.

The extreme complexity of the problem of interpreting the lunar highland rocks

A New Way of Exploration

The great explorations of history were unique, but in some ways all voyages of exploration are alike. There is a means of transportation; food, tools, and a way of living; methods of investigation, mapping, and sampling; and the collaboration of fellow explorers. What distinguishes one voyage from another is the place gone to and the discoveries made there. In our time, the places were on the Moon, and the discoveries have revealed another planet.

In Apollo, our ships were Saturns and CSMs, the latter more than just transport and resupply ships, for they were also orbiting laboratories, bearing spectrometers, lasers, and precision cameras. Our sketchbooks and notebooks were cameras and voice recordings. The alert experts of Mission Control were our guardian angels, in addition to those issued by Providence. The landed lunar module was our base camp, and the lunar rover was our steed. On the Apollo 17 mission, the LM *Challenger* could support us in its base-camp role for 75 hours, with a contingency margin of 12 hours more. In one-sixth gravity, its hammocks provided comfortable

sleeping positions, with none of the lumps of camp cots or pine boughs. The total absence of black flies or mosquitoes was wonderful; and the rehydrated food was better than the grub dished up by some camp cooks.

Out on the surface, the backpack could support us for seven to eight hours, depending on physical exertion. We carried enough extra oxygen, water, and batteries in *Challenger* to recharge the backpack twice, for three major excursions outside. We also had an emergency supply of oxygen in the backpack that could provide at least 30 minutes of suit pressure and oxygen at large rates of flow, in case the suit was holed. Only one aspect of work in the pressure suit was very difficult and that was the effort of gripping tools against pressure within the glove. Like squeezing a tennis ball repetitively for nine hours, this was very fatiguing to forearm muscles; we also separated our fingernails from the quick as the nails scraped against the glove fingers. The forearm fatigue disappeared after each night's sleep; sore fingernails did not disappear until days after we left the Moon.

The eyes of geochemical sensors peer through an opening in the sides of Apollo 15 CSM *Endeavour*. The instruments gave us broad-scale remote sensing of the lunar surface, allowing data from samples collected on the surface to be correlated across major areas of the Moon. Included were precision cameras and spectrometers that sensed x-rays, gamma rays, infrared radiation, and the chemical ions in the ultrathin lunar atmosphere. Apollo 15 cameras supplied much imagery used to plan our Apollo 17 exploration.

Adrift between the Earth and the Moon, Ron Evans retrieved the film canister of the mapping cameras on the day after Apollo 17 left lunar orbit. His space walk lasted an hour, and resulted in the successful retrieval of data from three experiments. Ron's oxygen was fed from the spacecraft through the umbilical hose, with an emergency supply on his back. I was in the open hatch to help in retrieval, which was necessary because the service module would be jettisoned before we reentered the Earth's atmosphere.

Apollo 17 photographed itself in the frame at right from its panoramic camera, which shows the valley of Taurus-Littrow after the landing of *Challenger*. The landing point (see inset below) is revealed by a bright spot, produced by the effects of the descent-engine exhaust. A reflection from and shadow of the LM are also visible in high-quality prints. The panoramic camera has an absolute resolution of about 1 meter in best prints.

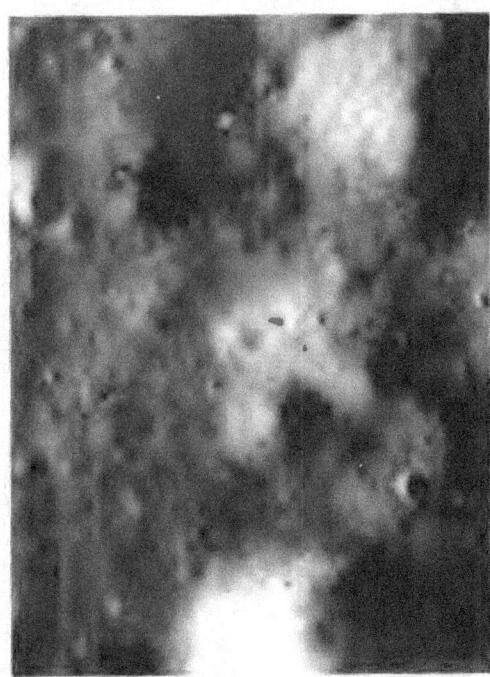

Detailed analysis of this panoramic photograph indicates that the light-colored avalanche, and many of the craters on the valley floor, are probably the result of the impact of material ejected some 50 million years ago from the crater Tycho 1300 miles to the southwest. Such orbital data have been invaluable in expanding the context of our interpretations of samples and data that were returned from explorations on the surface.

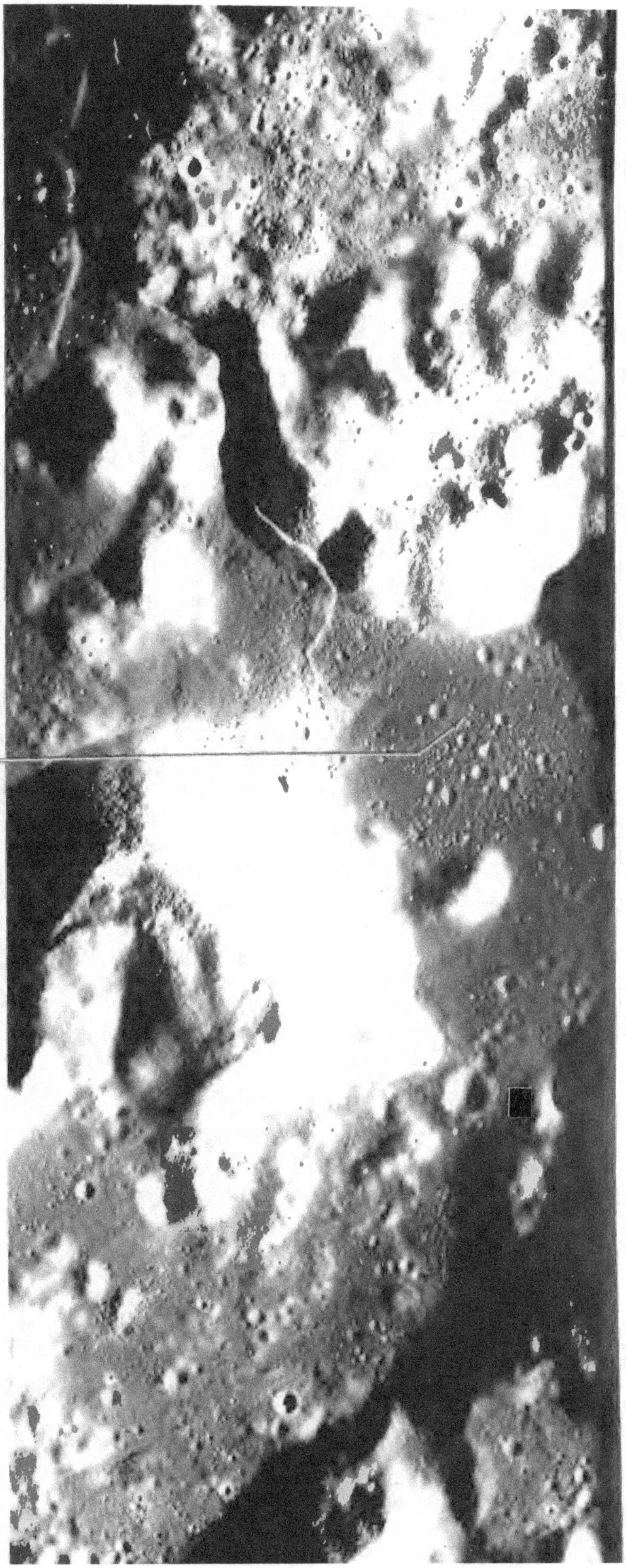

and processes became evident even as the Apollo 16 mission progressed. Rather than discovering materials of clearly volcanic origin as many expected, the men found samples that suggested an interlocking sequence of igneous and impact processes. A new chemical rock group known as "very high aluminum basalts" could be defined, although its ancestry relative to other lunar materials was obscured by later events that gave the cratered highlands their present form. The results of Apollo 16 have within them an integrated look at almost all previously and subsequently identified highland rock types. With this complexity comes a unique, as yet unexploited, opportunity to understand the formation and modification of the Moon's early crust and potentially that of the Earth.

The materials found in the Descartes region were similar to those sampled slightly earlier by Luna 20 in the Apollonius region. But there were significant differences in the aluminum content of debris representative of the two regions. Also there were differences in the abundance of fragments of distinctive crystalline rocks known as the anorthosite-norite-troctolite suite. After Apollo 15, this suite of rocks had been recognized as possibly being a much reworked leftover of at least portions of the ancient lunar crust. Luna 20 and Apollo 16 confirmed its great importance to the understanding of the ancient melted shell.

A MAJOR THERMAL EVENT?

The last crystallization age of some of the Apollo 16 rocks appeared to be about 3.9 billion years, and continued to indicate that this age is a major turning point in lunar history. This general age for the cooling of highland-like materials also was found to hold for the ejecta blanket of the Imbrium Basin at Fra Mauro, for the rocks of the Apennines, and later for some of the highland rocks at Taurus-Littrow. This limit suggested (1) a major thermal event associated with the formation of several large basins over a relatively short time, or (2) a major thermal event associated with the formation of the light-colored plains, or (3) the rapid cessation of the period of major cratering that continually reworked the highlands until most vestiges of original ages had disappeared and only the last local impact event was recorded. As we attempt to explain the absence of very old rocks on Earth, we should not forget these possibilities for resetting our own geologic clocks.

Apollo 16 continued the broad-scale geological, geochemical, and geophysical mapping of the Moon's crust from orbit begun by Apollo 15. This mapping greatly expanded our knowledge of geochemical provinces and geophysical variations, and has helped to lead to many of the generalizations it is now possible to make about the evolution of the lunar crust.

Apollo 17 carried Capt. Eugene A. Cernan, Capt. Ronald Evans, and me in December 1972 to the valley of Taurus-Littrow near the coast of the great frozen basaltic "sea" of Serenitatis. The unique visual character and beauty of this valley was, I hope, seen by most people on television as we saw it in person. The unique scientific character of this valley has helped to lessen our sadness that Apollo explorations ended with our visit. It would have been hard to find a better locality in which to synthesize and expand our ideas about the evolution of the Moon.

A landing site not visited, the crater Alphonsus is shown in this Apollo 15 mapping-camera picture. A point near the right edge of the crater floor in this southward-looking view was once the leading candidate for our landing site in Apollo 17. However, the need for greater geologic and geographic variety resulted in final selection of Taurus-Littrow instead. Our regret is that both could not have been explored.

Mysterious Aristarchus and Schröter's Valley also await some future explorers of the Moon to give us insight. Once considered for a mission, the sinuous rilles and volcanic features on the Aristarchus plateau have puzzling characteristics that are little understood.

A panorama of lunar history is captured in this view looking south over the Valley of Taurus-Littrow. A huge fragmented boulder had rolled almost a mile down the side of the North Massif to here, Station 6 on our traverse (see page 274). Our LM and its light area of surface altera-

tion can be seen on the photo about an inch to the right of the top point of the boulder. That's me at the left. Note the marks of my sampling scoop on the debris resting on a slanting surface of the boulder at left. Gene Cernan took the photos from which this mosaic was assembled.

At Taurus-Littrow we looked at and sampled the ancient lunar record ranging back from the extrusion of the oldest known mare basalts, through the formation of the fragmental rocks of the Serenitatis mountain ring, and thence back into fragments in these rocks that may reflect the very origins of the lunar crust. We also found and are now studying volcanic materials and debris-forming processes that range forward from the formation of the earliest mare basalt surface through 3.8 billion years of modification of that surface.

The pre-mare events in the Taurus-Littrow region that culminated in the formation of the Serenitatis Basin produced at least three major and distinctive units of complex fragmental rocks. The oldest of these rock units contains distinctive fragments of crystalline magnesium and iron-rich rocks that appear to be the remains of the crystallization of the melted shell. This conclusion is supported by a crushed rock of magnesium olivine with an apparent crystallization age of 4.6 billion years. The old fragmental rock unit containing these ancient fragments was intruded and locally altered by another unit which was partially molten at the time of intrusion about 3.9 billion years ago. Such intrusive fragmental rocks are probably the direct result of the massive impact event that formed the nearby Serenitatis Basin; however, an internal volcanic origin cannot yet be ruled out. The third fragmental rock unit seems to cap the tops of the mountains and it may be the ejecta from one of the several large old basins within range of the valley. This unit contains a wide variety of fragments of the lunar crust, including barium-rich granitic rock.

The valley of Taurus-Littrow and other low areas nearby appear to be a fortuitous window that exposes some of the oldest, if not the oldest, mare basalt extrusives on the Moon. They are about 3.8 billion years old, and are 50 to 100 million years older than the basalts at Tranquility Base. They also contain titanium oxide in amounts up to 13 percent by weight.

BEADS OF ORANGE GLASS

The modification of the surface of the valley basalt included the addition of mantles of beads of chemically distinctive orange glass (the "orange soil") and black devitrified glass. The beads appear to have been formed by volcanic processes having their origins in the deep interior of the Moon. The titanium-, magnesium-, and iron-rich nature of these silicate glasses surprised us in many ways. Their approximately 10- to 30-million-year age of exposure to the Sun is young and was expected for the dark mantling deposits seen in photographs; but their 3.5- to 3.7-billion-year age of cooling from a liquid was not expected. The explanation for this difference is not yet obvious.

All the Apollo missions left the Moon before the lunar sunrise had progressed into the vast regions of the lunar west: Mare Procellarum, where the mysterious features of that region's central ridge system still await the crew of a mission diverted after Apollo 13, and Mare Orientale, whose stark alpine rings have been viewed only in the subdued blue light of Earth. The promise of these regions and of the far side of the Moon has not diminished. They seemingly watch the progression of sunrises, awaiting the landing craft of another generation of explorers.

A parked Rover awaits our return from sleep before we set out on the third EVA of Apollo 17's surface exploration. We've parked it in an orientation that will minimize heating of its surfaces while we are eating and sleeping in the LM. Our activities around the LM have exposed dark soil beneath the lighter soil caused by our descent-engine exhaust. *Right*, a lonely Rover watches with its television eye as the last Apollo explorers depart from the Moon.

CHAPTER FIFTEEN

The Legacy of Apollo

By HOMER E. NEWELL

Because we live on it, the Earth is the center of things for us. Around Earth all other celestial bodies circle endlessly, or so it seemed to our forebears. For countless generations men who thought about such matters regarded the Earth as the center of the universe. So satisfying was this view, so entrenched in doctrine and dogma did it become, that when Copernicus and Kepler challenged the idea, they stirred up a hornet's nest. The concept of the Sun as the central stillness in the solar system around which Earth and all other planets revolve was considered too unsettling to be tolerated, and edict and persecution sought to suppress these dangerous new ideas. But in vain, for the Copernican revolution in human thought continues to this very day. In countless ways it colors the picture men draw of themselves and of man's place in the universe.

Apollo's greatest impact was to impress dramatically upon men's minds, more clearly than ever before, the significance of the Copernican view. The spectacle of a spacecraft leaving Earth with the incredible speed of almost 6 miles per second— thirteen times faster than a rifle bullet—traveling through space like a miniature planet, bearing men for the first time to another world, focused the attention of hundreds of millions of people. We saw Earth as only one of nine planets in the solar system, insignificant, except to us, among the unreachable stars in the vast expanse of the heavens. In cosmic perspective Earth is but a tiny object in a remote corner of space, companion to a modest star, one of a hundred billion stars making up one of billions of galaxies scattered over unimaginable distances to beyond the farthest reaches to which we have been able to peer with the most powerful telescopes.

But while helping to convey Earth's insignificance in the cosmic scale, Apollo dramatically displayed Earth's uniqueness and overwhelming significance on the human scale. Standing in imagination on the rocky rubble of a lunar plain, looking through an astronaut's eye and camera out over the vast arid wasteland of our inhospitable satellite, we saw above the horizon the beautiful, blue, fragile Earth. It

Our conceptions are altered when the point of view is shifted. When the Apollo 15 astronauts took this picture from lunar orbit, they saw the Earth as a thin crescent. At the same time, we on Earth were seeing a nearly full Moon.

awakened a heightened appreciation of and sense of responsibility toward our home in space. In the entire solar system, 6 billion miles across, only Earth so far as we now know nourishes the vast abundance of life that we so casually accept. Only by understanding thoroughly our planet and our place on it can we hope to learn how to use its resources wisely, to preserve for future generations our island in space in its pristine vigor and beauty. And that is where the true significance of Apollo lunar science comes in.

  To understand fully our own planet, it is essential that we study many planets,

Photography and Spacecraft

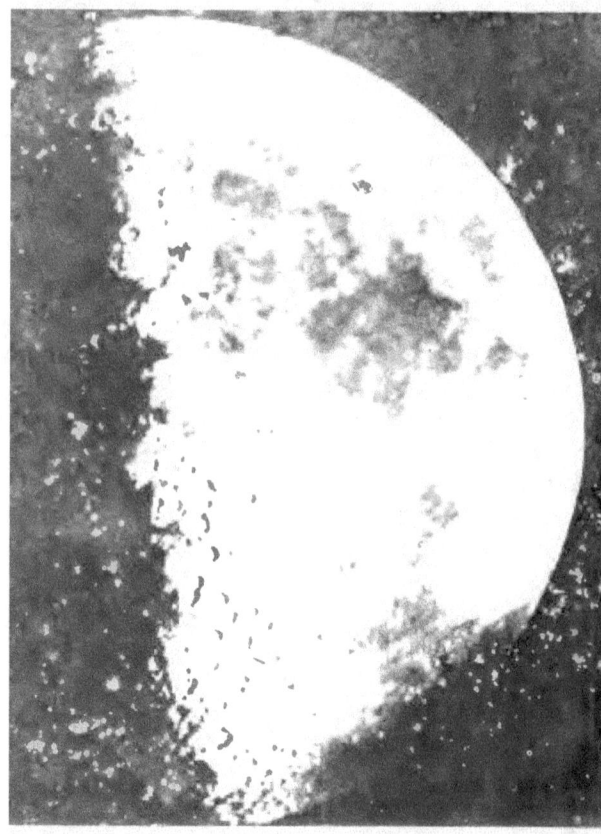

The oldest photograph of lunar details on record. The original of this picture was a "Crystalotype" print made from a daguerreotype taken in March 1851. As in all telescopic views of the half Moon, relief features are visible only near the terminator. The remainder of the image shows only the contrast in reflectivity between the maria and the highlands.

Space photography over a vast range of scales is providing the raw material for many kinds of study. This view of the almost-full lunar disk was taken when Apollo 17 was about 2000 miles from the Moon on its way home. More than a third of the area covered is never visible from the Earth. Mare Smythii, the dark circular area just to the right of the center, straddles the 90° east meridian. The crater Tsiolkovsky is near the terminator at the lower right.

making comparisons among them. An inevitable myopia interferes when we try to learn about planets from the study of only one. The Moon is a planet in its own right, by reason of its substantial size and mass, and what we learn of lunar science also advances Earth science.

As our nearest neighbor in space, the Moon has long been an object of wonder and study. Through the telescope the astronomer has seen myriads of craters on its surface, clear evidence of lava flows, and even some suggestion of current volcanic activity. The Moon is decidedly out of round. The sharpness of telescopic images, and

Brought the Moon Closer

This southward-looking oblique view from an orbital altitude of 70 miles places lunar features in a new perspective. The crater in the center is Aristarchus, the brightest large crater on the Moon. It is 25 miles in diameter and 2¼ miles deep. The low Sun angle has brought into bold relief the deposits of ejected material, as well as the strange sinuous rilles. When, finally, man walked the Moon, his footprints symbolized an enormous new gain in knowledge.

the suddenness with which stars disappear behind the Moon and later reappear show clearly that the Moon has virtually no atmosphere. No mountain systems like the Rockies or the Himalayas could be seen. Without atmospheric erosion and mountain-building activity, we supposed that the Moon would preserve on its face the record of solar system history to the very earliest days. But viewing the Moon from 239,000 miles away left much room for speculation and disagreement. So when rockets became available, plans were quickly laid for investigating the Moon close at hand, eventually by man himself.

IMPLANTING SENSORS ON THE MOON

Serious space probe studies of the Moon began with the Soviet Zond and Luna, and with the U.S. Ranger, Surveyor, and Lunar Orbiter. Armed with the information obtained by unmanned probes, the United States carried out the Apollo manned missions, implanting nuclear-powered geophysical laboratories at several landing sites, including seismometers, magnetometers, plasma and pressure gauges, heat flow instruments, and laser corner reflectors. These laboratories continue to operate long after the astronauts have returned to Earth. On two missions, the lunar surface laboratories were supplemented by satellites left in orbit to support geodetic studies. Detailed studies of lunar surface composition by X-ray fluorescence and radioactivity measurements were made from the Apollo spacecraft itself. Most importantly, the astronauts brought back from their six landing missions hundreds of pounds of lunar rocks and soil, which, together with the small amount returned by Soviet unmanned sample missions, have since been the subject of analysis and study by scientists around the world.

How old is the Moon? This was the first major question. Although a few thought the Moon would prove to be considerably younger than Earth, most felt it would turn out to be of the same vintage as our own planet. From radioactive dating of the lunar rocks and soil it is now established that the Moon formed about 4.6 billion years ago. The Moon is, indeed, very old, as old as Earth, and is telling us much about the earliest years of our planet and of the solar system. But to our great surprise the lunar rocks show that the Moon's surface is no longer in its original condition, having undergone considerable change in the first 1.5 billion years. A few of the highland rocks in the Apollo samples are older than 4 billion years—at least one is 4.6 billion years old, the probable time of formation and initial melting of the Moon—but most of them have been shocked and remelted by meteoroid impacts 4 billion years ago. Most of the mare lava flows are from 3 to slightly less than 4 billion years old, indicating that substantial melting and flow occurred on the Moon long after its formation. Gravitational energy released during the aggregation of the Moon, heat generated by radioactive elements in the lunar material, and thermal effects of the huge impacts that sculptured the lunar surface all undoubtedly contributed to the early melting of the outer layers.

What is the physical condition of the lunar surface? The question generated countless heated debates before the lunar landings. It was clear from the visible meteorite cratering that the surface must be thoroughly chopped up. How fine would the material be? Would it be loosely or densely packed? Could it hold a spacecraft landed upon it? How thick would the layer of rubble be? Some spoke of deep deposits of finely powdered dust into which a spacecraft would sink out of sight. Now we know that the Moon's surface is everywhere covered with many feet of fragmented material, or regolith. A sizable fraction of the soil is very fine dust, but the material is sufficiently cohesive and well packed that it easily supported the Apollo spacecraft, the astronauts, and their lunar rover.

Of what is the Moon made? This was a question about which one could only speculate from afar. It could be essentially meteoritic material. It might be like the Earth's crust. Or it might be something entirely different. Any proposition put forth, however, had to satisfy the constraint that the Moon's density is only 3.36, considerably lighter than Earth's average density of 5.5. One of the major questions was: Is there any water there? Water would be the most important constituent that one might hope to find on the Moon. It would be essential for the support of any microbial life there. It would be an invaluable resource for lunar bases that might be established in the remote future. No one expected to find water exposed on the surface—it would quickly evaporate in the airless, oven-hot environment of the lunar day—but many supposed that there might be substantial subsurface water in the form of ice as a permafrost. Even if free water were not found, it was fully expected that there would be water of crystallization in the lunar minerals.

Learning at firsthand about the stuff of which the Moon is made from the actual lunar material carried back by the Apollo astronauts has been one of the most exciting scientific undertakings of our day. More than 700 scientists, including several hundred from twenty other countries, have spent uncounted hours analyzing samples, conducting laboratory tests, and theorizing over the significance of what was being revealed for the first time. Annually at the Johnson Space Center they have assembled to share their findings and to try to explain them. No water was found. Lunar material is very dry, with practically no water of crystallization. The hydrous minerals so common on Earth are exceedingly rare in the lunar rocks and soil. Moreover, lunar material is depleted in most volatile elements, suggesting a very hot processing at some time in their history. The rocks in the lunar maria are similar to, though significantly different from, lavas on Earth. Highland samples show a considerable separation of lunar material into different minerals, showing a differentiation like that responsible for the wide variety of rocks and minerals on Earth. Moon material is neither exactly like the meteorites nor exactly like the Earth's crust, but all three could have had a common source with a different history.

HOT DEBATES ABOUT VOLCANOES

What have been the relative roles of impact cratering and volcanism? For the crater-ridden Moon the obvious first question is how were the craters formed: by meteorites hitting the surface at high velocity, or as the remains of once active volcanoes? Debates on this subject generated about as much heat as the volcanoes themselves. Many argued for, or gave the appearance of arguing for, one extreme or the other. More sober debaters recognized that both cratering and volcanism played a role and sought to discern their relative importance. It is clear that the Moon has changed considerably since it was first formed, but most of that change occurred more than three billion years ago. Apparently the first 1.5 billion years were a period of violent evolution, involving the Mare Imbrium event and other cataclysmic impacts, followed by the vast flooding of the mare plains with a series of lava flows. In contrast, the last three billion years have been relatively quiet, with occasional impacts like those that generated the craters Copernicus and Tycho, but no great lava events.

The implication of this for Earth is that during its first billion years Earth also must have been subjected to severe bombardment, generating huge craters like those

Space has brought dramatic new views of Earth . . .

At one point on its coasting path toward the Moon, Apollo 17 lined up with the Sun and the Earth, enabling the astronauts to take this full-disk view. Since it was December, the beginning of summer in the Southern Hemisphere, the icecap that covers the Antarctic continent is brightly illuminated. A most striking feature of the visible land mass of Africa and southwestern Asia is the transition from the tawny color of the Sahara, Libyan, and Arabian deserts, through the dark band of grass-covered savannah, to the cloud-strewn tropical rain forest.

still seen on the Moon, and this may also have been true for other planets as well. Certainly the Mariner 9 pictures of Mars and Mariner 10 pictures of Mercury show that both planets experienced substantial cratering, while terrestrial radars indicate the same for Venus. The evidence mounts that violent meteorite bombardment was wide-spread in the solar system in times past. On Earth, however, erosion, crumpling of the crust, and subsequent volcanism have erased most of the evidence of this early catastrophic period.

Is the Moon still active today? Infrequent observations of sudden localized

Shown us mysterious super-canyons on Mars . . .

A complex of huge valleys and tributary canyons, now named the Valles Marineris, can be traced across 2500 miles of the Martian surface. It is comparable in scale to the Red Sea or the east African rift valley system and probably originated, like them, in the pulling part of great plates of the planetary crust. Tectonic activity of this kind marks a planet that is still evolving. The tree-like tributary canyons in the picture (a 275-mile segment of Valles Marineris) may be the result of water erosion, even though the Martian atmosphere now contains little water.

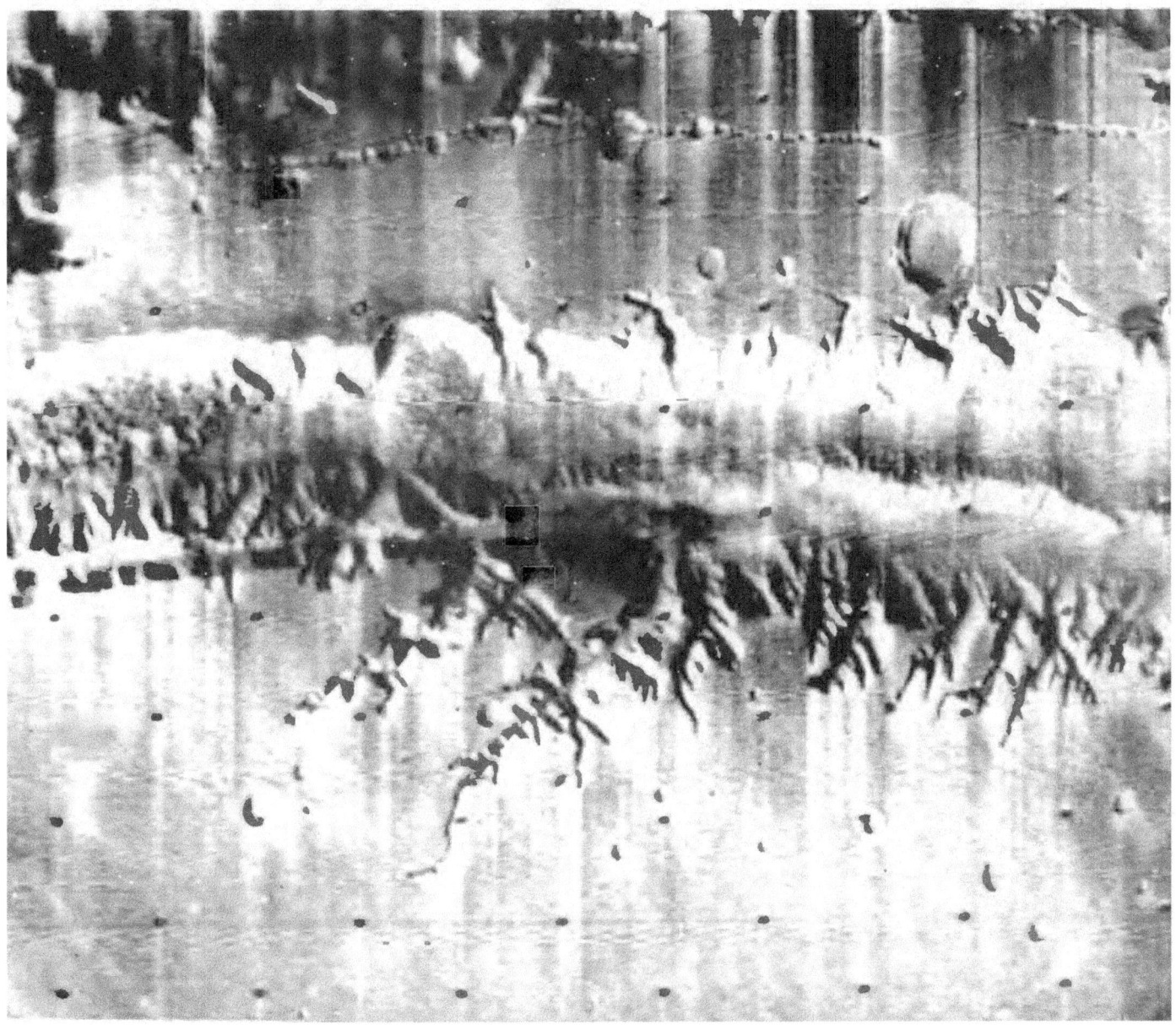

glows or hazes on the Moon have caused a stir when they occurred, and gave rise to speculation as to whether these were due to current volcanic activity, or were merely trapped gases shaken loose by moonquakes or meteorite impacts. There was speculation about whether our satellite was a dead planet or still a live one. Five Apollo seismometers were set up at different landing sites, and four of them are still working. At times of lunar perigee, these detect moonquakes of very deep foci centered at 500 to 620 miles below the lunar crust. The energy released over a year by these moonquakes, however, is a billion times less than that released by earthquakes

Given exciting views of giant, turbulent Jupiter . . .

Cloud tops high in the atmosphere form the giant planet's visible surface. This photograph, produced from the red and blue digital images of Pioneer 11's imaging photopolarimeter, shows the characteristic banding parallel to the equator and the elongated circles that mark regions of intense vertical convective activity. Most prominent among these is the Great Red Spot, a hurricane-like group of thunderstorms that has persisted through several centuries of observation. Jupiter's weather systems are long-lived because their heat comes mainly from the planet's liquid interior.

over a similar period. No evidence of current volcanic activity on the Moon has been found from either the unmanned or manned space missions. In most places the soil just below the surface appears to have been relatively undisturbed for millions of years, which is consistent with the rarity of large-scale meteorite impacts that Earth experiences today. At the very surface a slow erosion takes place by micrometeorite impacts and solar-wind particles, wearing away on the average a few molecular layers a year on exposed surfaces. Material does move around. Some material falls and slides down slopes, some may be moved around by electrostatic forces, and much of what

Revealed atmospheric circulations on overheated Venus . . .

Clouds in its very dense atmosphere completely hide Venus' solid surface. Although the clouds are nearly feature-less at visible wavelengths, they show a wealth of detail in the near ultraviolet. This ultraviolet picture, taken by the Mariner 10 television camera, is one of a post-encounter sequence that shows the cloud-bearing atmospheric layer in retrograde (right to left) rotation with a four-day period. The swirling currents of that rotation interact with con-vection cells rising from the subsolar point near the left edge of the disk.

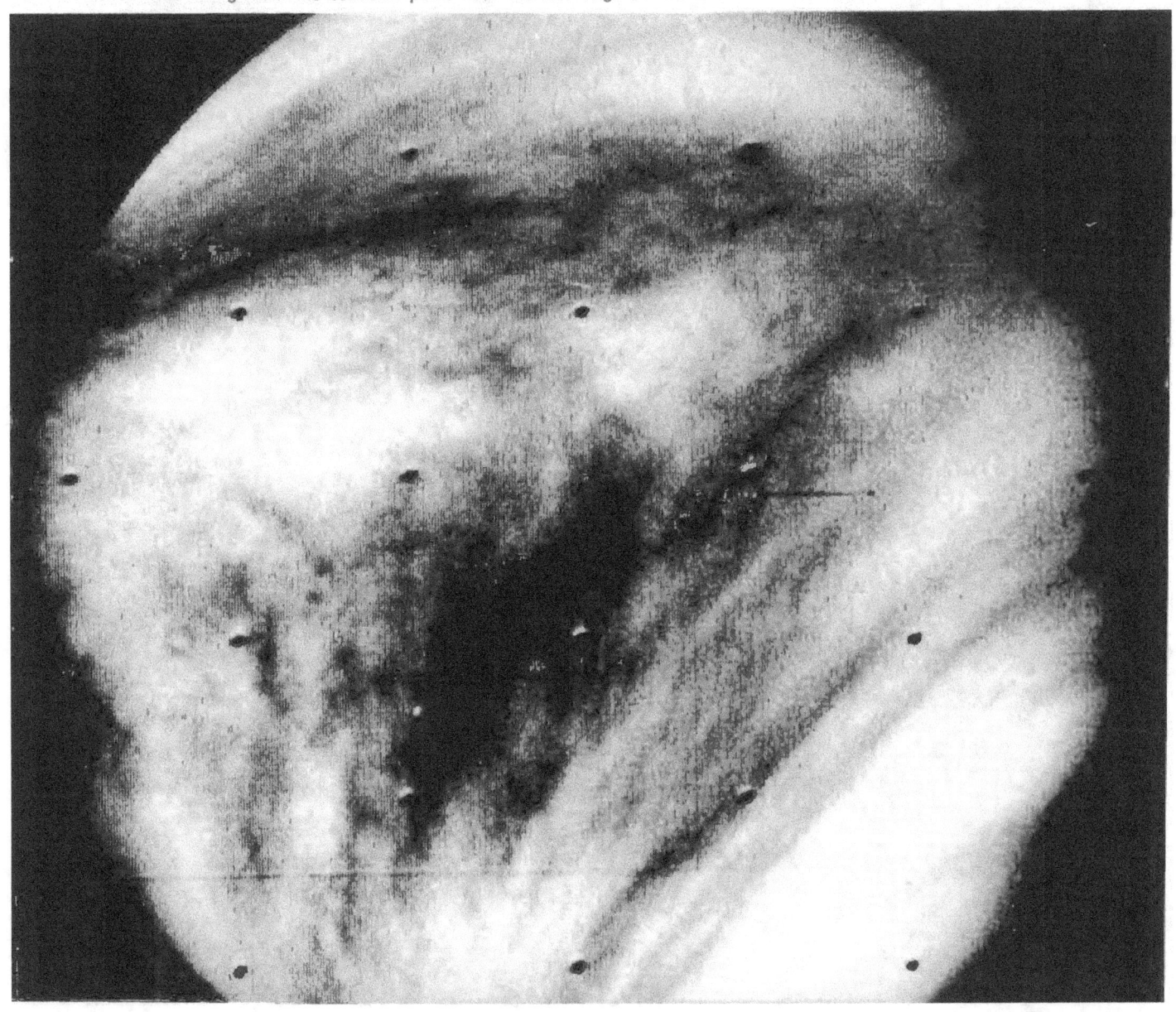

movement occurs is due to splashes from meteorite impacts. But all in all, the Moon appears to be extremely quiet now, in comparison with its earliest history or with Earth today.

Is the interior of the Moon hot or cold? Discussion of this subject once made the sparks fly. Assuming the Moon contains radioactive elements, as does Earth, then the heat generated by their decay should warm up the interior. But would this source provide enough heat to melt the Moon? Whatever the cause, the lava flows apparent on the surface of the Moon show that at least part of the Moon's interior actually

Unveiled the scarred, pocked face of Mercury . . .

The surface of Mercury was revealed to science on March 29, 1974, when Mariner 10 took over 1600 television frames just before and just after flying past the planet's dark side. The left and right halves of this picture are mosaics of the pre-encounter and post-encounter frames. A second encounter, six months later, provided additional coverage. Although the crater interiors on this airless planet closely resemble those on the Moon, study of the high-resolution frames shows external differences that may result from Mercury's stronger gravity field.

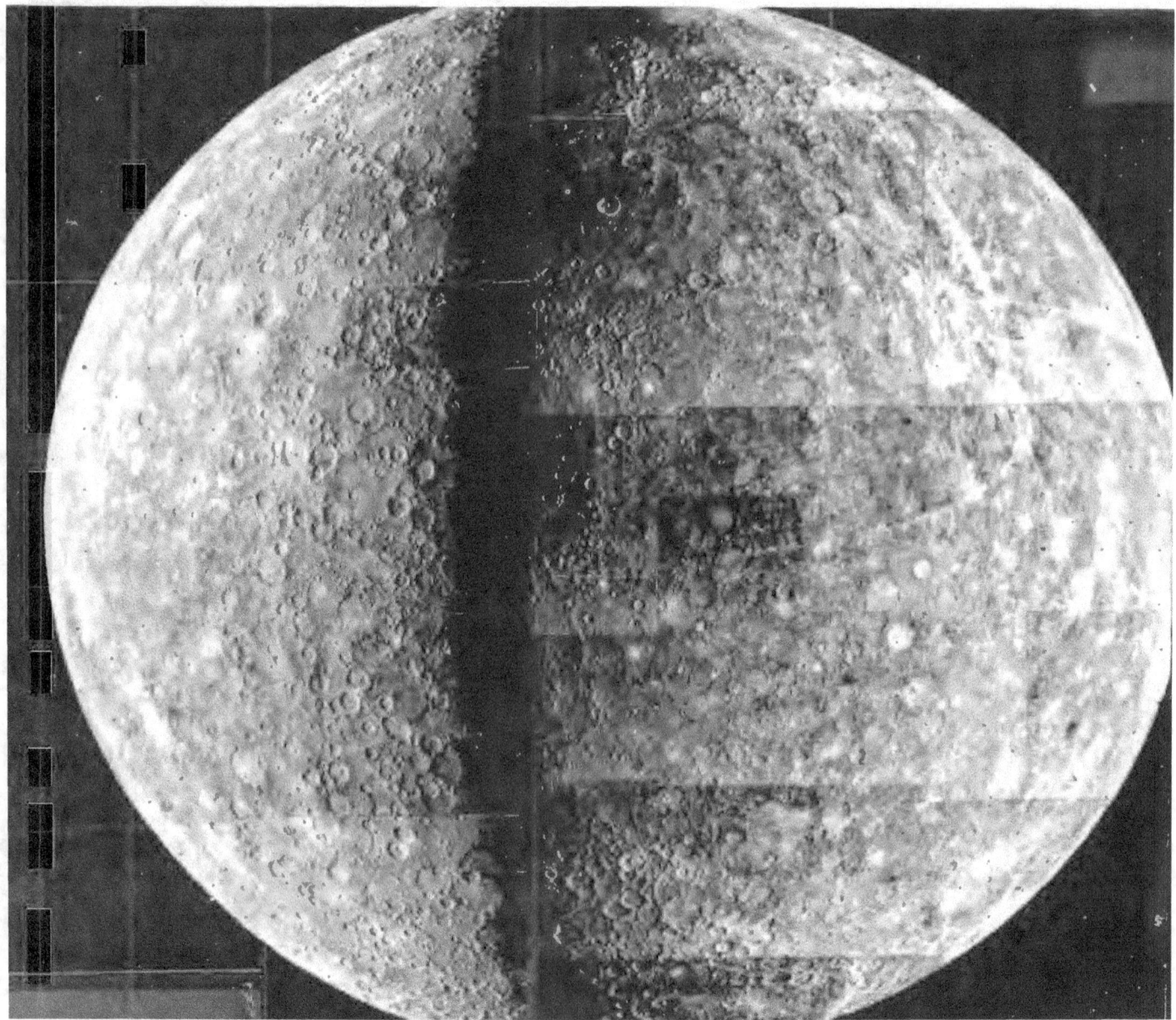

was molten at one time. But on the other side of the picture, the distinctly out-of-round shape implies a rigidity that a very hot and plastic Moon would not have. The Apollo measurements have added fuel to this flaming controversy. Anomalous concentrations of mass called mascons have been discovered in the great circular mare basins, detected by the way in which they distort the Moon's gravitational field. In a hot, plastic Moon, these mascons would have sunk until the gravity field was restored to equilibrium. The fact that they persist today indicates a rigid, cool Moon. Yet melting and lava flooding in the upper layers of the Moon are widespread. Electrical

And will soon visit Saturn's strange rings.

We shall have to wait a few more years for a closer view of Saturn and its rings. Pioneer 11 will encounter the planet in September 1979. Saturn is second in size only to Jupiter, with an average density less than that of water. Titan, the largest of its ten satellites, is the only one in the solar system known to have an appreciable atmosphere. Saturn's rings are thin layers of separate particles orbiting in its equatorial plane. The ring system is probably less than 60 miles thick, and its total mass is very small compared to that of Saturn's satellites.

conductivities inside the Moon, deduced from the Moon's reaction to the electrical charges and magnetic fields in the solar wind, indicate that the outer layers, down to 500 to 620 miles in depth, are now well below the melting point. Below those levels, however, is a region where seismic shear waves are markedly attenuated, indicating partial melting of the lunar material there. Moreover, while magnetic measurements do not reveal any dipole magnetic field at present like that of the Earth, remanent magnetization in rock samples, and substantial local magnetic fields, suggest that the Moon may well have had a dipole field in the past. Since it is believed that such a planetary magnetic dipole field is generated by circulation in a liquid core, this implies that at one time the core of the Moon was molten. This would further imply that the Moon was at one time quite hot throughout, which gets us right back to the difficulty of explaining how the mascons and the aspherical shape can continue to exist. There is a real puzzle here that needs sorting out.

Does the Moon have any atmosphere at all? Before Apollo it was already clear that the Moon could have very little atmosphere. It was expected that heavier gases like argon and krypton would be found clinging close to the surface, but in the lower gravitational field lighter gases would long since have escaped. In large measure this has been confirmed. Argon generated by radioactivity in the lunar crust, and hydrogen, helium, and neon from the solar wind, account for most of what little atmosphere there is, and that is over a billion times less than the Earth's atmosphere.

ANCIENT TRACKS OF RADIATION

What are the principal effects of the Sun upon the Moon, and how do they differ from the effects of the Sun on Earth? Protected by Earth's atmosphere, we are not directly exposed to the lethal part of the Sun's radiations. But the airless Moon is starkly exposed. By a neat twist, this situation gives us a way of checking up on the Sun's activity over the past few million years. Cosmic rays from the Sun and galaxy leave tracks in the soil grains and rock surfaces on which they impact; these tracks can be enhanced by etching, and counted. The track intensities indicate the intensity of the radiation to which the materials were exposed. By analyzing the rate at which lunar material is brought out onto the surface and then later buried again, it is possible to estimate the times when different layers in the lunar regolith were exposed, and for how long. Thus sample cores of the lunar regolith taken by the Apollo astronauts enable us to look back in time at the radiation environment experienced by the Moon. No significant changes in galactic cosmic radiation are seen, but there is a suggestion that there might have been variation in solar cosmic rays over the last one to ten million years. This would imply changes in solar activity, which in turn may have had effects on Earth.

Is there life on the Moon? Some of the bitterest exchanges took place over this question. If there were life, no matter how primitive, we would want to study it carefully and compare it with Earth life. This would require very difficult and expensive sterilization of all materials and equipment landed on the Moon, so as not to contaminate Moon life with Earth life. Also there was the question of contamination of Earth by Moon life, possibly a serious hazard to us on Earth. So difficult, time-consuming, and expensive quarantine procedures were urged. On the other side of this argument were those who pointed to the hostile conditions on the Moon: virtually no atmosphere, probably no water, temperatures ranging from 150° C during lunar night to more than 120 C at lunar noon, merciless exposure to lethal doses of solar ultraviolet, X-rays, and charged particles. No life, they argued, could possibly exist under these conditions. The biologists countered: water and moderate temperature conditions below ground might sustain primitive life forms. And so the argument went on and on, until finally Apollo flew to the Moon with careful precautions against back-contamination of Earth, but with limited effort to protect the Moon.

It turns out that there is no evidence that indigenous life exists now or has ever existed on the Moon. A careful search for carbon was made, since Earth life is carbon-based. In the lunar samples one hundred to two hundred parts per million of carbon were found; of this no more than a few tens of parts per million are indigenous to the lunar material, the rest being brought in by the solar wind. None of the carbon appears to derive from life processes. As a consequence, after the first few Apollo flights, even the back-contamination quarantine procedures were dropped.

THE NAGGING QUESTION OF ORIGIN

Where did the Moon come from? This is the big puzzler of them all. There were three major theories. The Moon came from Earth, possibly wrenched from what is now the Pacific Ocean. Or the Moon formed in the vicinity of Earth at the same time that Earth was forming. Or the Moon formed somewhere else in the solar system, and was later captured by Earth. Most students of the subject reject the first possibility, because it proves very difficult to explain all the steps that must have occurred to bring this about. That leaves various versions of the last two as principal contenders. Apollo has done little to favor one over the other. Indeed, some feel that we will never be able to say for sure. But the question is a nagging one, and scientists will continue to argue over it.

The deep significance of the Apollo investigations lies in the fact that these measurements and observations give us a detailed insight into a planetary body other than Earth, thereby helping us to understand better our own planet. Before space probes, lunar and planetary science had for a long time been inactive, due in part to lack of new data to spark serious thought. With the vast quantities of lunar data returned by Apollo and other lunar missions, together with rich new space-probe data from Mars, Venus, Mercury, and Jupiter, and new discoveries on Earth such as the slow spreading of ocean floors and the drifting of continents, a new field of comparative study of planets has virtually exploded into world science. In this comparative investigation of the planets, the Moon is an important link.

Already it is clear that bodies of planetary size will undergo considerable evolution after their formation. Most of this evolution takes place early, and it is probably less in the case of a Moon-sized planet, leaving the planet relatively quiet for most of its history. A planet the size of Mars, though substantially smaller than Earth, remains active for much longer than a Moon-sized body, as the Mariner 9 pictures clearly show. With its huge volcanoes, its giant canyon several times deeper than the Grand Canyon and long enough to span the United States, its variable polar caps, its suggestion of colliding crustal plates, Mars is clearly still an active planet. Venus, the size of the Earth, with its very hot surface and extremely dense and dynamic atmosphere, may well prove to be *more* active volcanically and tectonically than Earth. Mercury, intermediate between Mars and the Moon in size, is heavily cratered and seems to be much like the lunar highlands.

It will be fascinating when we can complete this perspective by studying the giant outer planets on the one hand, and the very small bodies like the comets and asteroids on the other. As Pioneer 10 and 11 have shown, Jupiter is extremely dynamic. We may expect the same to prove true of the other giant planets when we get a chance to see them close at hand. Moreover, the large planets, consisting, as they do, mainly of the lighter elements like hydrogen and helium, should provide a revealing insight into conditions in the solar nebula at the time the planets of the solar system were condensing out of the nebula. The same should be true of the comets and asteroids. The asteroids in particular are probably too small to have undergone any gravitational or radioactive melting, and therefore should in their interiors be as they were at the time of their formation, unless they, too, turn out to be fragments of what were once larger bodies.

LIFE IN AN OCEAN OF GAS

Our atmosphere is the breath of life to everything we do. It is an ocean of gas in which we live as fish live in an ocean of water. It is an integral part of our planet, and we cannot understand Earth's history without also understanding its atmosphere. In seeking to do this, the wide variation from the airless Moon to the exotic and turbulent atmosphere of Jupiter provides an invaluable context and perspective. As it happens, the solar system has several planetary atmospheres between these extremes.

Earth's atmosphere lies intermediate in surface pressure between that of Venus and Mars, Venus' atmosphere being about 100 times that of Earth, and Mars' about one one-hundredth. The composition of the atmospheres of Venus and Mars is predominantly carbon dioxide, while Earth's today is predominantly nitrogen and oxygen. On the other hand, the amount of carbon dioxide tied up in the carbonate of the Earth's crust is comparable to what we see free in the atmosphere of Venus today. Presumably on Earth carbon dioxide exhaled from volcanoes, in the presence of the abundant water, was soon taken up into the crustal rocks. Life on Earth doubtless accounts for the presence of so much oxygen in the atmosphere, most of which was probably generated by photosynthesis. From the Mariner observations of Mercury, it appears that Mercury is closer to the Moon than to Mars as far as atmosphere is concerned.

Much patient work remains to bring out the full significance of Apollo observations. Nevertheless it is already clear that Apollo has made a great contribution to the development of the new field of comparative planetary studies, a field that over the years will provide us priceless insights into our own planet. In days of deep concern over the Earth's environment and finite resources, comparative planetology is a new science of incalculable importance, and Apollo led the way.

The Contributors

Authoring the preceding fifteen chapters are these seventeen men who, along with thousands of others, were intensely involved in Apollo's expeditions to the Moon. Some have left the space program; all continue to make significant contributions to the world around them.

Edwin E. Aldrin, Jr. (b. 1930) A West Point graduate, Aldrin did his MIT doctoral thesis on manned orbital rendezvous. He flew with Lovell in Gemini 12 and set foot on the Moon as lunar module pilot on Apollo 11. A former combat pilot (66 Korean missions), Aldrin left the space program to command the Air Force Test Pilot School. Now retired from the military, he is a consultant to firms engaged in advanced technologies.

Michael Collins (b. 1930) A West Point graduate and test pilot, Collins flew on Gemini 12 and walked in space. On the historic Apollo 11 flight to the Moon, he commanded the Columbia in orbit above the lunar surface. A brigadier general in the Air Force Reserve, with some 266 hours of space flight, Collins is now the Director of the Smithsonian Institution's National Air and Space Museum in Washington, D.C.

Charles Conrad, Jr. (b. 1930) Princeton engineering graduate Conrad became a naval aviator and test pilot. He flew aboard Gemini 5 and Gemini 11, and was spacecraft commander in Apollo 12 and again in Skylab 2. Since his retirement from the Navy in 1974, Conrad has served as a vice president of the American Television and Communication Corporation.

Edgar M. Cortright (b. 1923) An aeronautical engineer, Cortright joined the NACA Flight Propulsion Laboratory in 1948. Ten years later he helped lay the groundwork for NASA's formation and continued at Headquarters to plan and direct space programs, including those for scouting the Moon. Since 1968 Cortright has been Director of Langley Research Center, and has led in the development of the Viking Mars landings.

Robert R. Gilruth (b. 1913) With two degrees in aeronautical engineering, Gilruth joined NACA's Langley Aeronautical Laboratory in 1937, becoming Assistant Director there in 1952. In 1958 he was named director of the Space Task Group and Project Mercury. When the STG became the Manned Spacecraft Center, Gilruth went to Houston as Director, and served from 1961 to 1972, leading the Gemini and Apollo programs.

Christopher C. Kraft, Jr. (b. 1924) came to NASA through NACA, joining the Langley laboratory in 1945 and becoming director of Project Mercury flight operations in 1959. His contributions to Mission Control continued in his service as Director of Flight Operations for the Gemini and Apollo programs. He was made Deputy Director of the Manned Spacecraft Center in 1969, and Director in 1972.

James A. Lovell (b. 1928) An Annapolis graduate and test pilot, Lovell has flown around the Moon twice, first in Apollo 8, man's initial lunar flight, and then in Apollo 13. Earlier he flew in Geminis 7 and 12. Lovell worked on space science applications from 1971 to 1972 at Manned Spacecraft Center; he is now president of the Bay-Houston Towing Co.

George M. Low (b. 1926) began his NASA career at Lewis Research Center in 1949. Named Chief of Manned Space Flight in 1958, and Deputy Director of Manned Spacecraft Center in 1964, he took over the Apollo spacecraft program in 1967. Low became NASA's Deputy Administrator in 1969.

George E. Mueller (b. 1918) came to NASA from industry in 1963. There he had directed ballistic missile and space probe programs. Noted for his work in microwave measurement and in telemetry, he served as NASA's Associate Administrator for Manned Space Flight from 1963 to 1969, when he returned to industry. He is now Chairman and President of the System Development Corporation in California.

Homer E. Newell (b. 1915) was the Naval Research Laboratory's Science Program Coordinator on Project Vanguard— America's first scientific satellite. He joined NASA in 1958, serving in Headquarters as Associate Administrator for Space Science and Applications from 1963 to 1967. From 1967 to his retirement in 1974, Newell was NASA Associate Administrator. He is the author of many works on space sciences.

Rocco A. Petrone (b. 1926) A West Point graduate, Petrone came to NASA in 1960. There he managed the activation of all Apollo launch facilities. He personally supervised the Apollo 11 launch. Named Director of the Apollo program in 1969 and of the Marshall Space Flight Center in 1973, Petrone became Associate Administrator of NASA in 1974. He now heads the National Center for Resource Recovery.

Samuel C. Phillips (b. 1921) An electrical engineer, Phillips became Deputy Director then Director of the Apollo program in 1964, serving till 1969. Before that he directed the Air Force's Minuteman program. A World War II fighter pilot and former Director, National Security Agency, Phillips, a four-star general, now heads the Air Force Systems Command.

Harrison H. Schmitt (b. 1935) A New Mexico geologist, Cal Tech and Harvard graduate, Schmitt became an astronaut-scientist in 1965, and went to the Moon as Apollo 17's lunar module pilot. Schmitt served as NASA's Assistant Administrator for Energy Programs during 1974 and 1975.

Alan B. Shepard, Jr. (b. 1923) America's first man to journey into space, flying a Redstone-boosted Mercury in 1961. He was also the spacecraft commander on the Apollo 14 flight and landed on the Moon in 1971. Shepard has served as chief of the Astronaut Office and as a delegate to the United Nations' General Assembly in 1971. An Annapolis graduate and former test pilot, Shepard holds the rank of rear admiral.

Robert Sherrod (b. 1909) Noted foreign correspondent and editor (*Time, Life, Saturday Evening Post*), Sherrod is the author of five books on the military, including his monumental *History of Marine Corps Aviation in World War II*. He covered the battles of Attu, Tarawa, Saipan, Iwo Jima, and Okinawa, and later, the Korean and Vietnamese conflicts. For several years he has been working on a book about Apollo.

Wernher von Braun (b. 1912) came to the United States in 1945 to conduct missile research for the Army. He directed the Pershing and Jupiter rocket programs before leading work on the powerful Saturn. Von Braun was Director of the Marshall Space Flight Center from 1960 to 1970 and NASA Deputy Associate Administrator for Planning, 1970 to 1972. He is now a vice president with Fairchild Industries.

James E. Webb (b. 1906) Prior to appointment as NASA Administrator in 1961 Webb, a lawyer, held major posts in government and industry. A Marine officer and pilot, he served as Under Secretary of State, as Budget Bureau Director, and as a corporate officer and director. Departing NASA in 1968, Webb resumed law practice in Washington, D.C.

Key Events in Apollo

1958   *July:* National Aeronautics and Space Act signed into law establishing NASA.

       October: Mercury program begun; six successful manned flights would occur.

       November: Space Task Group officially organized, at Langley Field, Va., to implement the manned satellite project, known later as Project Mercury.

1959   *April:* First group of astronauts is selected for manned spaceflight program.

       December: Saturn configuration and development program are approved.

1960   *February:* Ranger project for lunar hard landings is formally authorized.

       May: Lunar soft-landing program, later renamed Surveyor, commences.

       July: Apollo approved as name of advanced manned spaceflight program.

1961   *May:* First U.S. suborbital manned spaceflight, by Astronaut Alan B. Shepard, Jr., in a Redstone rocket-boosted Mercury capsule named *Freedom 7*.

       May: President Kennedy proposes broad, accelerated space program, including manned lunar landing within the decade, in a special message to the Congress.

       August: Cape Canaveral selected as the launch site for manned lunar flights.

       November: The Space Task Group redesignated as the Manned Spacecraft Center.

1962   *February:* First U.S. orbital manned spaceflight by Astronaut John H. Glenn, Jr., in an Atlas-boosted Mercury capsule named *Friendship 7*.

       July: Selection of lunar orbit rendezvous flight mode for Apollo missions.

1963   *May:* First U.S. long duration spaceflight, by Astronaut L. Gordon Cooper, Jr., aboard an Atlas-boosted Mercury, *Faith 7*; last flight in the Mercury program.

       August: Lunar orbiter program, to team with Ranger and Surveyor, approved.

1964   *May:* First flight of an Apollo-configured spacecraft with a Saturn vehicle (two-stage Saturn I), launched from Cape Kennedy; craft, upper stage orbited.

       July: Closeup television pictures of lunar surface are sent by Ranger VII.

1965   *March:* First U.S. two-man orbital spaceflight, by Astronauts Virgil I. Grissom and John W. Young in Titan II-boosted Gemini III.

       June: First U.S. extravehicular activity, Astronaut Edward H. White II maneuvering from Gemini IV piloted by Astronaut James A. McDivitt.

       December: First piloted rendezvous in space, by Astronauts Schirra and Stafford in Gemini VI with Gemini VII, Astronauts Borman and Lovell aboard.

1966   *February:* First launch of the Saturn IB (two stage) with an Apollo spacecraft.

March: First docking in space, made by Astronauts Neil A. Armstrong and David R. Scott aboard a Titan-boosted Gemini VIII.

June: Successful on first attempt, Surveyor I softlands on the lunar surface and reports back on soil bearing strength and surface temperatures.

August: First U.S. spacecraft to orbit the Moon, Lunar Orbiter I sends back Surveyor and Apollo landing site photos.

1967   *January:* Flash fire kills Apollo 1 Astronauts Grissom, White, and Chaffee in command module atop a Saturn I booster at Launch Complex 34.

November: First flight of the Saturn V is a success; Apollo 4 placed in orbit.

1968   *January:* First flight test of the lunar module, in Earth orbit, in Apollo 5.

October: First manned Apollo flight, by Astronauts Schirra, Eisele, and Cunningham, in the Saturn IB-boosted Apollo 7.

December: Astronauts Borman, Lovell, and Anders become first men to orbit the Moon, in the Saturn V-boosted CSM of Apollo 8.

1969   *March:* First manned flight of the lunar module, in Earth orbit, by Astronauts McDivitt and Schweickart with Astronaut Scott piloting CSM, in Apollo 9.

May: First lunar module orbit of the Moon, by Astronauts Stafford and Cernan with Astronaut Young manning the orbiting command module, in Apollo 10.

July: Astronauts Armstrong and Aldrin become first men to land on the Moon; Astronaut Collins pilots the command module orbiting the Moon, in Apollo 11.

November: Astronauts Conrad and Bean walk on the Moon, while Astronaut Gordon pilots command module, in Apollo 12.

1970   *April:* Ruptured oxygen tank aborts Apollo 13 mission. Astronauts Lovell, Swigert, and Haise return safely to Earth.

1971   *January:* Third successful manned lunar landing, Apollo 14, places Astronauts Shepard and Mitchell on Moon, while Astronaut Roosa pilots CSM.

July: Lunar Roving Vehicle used for first time on Moon, by Astronauts Scott and Irwin, while Astronaut Worden pilots the command module, in Apollo 15.

1972   *April:* Fifth successful lunar landing mission places Astronauts Young and Duke on Moon, while Astronaut Mattingly pilots command module, in Apollo 16.

December: Apollo program's longest mission, Apollo 17. Astronauts Cernan and Schmitt land on the Moon while Astronaut Evans pilots CSM.

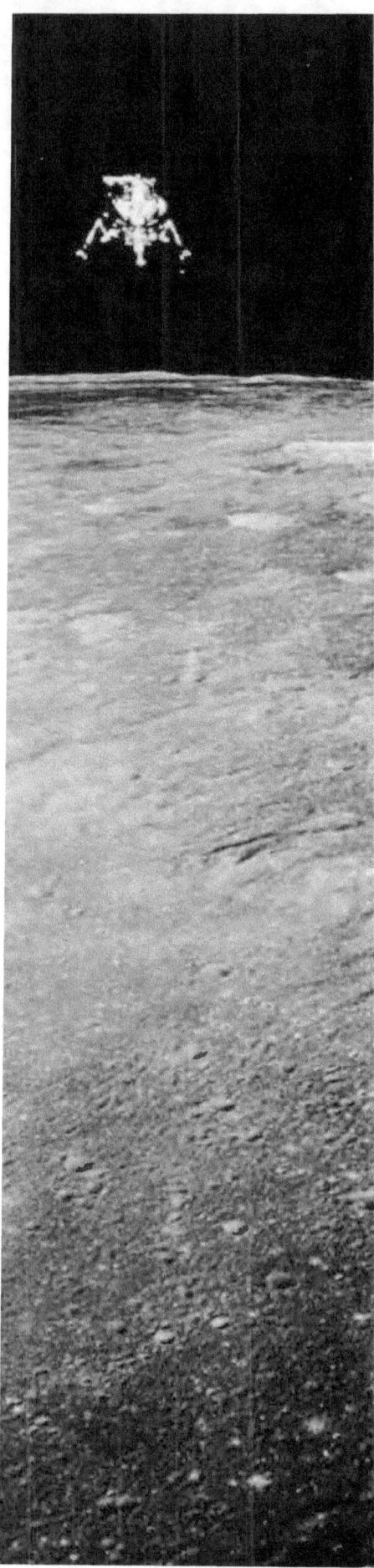

Index

Editor's Note

In planning this photo history we set out to record the story of Apollo before the colors fade and memories blur. At first we aimed to restrict ourselves to the actual expeditions to the Moon. But it soon became clear that this approach could not capture the scope and spirit of so far-reaching an enterprise. So we decided that the breadth of Apollo would be shown best from the differing perspectives of the people directly concerned. Each chapter author was encouraged to recount his part of the story as he remembered it. We refrained from homogenizing these contributions, although we recognized that they are necessarily personalized and slightly duplicative. But they do offer the viewpoints of some of the people who made Apollo happen, and thus may provide fresh insights into that incredible project.

To help develop the idea into a book, I turned to Frank Rowsome, NASA's technical publications chief, with whom I had collaborated on an earlier book, *Exploring Space with a Camera*. Others who helped were Sandra Scaffidi, photo editor; Kay Voglewede, copy editor; and Harry Samuels, art director. Special research was done by George Abbey, William R. Corliss, James Daus, Leon Kosofsky, Andrew Ruppel, and Ray Zavasky. Volta Torrey and David Anderton aided on Chapters 1, 10, and 12. Robert Sherrod not only wrote Chapter 8 but also drew on his extensive Apollo knowledge to give assistance on Chapters 6, 9, and 13. Harold Pryor of NASA's Scientific and Technical Information Office lent us people, facilities, and support.

We offer this book as a grateful memento to the hundreds of thousands of people who worked on Apollo, and as a fond tribute to our fellow Americans who gave the program such steadfast support. Apollo is worth remembering not only for what it did, but for what it taught us we can do.

EDGAR M. CORTRIGHT, *Director*
NASA Langley Research Center

JULY 28, 1975

313

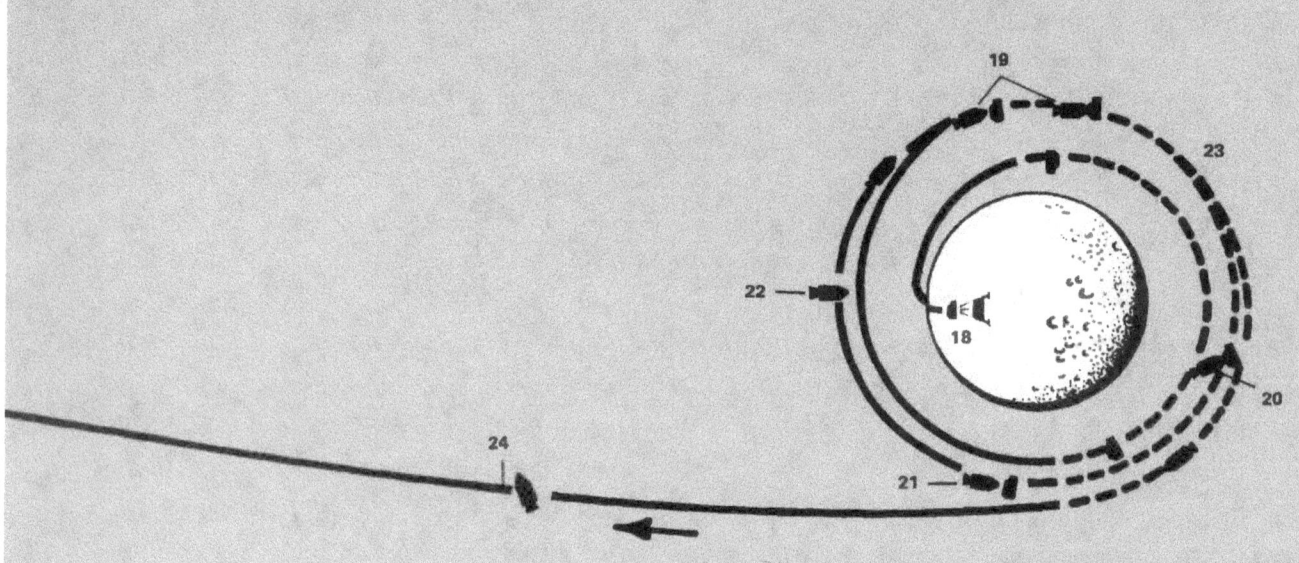

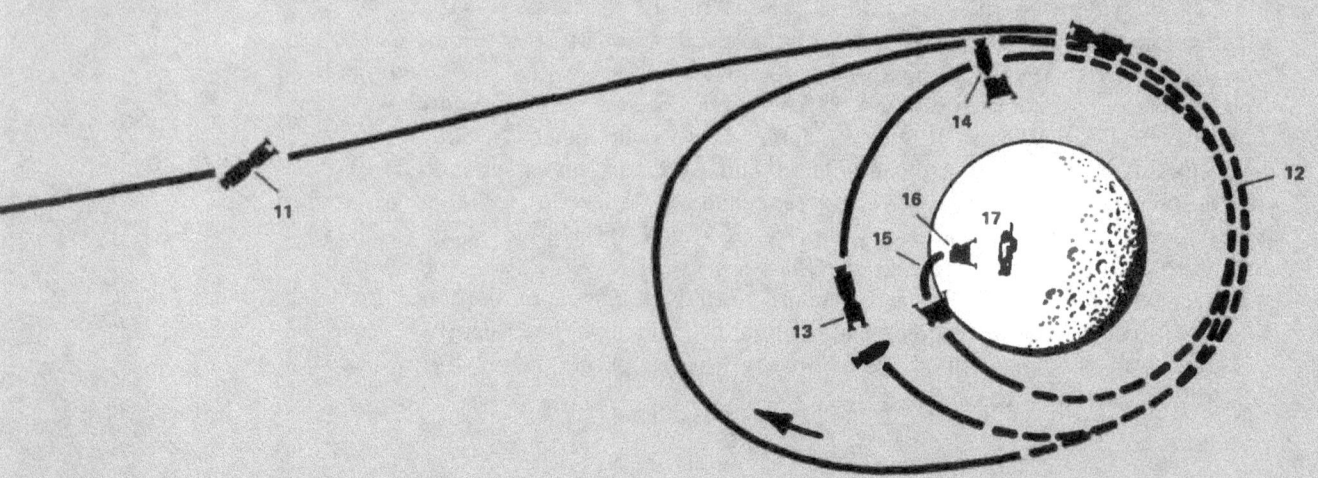

Broken trajectory lines indicate
loss of Earth communications.